Pests, Weeds and Diseases in Agricultural Crop and Animal Husbandry Production

Edited by Dimitrios Kontogiannatos, Anna Kourti and Kassio Ferreira Mendes

Published in London, United Kingdom

IntechOpen

Supporting open minds since 2005

Pests, Weeds and Diseases in Agricultural Crop and Animal Husbandry Production
http://dx.doi.org/10.5772/intechopen.87515
Edited by Dimitrios Kontogiannatos, Anna Kourti and Kassio Ferreira Mendes

Contributors
Wael ElSayed, Koji Nakamura, W. James Grichar, Peter A. Dotray, Ray Langham, Ricardo Alcántara-De La Cruz, Guilherme Oliveira, Leonardo Carvalho, Maria Fátima Da Silva, Sladjan Stankovic, Slobodan Krnjajic, Miroslav Kostic, Igor Kostic, Helen Karasali, Evangelia Tzanetou, Koperumselvan Karthika, Tamanreet Kaur, Mandeep Kaur, Nesreen Abd El-Ghany, Ivana Majić, Ankica Sarajlić, Emilija Raspudić, Zdenko Lončarić, Marko Josipović, Fei-Ying Yang, Weiyi He, Min-Sheng You, Pasquale Avino, Kassio Ferreira Mendes, Mamta Choudhary, Binod Kumar Choudhary, Ratan Chadra Ghosh, Shahenda Abu ElEla, Ivan Notardonato, Mario Vincenzo Russo, Alessandro da Costa Lima

First published in London, United Kingdom, 2021 by IntechOpen
IntechOpen is the global imprint of INTECHOPEN LIMITED, registered in England and Wales, registration number: 11086078, 5 Princes Gate Court, London, SW7 2QJ, United Kingdom
Printed in Croatia

British Library Cataloguing-in-Publication Data
A catalogue record for this book is available from the British Library

Additional hard and PDF copies can be obtained from orders@intechopen.com

Pests, Weeds and Diseases in Agricultural Crop and Animal Husbandry Production
Edited by Dimitrios Kontogiannatos, Anna Kourti and Kassio Ferreira Mendes
p. cm.
Print ISBN 978-1-78923-827-3
Online ISBN 978-1-78923-828-0
eBook (PDF) ISBN 978-1-83962-462-9

For EU product safety concerns:
IN TECH d.o.o., Prolaz Marije Krucifikse Kozulić 3, 51000 Rijeka, Croatia, info@intechopen.com or visit our website at intechopen.com.

Meet the editors

Dr. Dimitrios Kontogiannatos is a collaborative researcher at the Agricultural University of Athens (AUA) and the National Center for Scientific Research "Demokritos" (NCSR "Demokritos"), Athens, Greece. In 2013, he received his Ph.D. diploma in Insect Biotechnology and Insect Molecular Biology from the AUA's Department of Biotechnology, specializing in Pest Control Science and Insect Biotechnology. His research mainly focuses on the use of RNA interference (RNAi) technology in designing insecticidal formulations of low environmental cost. Additionally, he is studying the molecular and physiological impact of various xenobiotics on the developmental processes of terrestrial arthropods. From 2015 to the present he has been collaborating with the AUA's Biotechnology Department as a part-time Assistant Professor. In 2018 he founded and currently runs a Pest Control Company headquartered in Argyroupolis, Attica, Greece, named "PROPEST FHG".

Dr. Anna Kourti is a Professor of Insect Molecular Biology and Biotechnology at the Agricultural University of Athens (AUA), School of Applied Biology and Biotechnology, Department of Biotechnology, Laboratory of Molecular Biology, Athens, Greece. Her research mainly focuses on insect population genetics and biodiversity, insect molecular physiology, biology and genetics, molecular analysis and expression of environmentally related insect genes, transcriptomics, and metabolomics of insect diapause, circadian, and photoperiodic clocks. She is the author of many entomological publications in high impact factor peer-reviewed journals. She is a referee of a wide range of international peer-reviewed journals.

Kassio Ferreira Mendes is a Professor of Biology and Integrated Management of Weeds in the Department of Agronomy, Federal University of Viçosa, Brazil. He graduated from the State University of Mato Grosso in 2011 and received a Master's degree in Agronomy (Crop Science) from the Federal University of Viçosa in 2013. He received his post-doctorate degree (2019) and Doctorate degree (2017) in Sciences - Nuclear Energy in Agriculture (Chemistry in Agriculture and Environment)from the Center of Nuclear Energy in Agriculture, Campus "Luiz de Queiroz", University of São Paulo, Brazil with fellow University of Minnesota - USA (2016), Twin Cities Campus - College of Food and Agricultural Sciences in the Department of Soil, Water, and Climate and United States Department of Agriculture - Agricultural Research Service (USDA - ARS). Dr. Mendes is a member of the Brazilian Society of Weed Science.

Contents

Preface

This book highlights some of the most recent research and practical approaches with respect to emerging pest challenges in agricultural crop and animal husbandry production. The chapters cover the entire spectrum of plant and animal protection science focusing on insect pest control, weed management, herbicide application and food safety, animal pest control, and livestock disease identification. This book can be a useful resource for agricultural and horticultural science students, veterinarians and livestock professionals as well as agronomists and pest control scientists.

In the first section of this book, "Insect Science, Pest Management and Practical Approaches", we demonstrate the role of Integrated Pest Management in the incorporation of biological and behavioral approaches to preserve ecosystems and reduce the use of synthetic chemical pesticides. In addition, we present the use of natural products as an immeasurable source of bioactive insecticidal compounds; the importance of the high rate of diversification of herbivorous insects in agriculture as a consequence of their specialization to distinct host-plants (with a focus on the Orthoptera order), as well as the biological role of pheromones in the chemical communication of insect species. Finally, we analyze methodologies of Mass Spectrometry Imaging (MSI) in the visualization of the composition, abundance, and spatial distribution of molecules in tissues or cells of insects, as well as the role of irrigation and nitrogen fertilization in the biological control of the European Corn Borer, *Ostrinia nubilalis* (*Lepidoptera*: *Crambidae*).

In the second section of this book, "Herbicides: Scientific Background and Food Safety", we mainly focus on the most commonly used analytical methodologies for determining glyphosate and aminomethylphosphonic acid (AMPA) concentrations in various food products, soil, and air. Moreover, we analyze the herbicide resistance status of crops in Brazil with respect to glyphosate resistance and we present practical methodologies for herbicide application, dosing, and safety.

Finally, in the third section of this book, "Animal Pests and Diseases", the reader will find information on the diagnosis and pathoanatomy of Pasteurellosis in animals as well as the identification of ticks in dogs with Ehrlichiosis.

Pests, weeds, and diseases are the more destructive factors in agricultural crop and animal husbandry production, causing tremendous economic losses, influencing food production, and downgrading final products. Scientific progress and technological advances in pest and weed management and animal disease prevention constitute the most efficient weapons of the modern world in the fight against poverty and hunger.

This research is co-financed by Greece and the European Union (European Social Fund- ESF) through the Operational Programme «Human Resources Development,

Education and Lifelong Learning» in the context of the project "Reinforcement of Postdoctoral Researchers - 2nd Cycle" (MIS-5033021), implemented by the State Scholarships Foundation (IKY).

Dimitrios Kontogiannatos
School of Plant Sciences,
Department of Biotechnology,
Molecular Biology Laboratory,
Agricultural University of Athens

School of Applied Biology and Biotechnology,
Department of Crop Science,
Laboratory of General and Agricultural Microbiology,
Agricultural University of Athens

NCSR "Demokritos",
Institute of Biosciences and Applications,
Insect Molecular Genetics and Biotechnology,
Aghia Paraskevi, Athens, Greece

CEO "PROPEST FHG",
Argyroupolis, Attica, Greece

Anna Kourti
Agricultural University of Athens,
Greece

Kassio Ferreira Mendes
Universidade Federal de Viçosa,
Brazil

Section 1

Insect Science, Pest Management and Practical Approaches

Chapter 1

Integrated Pest Management: A Paradigm for Modern Age

Tamanreet Kaur and Mandeep Kaur

Abstract

Integrated pest management is an effective and environmentally sensitive approach for pest management. It plays an important role in sustainable agriculture and quality of food production by providing maximum economic yield to the farmer and also improving human health and environment. Recent developments in agricultural technology, modern communication tools, changing consumer trends, increased awareness for sustainably produced food systems, and globalization of trade and travel, have necessitated the need for the IPM paradigm as appropriate for modern times. Although the concept of integrated pest management originated almost 60 years ago, currently integrated pest management is a robust paradigm of pest control around the globe. This chapter reviews the history of integrated pest management, its main principles, and components of integrated pest management such as host plant resistance, cultural control, behavioral control, mechanical/physical control, biological control, and chemical control.

Keywords: pest management, global losses, sustainable agriculture, new model, control measures

1. Introduction

One of the major challenges of the twenty-first century is to provide food for its ever-growing population. It has pushed food production systems to maximum efficiency and the demand requires farmers to produce more crops on existing farmland that needs continuous improvement of agricultural technologies to minimize crop losses. Although chemical pesticides have played a vital role in providing an abundant and inexpensive food source [1], its persistent overuse has resulted in a number of adverse environmental impacts such as pesticide resistance, resurgence of insect pests, pesticide poisoning, environmental toxicity, elimination of predator species, negative outcomes for other nontarget organisms, disruption in the food web, accumulation of toxins in the food webs, and reduced crop yields [2, 3]. Thus, to feed the future generations and to meet increasing demand for wide spectrum of high-quality fresh products without degrading the resources, strategy must be economically viable and ecologically sustainable. Integrated pest management (IPM) strategy being environmentally friendly pest management is increasingly being adopted in both developed and developing countries for adequate safe and quality food production, improves farmer's livelihood and conserves nonrenewable resources.

1.1 Definition of integrated pest management

Although multiple sources define IPM in diverse ways, previous models primarily focused on the ecological, and to some extent on the evolutionary, aspects of pest management [4]. IPM is a holistic "approach" or "strategy" to combat plant pests and diseases using all available methods, while minimizing applications of chemical pesticides [5]. The basic aim of IPM is not to eradicate pests, but to manage them, maintaining their populations below economic injury levels [6, 7]. IPM is a combination of methods to manage the pest population with considerations of economic efficiency and environmental effects rather than an eradicative method, which was used in traditional practices [8]. The Food and Agriculture Organization of the United Nations (FOA) defines integrated pest management as careful consideration of all available pest control methods and subsequent integration of appropriate measures that deter the development of pest populations and keep pesticides and other interventions to levels that are economically justified and minimize risks to human health and the environment. The United States Department of Agriculture-Agricultural Research Service [9] defines integrated pest management as a sustainable, decision-making process that aims at keeping pest population at below economic threshold levels by employing pest control techniques such as biological, cultural, physical, and chemical methods to identify, manage, and reduce risk from pests and pest management tools and strategies in a way that minimizes overall economic, health, and environmental risks. This strategy avoids undesirable short-term and long-term ripple effects and will ensure a sustainable future [10]. IPM differs from organic agriculture as it allows the judicious use of pesticides, fertilizers, and other materials made from synthetic materials when necessary whereas organic agriculture largely restrict to allowable pesticides made from natural materials only [11].

1.2 Integrated pest management: a historical perspective

In the 1940s with development of synthetic pesticides, the whole scenario of crop pest management changed. Pesticides played a major role in crop production due to their efficacy, convenience, flexibility, and economy. It began with the introduction of alkyl thiocyanate insecticides, and then the discovery of remarkable insecticidal properties of DDT (dichlorodiphenyltrichloroethane) in 1939 by Paul Muller. DDT was followed by the manufacture of other chlorinated hydrocarbons, including aldrin, endrin, heptachlor and recognition of the herbicidal activity of the phenoxyacetic acids-MCPA (2-methyl-4-chlorophenoxyacetic acid) and 2,4-D (2,4-dichlorophenoxyacetic acid). A number of synthetic inorganic insecticides containing arsenic, mercury, tin, and copper were also developed in nineteenth century. By the 1950s, overuse of insecticides had generated numerous well-recognized cases of pest resistance and destruction of natural enemies of pests [12]. Due to over reliance on synthetic pesticides from the late 1940s to mid-1960s, the period has been called "the dark ages" of pest control. However, in the late 1950s, entomologists began to identify the problems associated with extensive and intensive use, misuse, and abuse of insecticides and pesticide resistance, secondary pest outbreaks, hazards of toxic residues in food commodities and biomagnifications, environmental pollution, and killing of nontarget beneficial organisms. Although many components of IPM were developed long time back through trial and error experiences, farmers had developed a number of mechanical, cultural, and physical control measures of different pests; however, the concept of IPM came into existence only after realizing the harmful effects of chemical pesticides. The term Integrated Pest Control was first used as "integrated control" by Barlett [13] for the integrated use of biological and chemical control to manage insect pests of agricultural crops. The first integrated control program was devised for managing spotted alfalfa aphid, *Therioaphis maculata* (Buckton), on alfalfa grown for hay purposes. It was further elaborated as an approach that applies to the concept of integrating the biological and other controls in complementary ways [7]. The concepts of economic threshold level and economic injury level were also introduced by these authors. Subsequently, it was broadened to include all control methods and all classes of pests (insects, plant pathogens, nematodes, weeds, vertebrate pests, etc.). Shortly after IPM concept first appeared, Rachel Carson's book "Silent Spring" in 1962 was published, which explored the effects of pesticide overuse on environment and nontarget species [14–16]. Hence, the public awareness was raised and thereby the concept "integrated control" became popular in both scientific literature and practice [15, 17]. From past 30 years, IPM has been a valuable paradigm for organizing research and extension efforts worldwide and since then numerous IPM programs are being implemented worldwide. The future aim of IPM programs should not be restricted to only efficient use of pesticides and product substitution; rather these programs should aim at fundamental structural changes through effective understanding of ecological processes and synergy between crops.

2. General principles of integrated pest management

Main proponents of IPM suggest six basic strategies to improve insect management strategies:

2.1 Prevention

The cheapest and most reliable way to avoid many pest problems is to provide an environment that discourages pest activities/infestation. These types of methods include suppression of harmful organisms from becoming problems by planning and managing various options such as:

1. Crop rotation and intercropping.

2. Use of adequate cultivation techniques like seedbed sanitation, management of sowing/planting time and plant densities, under-sowing, conservation tillage, pruning, and direct sowing.

3. Use of resistant cultivars, standard/certified seeds and planting material.

4. Well-balanced nutrient supply and optimal water management.

5. Preventing the spread of harmful organisms through field sanitation and hygiene measures.

6. Protecting and enhancing beneficial organisms.

2.2 Pest identification

Pest identification is one of the foremost strategies to control the pest population. Moreover, when the identity of a pest is not known, then, a strategy built to control the pest at a particular site cannot be transferred to another, as pest species or strain at another site might behave differently. Thus, a solid foundation must be built on pest's systematic, taxonomy, etiology, and spatial distribution [18].

2.3 Establishing a periodic inspection and monitoring system

A pest is an anthromorphic categorization which is beneficial and harmful at the same time. For instance, termites considered beneficial organisms in forests converting dead trees to organic matter are also considered as pests as they on wood having high economic value [19]. Pest inspection includes regular site inspections and trapping to determine the extent of infestation levels and types of pests at particular site. It also includes regular check on the occurrence of species identified properly and considered to be pests or beneficial organisms, the damage caused by the pests, the crop characteristics, and the environmental factors. Monitoring procedure is a key element of IPM programs as it helps early detection, timely information on pest activity, ranking of the severity of infestations, identify its causes, and estimation of future populations. Environment monitoring methodologies must be designed for assessing instantaneous and dynamic aspects of the pest's density, activity, or incidence [18]. Understanding these environmental interactions allows crop advisors to react to changing environmental conditions and helps to determine acceptable pest population levels, effective reduction measures, and breach of the action threshold.

2.4 Determine economic injury and action threshold level of pest activity

The primary objective in integrated pest management is not to eliminate a pest but to bring it into acceptable boundary. FOA defines pest as any species/strain/biotype of plant, animal, or pathogenic agent injurious to plants and plant products, materials, or environments and includes vectors of parasites or pathogens of human and animal disease and animals causing public health nuisance. In IPM, a few pests can be tolerated and it is compulsory to take action when pest numbers reach a certain level, this level is known as threshold. The lowest level of injury to crop plants where the damage can be measured is called the damage boundary and the lowest number of insects that will cause economic damage is referred to as economic injury level (EIL). Economic threshold level (ETL) is defined as the pest density at which control measures should be applied to prevent an increasing pest population from reaching economic injury level. An action threshold level (ATL) is the pest population size that requires remedial action for human-health, economic, or aesthetic reasons and it will vary depending on the site structure and how it is being used (**Figure 1**). As ETL and ATL are pest and site specific, meaning that it may be acceptable at one site but at another site it may

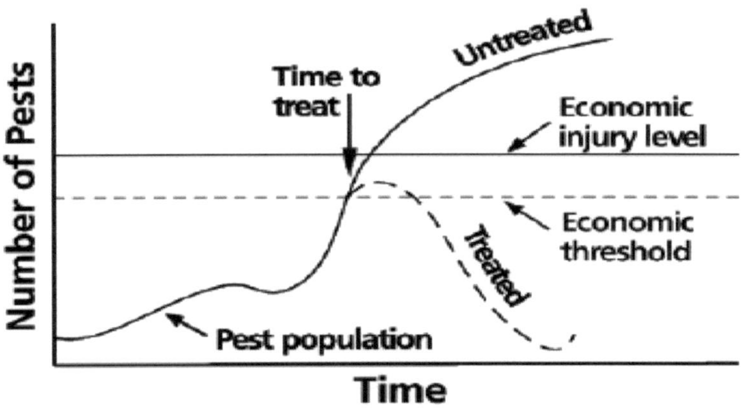

Figure 1.
To make control practice profitable, or at least break even, it is necessary to set the economic threshold (ET) below the economic injury level (EIL). Graphic: National Pesticide Applicator Certification Core Manual, NASDARF.

not be acceptable. Next step involves decision-making process that draws on accurate, timely information to make pest prevention and its management decisions.

2.5 Developing management strategies

In IPM, implementation of treatment strategy involves mechanical, cultural, biological, or chemical controls, or a combination of these strategies. Although using a single strategy may be successful for a short duration but by integration of these practices may provide safe guards against ecological disruptions (pest resistance or destruction of natural enemies) that often develop because of reliance on a single strategy [20]. If all methods have failed and the monitoring system shows that pest population is still beyond action thresholds, then the use of synthetic chemicals should be last resort only, but when used, the least toxic materials should be chosen to minimize exposure to all nontarget organisms. Ultimately, the goal is to control pests with little impact on the environment.

2.6 Evaluating and record keeping

Lastly, evaluation is often considered as one of the most important steps in integrated pest management [21]. A regular evaluation program is essential step to determine the success of the pest management strategies. It is the process of reviewing an IPM program and the results it has generated. Moreover, understanding the effectiveness of the IPM program allows making necessary modifications to the IPM plan prior to pests reaching the action threshold and requiring action again. A record keeping system is essential to establish trends and patterns in pest outbreaks. Information recorded at every inspection or treatment should include pest identification, its population size, distribution, recommendations for future prevention, and complete information on its treatment.

3. Pest management tactics under integrated pest management

The different pest management tactics to suppress pests includes host plant resistance, cultural control, behavioral control, mechanical/physical control, biological control, and chemical control. Each category as discussed below employs a different set of mechanisms for suppressing pest populations.

3.1 Host plant resistance

Host plant resistance approach is the first line of defense in IPM. During domestication of crops many resistance traits have been lost [22, 23]. It involves the use of pest-resistant and pest-tolerant cultivars developed through traditional breeding/genetic engineering [24–26]. The cultivars produced possess physical, morphological, or biochemical characters that reduce the plant's attractiveness for the pest to feed, develop, or reproduce successfully and thus reduce the yield losses. Moreover, it also involves withstanding the infestation/infection of pests to reduced level that they are not large numbers during the plant growth period [27].

3.2 Cultural control

Adopting good agronomic practices that avoid/reduce pest infestations and damage is referred as cultural control. The various cultural practices have been grouped as below:

1. Preparation of nurseries/main fields free from pest infestation by following practices such as removing plant debris, trimming of bunds, treating of soil and deep summer plowing, which kills various stages of pests. Plowing is considered an important control option to destroy the crop residue and expose the soil-inhabiting stages of several vegetable pests [28]. Proper drainage system in field is also to be adopted.

2. Testing of soil for nutrient deficiencies for application of appropriate fertilizers. Use of farm yard manure (FYM) and biofertilizers should be encouraged. High or low nitrogen [29, 30] content in the plant can also contribute to some disease problems.

3. Selection of clean, certified, pest-resistant/tolerant seeds and treating seeds with fungicide/biopesticides before sowing for seedborne disease control.

4. Proper adjustment of time of sowing and harvesting to escape peak season of pest attack and rotation of host crops with non-host crops.

5. Accurate plant spacing, which makes plants healthier and less susceptible to pests.

6. Proper water management as the high moisture in soil for prolonged period is conducive for development of pests, especially soilborne diseases.

7. Proper weed management as most of weeds besides competing with crop for micronutrients also harbor many pests.

8. Community approach is required for synchronized sowing the crops simultaneously in vast area so that pest may not get different staged crops suitable for its population buildup. If pest appears in damaging proportion, control operation could also be applied effectively in whole area.

9. Crop rotation with nonhost or tolerant crops will break the pest cycles and reduce their buildup year after year. Crop rotation tactic has been used for insect, disease, and weed management in many cropping systems [31–34].

10. Growing trap crops [35] on the borders or peripheries of fields as by growing such crops on the border of the fields develops pest population that can be source of natural enemies providing top-down control [36]. Intercropping/multiple cropping wherever possible as certain crops act as repellents, thus keeping the pest species away from preferred crops results in reduction of pest incidence [37, 38]. For instance, significant disease reduction was seen by interspacing a rice cultivar susceptible to *Magnaporthe oryzae* (causing rice blast) with a resistant one [39, 40].

11. For excellent fruit set in orchards, pollinizer cultivars should be planted in required proportion.

12. Harvesting should be done close to ground level as certain developmental stages of insect pests/diseases remain on the plant parts, which act as primary inoculum for the next crop season.

13. While pruning fruit trees, it is advised to remove crowded/dead/broken/diseased branches and destroy them and large pruning wounds should be covered with bordeaux paste/paint to protect the plants from pest/disease attack.

3.3 Behavioral control

The behavior of a pest can be exploited and controlled through baits, traps, and mating disruption techniques [41–43]. Use of baits containing poisonous material will attract and kill the pests when distributed in the field or placed in traps. Pests are attracted to certain colors, lights, odors of attractants or pheromones. These devices one or more can be used to attract, trap, or kill pests. For instance, pheromone traps involves dispensing large amounts of sex pheromones in plantation area, thereby suppressing the male's abilities to find female conspecifics for mating [44]. Thus, pheromone lures confuse adult insects and disrupt their mating potential, monitoring pest levels, mass trapping, and thus reduce their offspring.

3.4 Mechanical/physical control

This approach refers to the use of a variety of physical/mechanical techniques for pest exclusion, its trapping, removal, or destruction [45–47]. These treatments use equipment, devices, barriers, or extreme temperatures to reduce pests. Mechanical/physical controls include:

1. Agricultural practices like tillage, slash and burn, and hand weeding.
2. Pruning of infested parts of fruits and forest trees and defoliation in certain crops.
3. Mechanical cultivation of soil to kill weeds or overwintering insects.
4. Mowers and brushing equipment for plant control.
5. Setting up of traps for insects, rodents, mollusks, or other pests.
6. Pest exclusion with screens, plants collars, netting, handpicking, or vacuuming.
7. Freezers to control pests in stored products.
8. Flame, hot water, or infrared light for weed control.
9. Noisemakers or other pest repelling devices.
10. Modifying environmental conditions such as heat or humidity in greenhouses, steam sterilization, or solarization.
11. Installation of bird perches in the field for allowing birds to sit and feed on insect pests and their immature stages viz., eggs, larvae, and pupae.
12. Installation of visual or physical bird deterrents such as reflective material or sonic devices or bird scarer in the field as per requirement.

3.5 Biological control

Biological control/biocontrol involves the use of living organisms to manage crop-damaging pests. It is one of the oldest nonchemical control methods used in agriculture [48], and is probably the most well-researched part of the IPM concept. In biological control, arthropod pests are mainly controlled using biological control agent's viz. predators, parasitoids, and pathogens. Biological control agents may provide good control option under certain conditions (temperature, humidity, length of day) or on certain crops. Most biological control agents are highly perishable, so they need to be handled with care and must be released soon they are received. Its release must be planned for the right time and biology must be thoroughly understood as most species are effective on one or a few species of pests. Beneficial insects have been successfully used to control pests in greenhouses [49] and outdoor specialty crops such as strawberries [50]. Most of the intrinsic problems associated with biological control appear mainly in open areas with arthropod agents, which might emigrate from the plantation leaving the pest behind and attack each other (intraguild predation) rather than the target pest [51], or attack nontarget prey [52]. Biological control can be classified into three basic categories namely classical, conservation, and augmentation [53, 54]:

1. Classical biological control involves collection of natural enemies from their native region and releasing them in the new area where their host pest was introduced accidentally [55, 56]. Natural enemies such as predatory arthropods and parasitic wasps can cause significant reductions in pest populations at certain circumstances [57]. In microbial control, disease microorganisms are used to control pests/weeds. For instance, *Bacillus thuringiensis* (Bt), a soil bacterium that contains a chemical toxic to larval insect pests, acts by blocking the larvae from absorbing nutrients in their digestive systems.

2. Conservation biological control is aimed at promoting the survival and activity of natural enemies at the expense of pest populations [48]. For instance, ecological strips can be deliberately created consisting of selected non-crop plants to provide food sources, overwintering shelters, and protection of local natural enemies from pesticide disturbances [58, 59].

3. Augmentative biological control is the periodic release of large numbers of mass reared natural enemies with the aim of supplementing natural enemy population/inundating pest population with natural enemies [54, 60]. The practice of augmentation is based on the knowledge or assumption that in some situations there are not adequate numbers or species of natural enemies to provide optimal biological control, but that the numbers can be increased (and control improved) by releases. This relies on an ability to mass-produce large numbers of the natural enemy in a laboratory or by commercial companies.

3.6 Chemical control

Chemical pesticides are the last resort when all other methods fail to keep the pest population below economic level. The four major problems encountered with chemical pesticides are pest resistance, toxic residues, secondary pests, and pest resurgence [10]. Chemical control includes synthetic chemicals as well as chemicals of microbial (avermectin and spinosad) or botanical origin (azadirachtin and pyrethrins). Pesticides that are generally highly toxic and are known to have toxic residual effects should not be recommended off hand. The use of natural pesticides and organophosphates being more environmentally friendly is encouraged and synthetic pesticides should only be used as a last resort or only used as required and only at specific times in a pest's life cycle. Chemical pesticides are categorized into different groups based on their mode of action [61] and rotating chemicals from different mode of action groups is essential to reduce the risk of resistance development [62]. Pests can also develop resistance to botanical and microbial pesticides if they are overused [63]. Thus, use of pesticides should be judicious, based on pest surveillance and economic threshold level. While going for chemical control, we must understand thoroughly what to spray, when to spray, where to spray, and how to spray.

4. Integrated pest management certification

Pest control operators, farmers, grounds managers, crop consultants, wildlife management specialists, and others can have their products certified under a variety of programs that use IPM as a requirement. Certification means that a product/service meets a well-defined standard.

5. Pros and cons of an integrated pest management program

The key benefits of integrated pest management to farming and society include:

- IPM emphasizes understanding the agroecosystem, integration of new management skills and the new concepts for pest management to protect our environment and make sure the uninterrupted safe and nutritious food supply for the growing world population.

- IPM, besides sustaining biodiversity, slows the development of resistance of pests to synthetic pesticides.

- It improves profitability to farmer as pest management costs are reduced.

- It reduces risk of crop loss by a pest and long-term answers to pest problem.

- It protects environmental and human health by restricting broad spectrum pesticide use.

In spite of benefits of IPM stated so far, there are also some drawbacks to it:

- IPM involves more technicalities and decision-making.

- An IPM program requires a higher degree of planning and management.

- It is more time and energy consuming.

- It requires more resources as a substitute to pesticides.

6. Conclusion

In agriculture sector, increased pest resistance and ecological backlash can only be corrected by effective, safe, and sustainable pest management strategies. IPM can be expected to continue to be dominant theme in the future as it can better exploit the modern science and the traditional agricultural systems based on indigenous farming practices. Overall, IPM addresses all the economic, environmental, and social aspects and provides safe and affordable food to the consumers and profits to producers and sellers along with maintaining environmental health. Moreover, to develop IPM programs for the twenty-first century, further research and, more importantly, field studies and on-farm validation are needed to advance these approaches for a pest-free crop and pesticide-free product.

Conflict of interest

No conflict of interest is indulged.

Abbreviations

IPM	integrated pest management
FOA	Food and Agriculture Organization of the United Nations
DDT	dichlorodiphenyltrichloroethane
MCPA	2-methyl-4-chlorophenoxyacetic acid
2,4-D	2,4-dichlorophenoxyacetic acid
FYM	farm yard manure
EIL	economic injury level
ETL	economic threshold level
ATL	action threshold level
Bt	*Bacillus thuringiensis*

Author details

Tamanreet Kaur* and Mandeep Kaur
Department of Zoology, Kanya Maha Vidyalaya, Jalandhar, India

*Address all correspondence to: tamanreetkaur@gmail.com

References

[1] USEPA (United States Environmental Protection Agency). Agricultural Pesticides: Management Improvements Needed to Further Promote Integrated Pest Management [Internet]. 2015. Available from: http://www.epa.gov/agriculture/tipm.html [Accessed: 12 March 2020]

[2] Aktar MW, Sengupta D, Chowdhury A. Impact of pesticides use in agriculture: Their benefits and hazards. Interdisciplinary Toxicology. 2009;**2**:1-12. DOI: 10.2478/v10102-009-0001-7

[3] Sharma R, Peshin R, Shankar U, Kaul V, Sharma S. Impact evaluation indicators of an integrated pest management program in vegetable crops in the subtropical region of Jammu and Kashmir, India. Crop Protection. 2015;**67**:191. DOI: 10.1016/j.cropro.2014.10.014

[4] Peterson RKD, Higley LG, Pedigo LP. Whatever happened to IPM? American Entomologist. 2018;**64**:146-150. DOI: 10.1093/ae/tmy049

[5] Stenberg JA. A conceptual framework for integrated pest management. Trends in Plant Science. 2017;**22**:749-769. DOI: 10.1016/j.tplants.2017.06.010

[6] Smith RF, van den Bosch R. Integrated control. In: Kilglore WW, Doutt R, editors. Pest Control: Biological, Physical and Selected Chemical Methods. Academic Press: Elsevier; 1967. pp. 295-340

[7] Stern VM, Smith RF, van den Bosch R, Hagen KS. The integrated control concept. Hilgardia. 1959;**29**:81-101. DOI: 10.3733/hilg.v29n02p081

[8] Galea V. Principles of Integrated Plant Protection [Learning Guide]. Queensland: The University of Queensland; 2010

[9] USDA-ARS (United States Department of Agriculture-Agricultural Research Service). A National Road Map for Integrated Pest Management [Internet]. 2018. Available from: https://www.ars.usda.gov/ARSUserFiles/OPMP/IPM%20Road%20Map%20FINAL.pdf [Accessed: 10 March 2020]

[10] Lewis WJ, van Lenteren JC, Phatak SC, Tumlinson JH. A total system approach to sustainable pestmanagement. Proceedings of the Natural Academy of Sciences of the United States of America. 1997;**4**:12243-12248. DOI: 10.1073/pnas.94.23.12243

[11] US EPA (United States Environmental Protection Agency). Key Factors in Implementing Integrated Pest Management [Internet]. 2018. Available from: https://www.epa.gov/managing pests-schools/key-factors-implementing-integratedpest-management [Accessed: 14 March 2020]

[12] Perkins JH. Insects, Experts and the Insecticide Crisis: The Quest for New Pest Management Strategies. New York: Plenum Press; 1982

[13] Barlett BR. Natural predators. Can selective insecticides help to preserve biotic control. Journal of Agricultural and food Chemistry. 1956;**11**:42-44

[14] Briggs SA. Silent Spring: The view from 1990. Ecologist. 1990;**20**:54-60

[15] van Emden HF, Peakall DB. Beyond Silent Spring: Integrated Pest Management and Chemical Safety. London: Chapman and Hall; 1996. xviii + 322 p

[16] Pollock CG. Silent spring revisited: A 21st-century look at the effect of pesticides on wildlife. Journal of Avian

Medicine and Surgery. 2001;**15**:50-53. DOI: 10.1647/1082-6742

[17] Kogan M. Integrated pest management: Historical perspectives and contemporary developments. Annual Review of Entomology. 1998;**43**:243-270. DOI: 10.1146/annurev.ento.43.1.243

[18] Irwin ME. Implications of movement in developing and deploying integrated pest management strategies. Agricultural and Forest Meteorology. 1999;**97**:235-248. DOI: 10.1016/s0168-1923(99)00069-6

[19] Capinera JL. Handbook of Vegetable Pests. Orlando: Academic Press; 2001

[20] Pedigo LP, Rice ME. Entomology and Pest Management. 6th ed. Illinois: Waveland Press Long Grove; 2009

[21] Bennett GW, Owens JM, Corrigan RM. Truman's Scientific Guide to Pest Management Operations. 6th ed. Purdue University: Questex; 2005. pp. 10-12

[22] Chen YH, Gols R, Brevik KA, Benrey B. Complex tritrophic interactions in response to crop domestication: Predictions from the wild. Entomologia Experimentalis et Applicata. 2015;**157**:40-59. DOI: 10.1111/eea.12344

[23] Tamiru A, Khan ZR, Bruce TJA. New directions for improving crop resistance to insects by breeding for egg induced defence. Current Opinion in Insect Science. 2015;**9**:51-55. DOI: 10.1016/j.cois.2015.02.011

[24] Douglas AE. Strategies for enhanced crop resistance to insect pests. Annual Review of Plant Biology. 2018;**69**:637-660. DOI: 10.1146/annurev-arplant-042817-040248

[25] Kennedy GG. Integration of insect-resistant genetically modified crops within IPM programs. In: Romeis J, Shelton A, Kennedy G, editors. Integration of Insect-resistant Genetically Modified Crops within IPM Programs. Dordrecht: Springer; 2008. pp. 1-26

[26] Nelson R, Wiesner-Hanks T, Wisser R, Balint-Kurti P. Navigating complexity to breed disease-resistant crops. Nature Reviews Genetics. 2018;**19**:21-33. DOI: 10.1038/nrg.2017.82

[27] Sharma HC. Host plant resistance to insects: Modern approaches and limitations. Indian Journal of Plant Protection. 2007;**35**:179-184

[28] Kunjwal N, Srivastava RM. Insect pests of vegetables. In: Omkar, editor. Pests and Their Management. Singapore: Springer; 2018. pp. 163-221

[29] Mitchell CE, Reich PB, Tilman D, Groth JV. Effects of elevated CO_2, nitrogen deposition, and decreased species diversity on foliar fungal plant disease. Global Change Biology. 2003;**3**:438-451. DOI: 10.1002/ecs2.2111

[30] Snoeijers SS, Perez-Garcia A, Joosten MHAJ, De Wit PJGM. The effect of nitrogen on disease development and gene expression in bacterial and fungal plant pathogens. European Journal of Plant Pathology. 2000;**106**:493-506. DOI: 10.1023/A:1008720704105

[31] Curl EA. Control of plant diseases by crop rotation. The Botanical Review. 1963;**29**:413-479

[32] Wright RJ. Evaluation of crop rotation for control of Colorado potato beetles (Coleoptera: Chrysomelidae) in commercial potato fields on Long Island. Journal of Economic Entomology. 1984;**7**:1254-1259

[33] Liebman M, Dyck E. Crop rotation and intercropping strategies for weed management. Ecological Applications. 1993;**3**:92-122. DOI: 10.2307/1941795

[34] Mohler CL, Johnson SE. Crop Rotation on Organic Farms [Planning Manual]. Ithaca: Natural Resource, Agriculture, and Engineering Service; 2009

[35] Aluja M, Jimenez A, Camino M, Pinero J, Aldana L, Caserjon V, et al. Habitat manipulation to reduce papaya fruit fly (Diptera: Tephritidae) damage: Orchard design, use of trap crops and border trapping. Journal of Econonic Entomology. 1997;**90**:1567-1576. DOI: 10.1093/jee/90.6.1567

[36] Paredes D, Cayuela L, Campos M. Synergistic effects of ground cover and adjacent vegetation on natural enemies of olive insect pests. Agriculture, Ecosystem and Environment. 2013;**173**:72-80. DOI: 10.1016/j.agee.2013.04.016

[37] Pretty J, Bharucha ZP. Integrated pest management for sustainable intensification of agriculture in Asia and Africa. Insects. 2015;**6**:152-182. DOI: 10.3390/insects6010152

[38] Nielsen AL, Dively G, Pote JM, Zinati G, Mathews C. Identifying a potential trap crop for a novel insect pest, *Halyomorpha halys* (Hemiptera: Pentatomidae), in organic farms. Environmental Entomology. 2016;**45**:472-478. DOI: 10.1093/ee/nvw006

[39] Zhu YY, Chen HR, Fan JH, Wang YY, Li Y, Chen JB, et al. Genetic diversity and disease control in rice. Nature. 2000;**406**:718-722. DOI: 10.1038/35021046

[40] Raboin LM, Ramanantsoanirina A, Dusserre J, Razasolofonanahary F, Tharreau D, Lannou C, et al. Two-component cultivar mixtures reduce rice blast epidemics in an upland agrosystem. Plant Pathology. 2012;**61**:1103-1111. DOI: 10.1111/j.1365-3059.2012.02602.x

[41] Vlades EMAE, Aldana LLL, Figueroa BR, Gutierrez OM, Hernandez RMC, Chavelas MT. Trapping of *Scyphophorus acupunctatus* (Coleoptera: Curculionidae) with two natural baits in a field of *Polianthes tuberosa* (Liliales: Agavaceae) in the state of Morelos, Mexico. Florida Entomologist. 2005;**88**:338-340. DOI: 10.1653/0015-4040(2005)088[0338:TOSACW] 2.0.CO;2

[42] El-Sayed AM, Suckling DM, Byers JA, Jang EB, Wearing CH. Potential of "lure and kill" in long-term pest management and eradication of invasive species. Journal of Economic Entomology. 2009;**102**:815-835. DOI: 10.1603/029.102.0301

[43] Morrison WR, Lee DH, Short BD, Khrimian A, Leskey TC. Establishing the behavioral basis for an attract-and-kill strategy to manage the invasive Halyomorpha halys in apple orchards. Journal of Pest Science. 2016;**89**:81-96. DOI: 10.1007/s10340-015-0679-6

[44] Witzgall P, Kirsch P, Cork A. Sex pheromones and their impact on pest management. Journal of Chemical Ecology. 2010;**36**:80-100. DOI: 10.1007/s10886-009-9737-y

[45] Gamliel A, Katan J. Solarization: Theory and Practice. Saint Paul: The American Phytopathological Society; 2012

[46] Gogo EO, Saidi M, Ochieng JM, Martin T, Baird V, Ngouajio M. Microclimate modification and insect pest exclusion using agronet improve pod yield and quality of French bean. Horticultural Science. 2014;**49**:1298-1304. DOI: 10.21273/HORTSCI.49.10.1298

[47] Dara SK, Peck D, Murray D. Chemical and non-chemical options for managing two spotted spider mite, western tarnished plant bug and other arthropod pests in strawberries. Insects. 2018;**9**:156. DOI: 10.3390/insects9040156

[48] Smith HS. On some phases of insect control by the biological method. Journal of Economic Entomology. 1919;**12**:288-292

[49] van Lenteren JC. Biological and integrated pest control in greenhouses. Annual Review of Entomology. 1988;**33**:239-269

[50] Zalom FG, Bolda MP, Dara SK, Joseph S. UC IPM Pest Management Guidelines: Strawberry (Insects and Mites) [Statewide IPM Program]. Oakland: University of California; 2018

[51] Frago E. Interactions between parasitoids and higher order natural enemies: Intraguild predation and hyperparasitoids. Current Opinion in Insect Science. 2016;**14**:81-86. DOI: 10.1016/j.cois.2016.02.005

[52] Lu X, Siemann E, He M, Wei H, Shao X, Ding J. Climate warming increases biological control agent impact on a non-target species. Ecology Letters. 2015;**18**:48-56. DOI: 10.1111/ele.12391

[53] Hatano E, Saveer AM, Borrero-Echeverry F, Strauch M, Zakir A, Bengtsson M, et al. A herbivore-induced plant volatile interferes with host plant and mate location in moths through suppression of olfactory signalling pathways. BMC Biology. 2015;**13**:1-15. DOI: 10.1186/s12915-015-0188-3

[54] Eilenberg J, Hajek A, Lomer C. Suggestions for unifying the terminology in biological control. Biocontrol. 2001;**46**:387-400. DOI: 10.1023/A:1014193329979

[55] Kenis M, Hurley BP, Hajek AE, Cock MJW. Classical biological control of insect pests of trees: Facts and figures. Biological Invasions. 2017;**19**:3401-3417. DOI: 10.1007/s10530-017-1414-4

[56] Heimpel GE, Cock MJW. Shifting paradigms in the history of classical biological control. BioControl. 2018;**63**:27-37. DOI: 10.1007/s10526-017-9841-9

[57] Hajek AE, Eilenberg J. Natural Enemies: An Introduction to Biological Control. Cambridge: Cambridge University Press; 2018

[58] Mommaerts V, Put K, Smagghe G. *Bombus terrestris* as pollinator-and-vector to suppress *Botrytis cinerea* in greenhouse strawberry. Pest Management Science. 2011;**67**:1069-1075. DOI: 10.1002/ps.2147

[59] Huang F, Andow DA, Buschman LL. Success of the high-dose/refuge resistance management strategy after 15 years of Bt crop use in North America. Entomologia Experimetalis et Applicata. 2011;**140**:1-16. DOI: 10.1111/j.1570-7458.2011.01138.x

[60] Agren GI, Stenberg JA, Bjorkman C. Omnivores as plant bodyguards—A model of the importance of plant quality. Basic and Applied Ecology. 2012;**13**:441-448. DOI: 10.1016/j.baae.2012.07.005

[61] IRAC (Insecticide Resistance Action Committee). IRAC Mode of Action Classification Scheme [Internet]. 2018. Available from: https://www.irac-online.org/documents/moa-classification/?ext=pdf [Accessed: 16 March 2020]

[62] Sparks TC, Nauen R. IRAC: Mode of action classification and insecticide resistance management. Pesticity Biochemistry and Physiology. 2015;**121**:122-128. DOI: 10.1016/j.pestbp.2014.11.014

[63] Dara SK. Insect Resistance to Biopesticides. University of California Agriculture and Natural Resources eJournal Strawberries and Vegetables [Internet]; 2017. Available from: https://ucanr.edu/blogs/blogcore/postdetail.cfm?postnum=25819 [Accessed: 16 March 2020]

Chapter 2

Pheromones and Chemical Communication in Insects

Nesreen M. Abd El-Ghany

Abstract

Chemical communication is an essential item for insects' survivals that qualify them to adapt their behavior depending on the surrounding environment. Semiochemicals defined as informative molecules (M) mainly play an important role that conveys specific chemical messages between insect and insect and plant and insect. Olfaction mechanism in insects is a key point of chemical communication between the same and different insect species. Discrimination of various odors through the olfaction system depends only on the evolutionary pressures of the molecules which stimulate the development of specific binding proteins (BPs) and specific receptor sites present on individual chemosensory neurons. Pheromones are defined as species-specific chemical signals which enable communication between life-forms of the same species. Recently, semiochemicals become as alternative or complementary components to insecticide approaches in integrated pest management (IPM) strategies. Pheromones are secreted by insects causing a specific reaction, for example, either a definite behavior or a developmental process. Pheromones have been classified into eight various types: aggregation pheromones, alarm pheromones, oviposition-deterrent pheromones, home recognition pheromones, sex pheromones, trail pheromones, recruitment pheromones, and royal pheromones. Pheromones are promising and can be used singly or in integration with other control strategies for monitoring and controlling insect pests in agricultural systems.

Keywords: chemical communication, semiochemicals, pheromones, insect olfactory

1. Introduction

Chemical communications in insects are exploited for many features as food seeking and preference, orientation, recruitment, defense, reproductive habitats, predator recognition, and mate attraction [1]. Chemical communication is distinguished by its effectiveness over long distances than others as mechanical and visual communications. Various active compounds were isolated and identified from different plant species that explore several activities toward other organisms [2–5]. The wide range of these compounds affects different insect pests in different ways. Herbivorous insects may use host plant volatiles for determination of food, mates, and/or oviposition and hibernation sites by stimulation of insect chemoreceptor cells in taste sensilla present on antennae, tarsi, and mouthparts [6]. The manipulation of insect behavior occurs by detection of the chemical stimuli known as semiochemicals [7] or infochemicals [8]. Semiochemicals are defined as informative molecules released from one organism that evokes either a behavioral or physiological response between members of the same or different species. They are mainly used in plant-insect or insect-insect interactions as alternative or complementary components to insecticide approaches in different integrated pest management (IPM) strategies. Such compounds are mainly affecting the behavior of various insect pests via chemical signals which occur between insect and insect or plant and insect. Semiochemicals considered a promising component in IPM programs for controlling insect pests. They are involved in different control strategies such as monitoring, mass trapping, mating disruption, and attract-and-kill and push-pull strategies [1].

2. Insect-insect and plant-insect interactions

In insects, the interaction of chemical signals can either stimulate or inhibit the behavior of the pest and so change its response. The response of insects to plant volatiles differs and is either attractive (adapted herbivore) or repellent (non-adapted herbivore). The classification of plant volatiles as attractants and repellents is not standardized due to fluctuation of insect behavior responses to such volatiles depending on their concentration. Herbivorous insects develop host plant compounds and use them as sex pheromone precursors or sex pheromones [9]. For example, male orchid bees assemble terpenoid mixtures from orchids and transfer them as an aggregation pheromone to stimulate leakage in mating [10]. Furthermore, moths, butterflies, grasshoppers, beetles, and aphids utilize pyrrolizidine alkaloids as feeding deterrents against their parasites and/or predators [11]. The interactions which occur between different organisms are divided into two main categories, intraspecific and interspecific, depending on how the interactions occur. An intraspecific communication passes between individuals of the same species, while an interspecific communication involves an interaction between members of different species. Based on the communication signal and subsequently the relation between the receiver and the emitter, semiochemicals are classified into two main functional groups: pheromones and allelochemicals [1].

2.1 Pheromones

Pheromones are defined as species-specific chemical signals which enable communication between life-forms of the same species. Pheromones are secreted by insects which caused a specific reaction, for example, either a definite behavior (immediate effect on the behavior of the receiver) which is called a releaser pheromone or a developmental process (physiological effects on the receiver) which is called a primer pheromone [12]. Pheromones have been classified into eight various types: aggregation pheromones, alarm pheromones, oviposition-deterrent pheromones, home recognition pheromones, sex pheromones, trail pheromones, recruitment pheromones, and royal pheromones. Primer pheromones stimulate the olfactory sensory neurons that emit signals to the insect's brain which stimulate hormones released by the endocrine system [13]. Caste determination in social insects (bees, wasps, ants, termites, locusts) resembles the most famous example for primer pheromone in **Figure 1** [14]. Releaser pheromones are divided by function into sex pheromones, trail pheromones, alarm pheromones, etc. Sex pheromone is the most commonly known which species specific that attract opposite sexes for mating is highly. Concerning trail pheromones, these are commonly known in social insects for orientation and also for recruit nest mates toward a suitable food source. For example, ants and termites deposit these pheromones as they navigate their territory, thus promoting extensive nets of chemical routes [15].

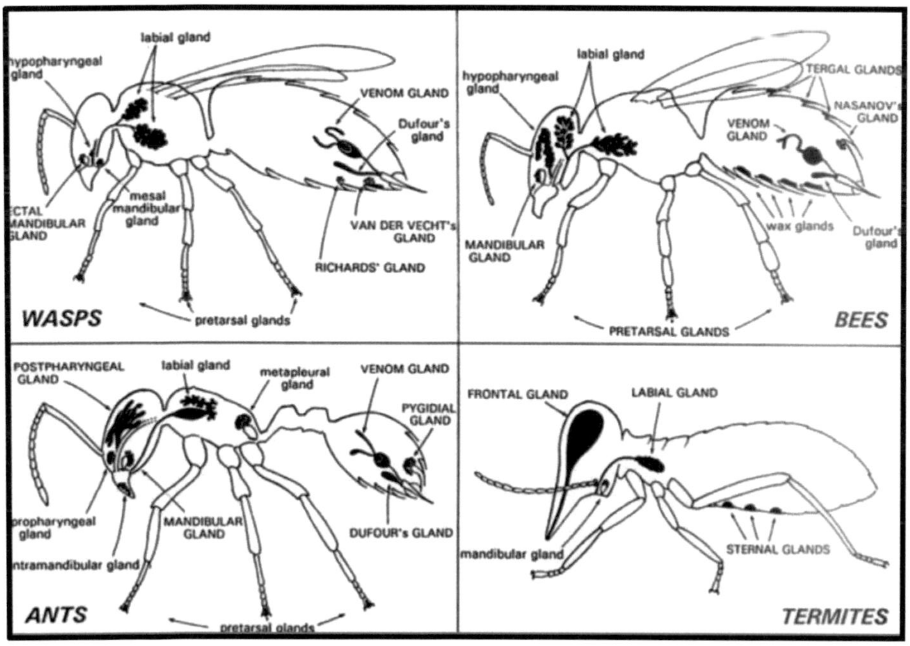

Figure 1.
Schematic profile drawings for exocrine glands of some social insects with a pheromonal function (capital lettering) [14].

On the other hand, bees release airborne orientation pheromones including forage marking, nest entrance finding, and swarming from the Nasonov gland. These pheromones are composed of mixtures of geraniol, farnesol, citral, and other minor compounds [16]. Alarm pheromones are well-developed pheromones in social insects for defensive function and are composed of multicomponent volatiles as mono- and sesquiterpenes and acetates [16–18]. The aggregation pheromones attract conspecifics of both sexes, e.g., bark beetles. The beetles start digging up into the bark of the host tree, thus releasing a mixture of terpenoids which are long-range aggregation pheromones that synthesized de novo, and others produced terpenoids via gut symbiotic bacteria or sequestered from the host tree. Depending on evoke aggregation pheromones, a great number of beetles attack, leading to killing of the host tree [14, 19].

2.2 Allelochemicals

The second subclass of semiochemicals is allelochemicals which includes substances that transmit chemical messages between different species. Fundamentally, these substances resemble an interspecific communication which are emitted by individuals of one species and are understood by individuals of a different species. Allelochemicals are divided depending on the benefits and costs to the signaler and receiver. They have been divided into five categories according to [1, 20] as follows.

2.2.1 Allomones

Allomones (from Greek "allos + hormone" = excite others): released from one organism that stimulate a response in an individual of another species. The response is beneficial to the emitter, e.g., poisonous allelochemicals. They can also be seen as a deterrent emitted by insects against their predators as a defense mechanism. Granular trichomes which cover plant leaves and stems release herbivore-deterring allomones under stress conditions as a defense process. These allomones are toxic for the herbivorous insect pests, e.g., nicotine from a tobacco plant. Moreover, bolas spiders can deceit, lure, and capture male moths by synthesizing and mimicking moth pheromones [14].

2.2.2 Kairomones

Kairomones (from a Greek word "kairos" = opportunistic or exploitative): emitted by one organism that stimulate a response in an individual of another species. The response is beneficial to the recipient, e.g., orientation of predaceous checkered beetles (Coleoptera, Cleridae) toward the aggregation pheromone of their prey and bark beetle (Coleoptera, Curculionidae, Scolytinae) [14, 21]. Kairomones may be allomones or pheromones depending on the circumstances. For example, American bolas spiders attract their prey (male moths) by releasing attractant allomones which serve as sex pheromones emitted by female moths. Also, exudates of warm-blooded animals that pull blood-sucking insects toward their hosts serve as kairomones.

2.2.3 Synomones

Synomones: beneficial to both the releaser and receiver. Examples include scents used by flowers to attract pollinating insects. Moreover, herbivore-induced plant volatiles are considered to be active synomones which recruit natural enemies of insect pests toward the affected plants [22]. Also, synomones play an essential role in mate-finding communication. This role relies on the reduction of competition in the olfaction communication channel between closely related species with overlapping pheromone components. This advisable action is important in preventing exhaustion from the time and energy required for orientation toward heterospecifics [23]. In termites, hydroquinone is a phagostimulant compound secreted by labial glands distinguished as pheromones and synomones when different species are partaking the same foraging territory. It acts as a pheromone when recognized by nest mates of the same species and as a synomone when perceived by another termite species [24].

2.2.4 Antimones

Antimones: maladaptive for both the releaser and receiver. These substances are produced or acquired by an organism that, when encountered by another individual of a different species in the natural environment, activate in the receiving individual a repellent response to the emitting and receiving individuals [1].

2.2.5 Apneumones

Apneumones (from a Greek word "a-pneum" = breathless or lifeless): emitted by a non-living source, causing a favorable behavioral or physiological reaction to a receiving organism, but harmful to other species that may be found either in or on the non-living material. Apneumones were suggested by [7]. Rare cases of these allelochemicals have been found later in the literature, e.g., hexanal and 2-methyl-2-butanol released from rabbit stools attract sandfly females for oviposition [25].

3. Mechanisms of chemical communication in insects

Chemical communication is an essential item for insects' survivals that qualify them to adapt their behavior depending on the surrounding environment [1]. In insects, chemical communication is based on a mixture of one or several semiochemical substances which stimulate various receptor organs. The efficiency of semiochemicals in chemical communication is mainly based on various physical properties such as chemical nature, solubility volatility, and its lifetime in the environment. Also, the stability of such volatiles affects their efficiency in IPM programs [1]. Dispersal is a natural activity of insect where the movement is directed (taxes) or random (kineses) which is motivated by chemical or visual stimuli. There are three mechanisms of insect behavioral responses for finding an odor source. In the first mechanism called true chemotaxis, the insects align their body directly toward the odor source due to sensing the gradient of odor molecules. For the second mechanism, the insect does not discover the odor direction but becomes stimulated either for moving at different rates which is called orthokinesis or turning at various frequencies depending on changes in odor concentration (klinokinesis). The third mechanism depends on the odor of molecules impulse insect toward some other stimulus. Anemotaxis is the most common example for this mechanism where the molecules of an attractive chemical stimulate the receptive insects to fly upwind [26].

4. Chemosensory stimulation in insects

In insects, chemosensory stimulation occurs in various receptor organs via constant bombardment of chemical signals which improved the insect's ability to detect, discriminate, and distinguish innumerable different molecules as different odors. The insect receptor organs include antennae, mouthparts, and ovipositors. These receptors are very sensitive even for a few molecules of specific semiochemicals. Attraction (directed movement toward stimuli) and repulsion (directed movement away from stimuli) are the main insect responses to various odors. For field traps, insect catches not only occur via taxes but also via kineses (random movement). The insect can detect any odor by olfactory receptors located in the sensory organs including antennae, mouthparts, and ovipositors [27]. Various types of sensilla are recorded including trichodea, basiconica, styloconica, chaetica, etc. Knowledge of the types of sensilla on the antennae and mouthparts provides a foundation for understanding the olfaction and feeding preferences of herbivorous insect pests and subsequently can be useful for improving new control strategies for the target pests [28, 29]. The basic structure of sensillum is explored by [30] in **Figure 2**. The sensillum formed from the sensory neuron attached to branched cuticular pores (P) which allows odor passage. Sensillum pores act to filter molecules received from the airstream and concentrated it in the lumen of the sensillum and passed to branched neurons which convey impulses from and to the central nervous system.

In insects, the ability to discriminate different odors depends only on the evolutionary pressures of the molecules which stimulate the development of specific binding proteins (BPs) and specific receptor sites present on individual chemosensory neurons. This selectivity bestowed upon chemosensory neurons by the receptor types expressed represents one level of signal filtering in the insect's olfactory system. The olfaction mechanism in insects is summarized by [14] in **Figure 3**. In brief, a chemical signal crosses the sensillum lymph (SL) through a pore and then binds to highly specific binding proteins: pheromone binding proteins/

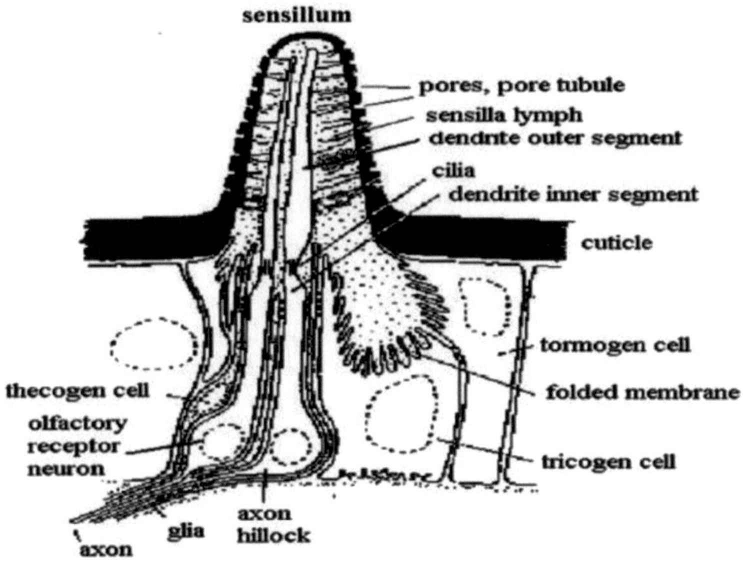

Figure 2.
Basic structure of sensillum [30].

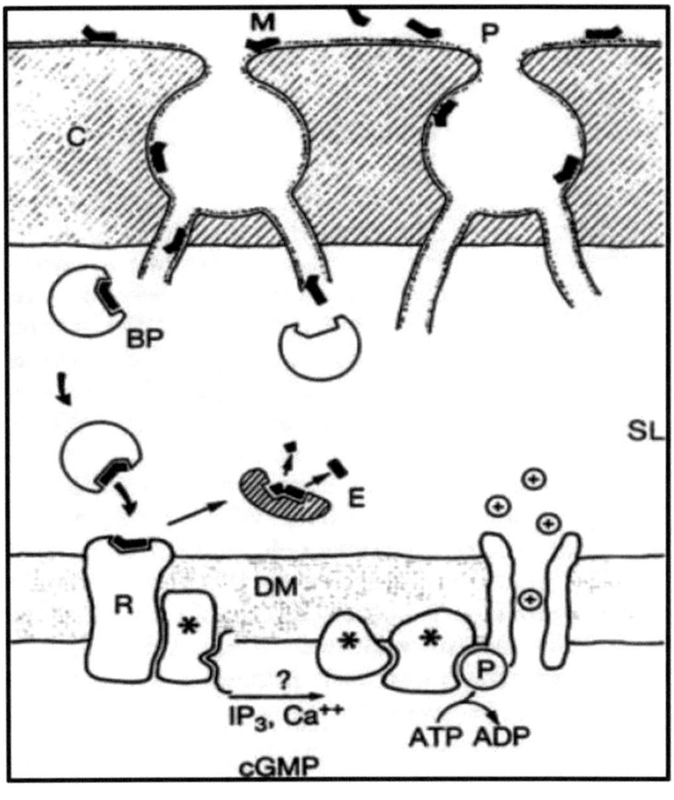

Figure 3.
Simplified schematic concept of perireceptor events in the insect's chemosensory sensilla. Absorbed stimulus molecules diffuse from the sensillum surface through pores in the cuticle (C) into the sensillum lymph. There, they are taken up by odorant- or pheromone-binding proteins and are transported through the aqueous lymph until they reach a specific receptor molecule (R) on the outer dendritic membrane (DM). This activates dendritic ion channels via membrane-bound proteins (*) and intracellular second messenger cascades such as cyclic guanosine monophosphate (cGMP), inositol trisphosphate (IP3), and Ca ions. Also, the stimulus molecule could degrade in the sensory lymphatic room by specific enzymes (E) into inactive metabolites so that it can no longer activate the receptor [35].

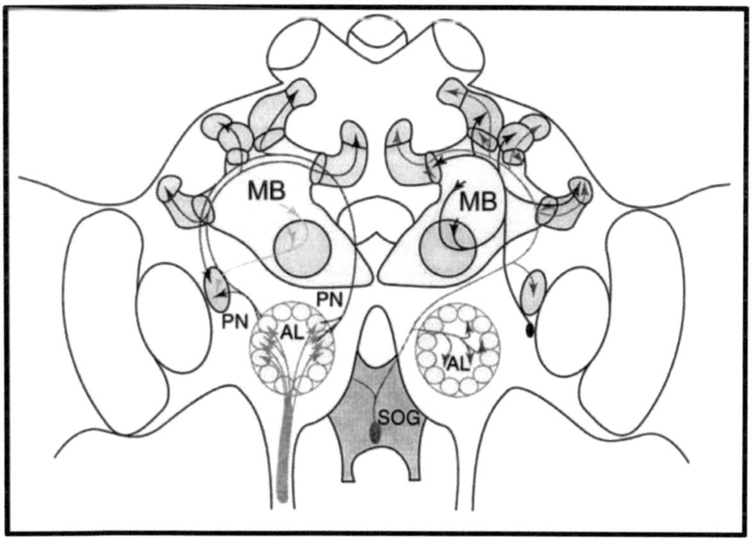

Figure 4.
Schematic view of the central brain area of the honeybee showing the antennal lobes with their specific glomeruli (small circles). From the AL projection neurons (PN) send olfactory information into the mushroom bodies. The MBs are higher-order integration centers of olfactory, visual, and mechanosensory information and are believed to play a role in the control of complex behaviors as well as learning and memory. SOG: Sub-esophageal ganglion [36].

odorant binding proteins (PBPs/OBPs). The signal-PBP/OBP-complex passes or is transported to the chemosensory neuron, where it binds to a specific olfactory receptor protein (OR or R) in the neuron membrane. These receptor proteins were identified in 1999 by [31, 32]. They all belong to the same "seven-transmembrane-domain" protein family; however, they differ between taxa a great deal [33]. From a molecular perspective, binding to the OR activates so-called G-proteins, which are also located in the neuron membrane and part of a phosphorylation-dependent energy exchange, triggering a cascade of signaling reactions. These eventually lead to electrical impulses being sent down from the axon of the neuron to the antennal lobe (AL) (**Figure 4**). The AL is structured into a number of neuron groups (glomeruli) that are innervated separately and only in response to specific individual odors or classes of chemically similar ones [34]. Filtering of these signals is accomplished after reaching the AL glomeruli depending on their quality, quantity, and temporal and spatial characteristics. From the AL, specific patterns of neural activity are processed to higher integrative centers of the brain, such as the mushroom bodies (MBs; **Figure 4**), which are believed to be involved in the control of complex behaviors.

5. Utilization of olfactory communication in IPM

Olfactory/chemical signals represent essential components in different insect management strategies including monitoring, mass trapping, luring and killing, mating disruption, and push-pull strategy (stimulo-deterrent diversion). Also, host plant volatiles play an important role in IPM strategies as the main olfactory response of insect pests for determination of food, mates, and/or oviposition and hibernation sites [1]. Host plant volatiles are often induced by different environmental factors. For instance, the feeding process of herbivore may increase emission of volatiles in plants; these volatiles are referred to as herbivore-induced plant volatiles that stimulate natural enemies to find their prey as illustrated in **Figure 5** [37].

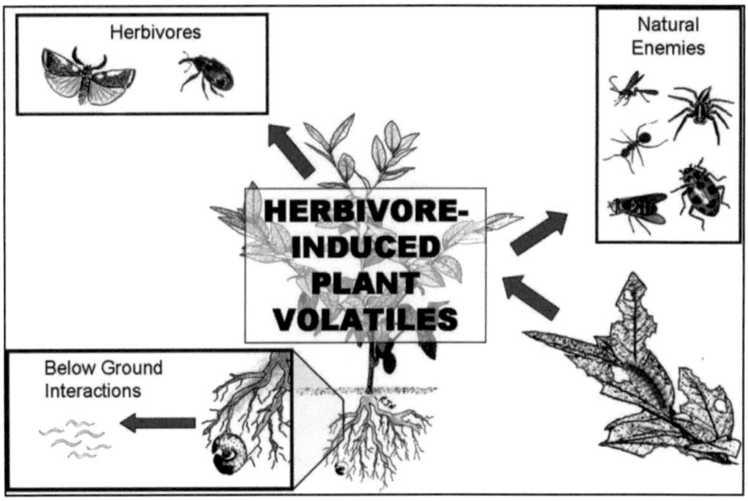

Figure 5.
Herbivore-induced volatile effects on herbivores and their natural enemies [37].

Moreover, isolation and identification of such molecules are essential for consideration as new substances involved in IPM programs.

Recently, the application of different semiochemicals has become an important category of integrated pest management. Various semiochemical compounds are widely applied not only for controlling insect pests [38–41] but also for conservation of rare and threatened insects [42]. Semiochemical substances provide prospective interest in IPM programs depending on the outcome advantages of using such substances. For instance, these substances are distinguished by high volatility that allows diffusion for long distances, application in low concentrations, and rapid dissipation that reduces health and environmental risks compared with chemical pesticides. The efficacy of such molecules mainly depends on their physical properties, i.e., molecular structure, volatility, solubility, and lifetime in the environment. Also, the environmental factors are an important parameter that affect the activity of semiochemical compounds. For example, temperature affects the stability of such compounds by increasing the diffusion of volatile compounds, leading to decreased molecule lifetime in the environment [1].

Control strategies of herbivorous insects are mainly based on semiochemicals which include monitoring, mass trapping, lure-and-kill (attract-annihilate), mating disruption, and push-pull strategy (stimulo-deterrent diversion) tactics. Pheromones are considered as a promising and important component in IPM programs. It can be applied singly or in integration with other control strategies in the agricultural system management for monitoring and controlling various insect pests [1]. The pheromone application is performed in two ways: indirect control and direct control strategies. The direct control involved mass trapping and area-wide dissemination which includes disruption, attractant, and attract-and-kill (lure-and-kill). However, the indirect control involves monitoring for quarantine and spray timing strategy. Pheromone traps are widely used commercially for different purposes in IPM strategies. For example, pheromone-baited traps are used as attract-and-kill or mating disruption techniques to prevent males from reproducing. Furthermore, pheromone can play an important role for detection of information about insect populations. It represents an overview for sex ratio and the mating status which are serious data for the detection of the population phase which is subject to cyclical changes in population density [43, 44]. Interestingly, strategies depending on pheromone application are useful for measuring the genetic diversity of insect pests. For example, the genetic diversity of the Asian long-horned beetle in Asia, North America, and Europe is reported to be based on pheromone traps [45].

6. Combination of chemical and other communication signals in IPM

Combinations of different communication signals are extremely more efficient in attracting insects than a single stimulus for controlling insect pests. The most successful strategies for insect management were recorded for a combination between different communication signals as visual (color, shape, or size) and olfactory stimuli [1]. Lure-and-kill strategy is an important and widespread tactic which used sticky materials to prevent captured insect from escaping and/or baited with insecticide. Also, combining an insecticide and/or a food stimulant can further enhance the efficacy of visual-depending traps for field applications. The chemical and visual stimuli that attract insects to their host plants have been incorporated into a wide range of insect traps that work better than using a single stimulus [46–50]. Many examples exist where visual stimuli enhance insect responses to semiochemical-based traps [51–53]. Using spheres with red color attractant coated with a non-drying adhesive combined with attractants with odors resembling ripening apples results in an excellent control of the apple maggot fly, *Rhagoletis pomonella* (Walsh) [47]. Also, the choice for suitable places for female mosquitoes to lay eggs is a key factor for the survival of immature stages (eggs and larvae). This knowledge stands out in importance concerning the control of disease vectors. The selection of a place for oviposition requires a set of chemical, visual, olfactory, and tactile cues that interact with the female before laying eggs, helping the localization of adequate sites for oviposition [54].

Author details

Nesreen M. Abd El-Ghany
Department of Pests and Plant Protection, Agricultural and Biological Division, National Research Centre, Dokki, Giza, Egypt

*Address all correspondence to: nesreennrc@gmail.com

References

[1] Abd El-Ghany NM. Semiochemicals for controlling insect pests. Journal of Plant Protection Research. 2019;**59**(1): 1-11. DOI: 10.24425/jppr.2019.126036

[2] Hassan EM, Shahat AA, Ibrahim NA, Vlietinck AJ, Apers S, Pieters L. A new monoterpene alkaloid and other constituents of *Plumeria acutifolia*. Planta Medica. 2008;**74**(14):1749-1750. DOI: 10.1055/s-0028-1088317

[3] Salib JY, El-Toumy SA, Hassan EM, Shafik NH, Abdel-Latif SM, Brouard I. New quinoline alkaloid from *Ruta graveolens* aerial parts and evaluation of the antifertility activity. Natural Product Research. 2014;**28**(17):1335-1342. DOI: 10.1080/14786419.2014.903395

[4] Hassan EM, Matloub AA, Aboutabl ME, Ibrahim NA, Mohamed SM. Assessment of anti-inflammatory, antinociceptive, immunomodulatory, and antioxidant activities of *Cajanus cajan* L. seeds cultivated in Egypt and its phytochemical composition. Pharmaceutical Biology. 2016;**54**(8):1380-1391. DOI: 10.3109/13880209.2015.10783 83

[5] Hassan RA, Hassan EM, Ibrahim NA, Nazif NM. Triterpenes and cytotoxic activity of *Acokanthera oblongifolia* Hochst. growing in Egypt. Research Journal of Pharmaceutical, Biological and Chemical Sciences. 2015;**6**(1):1677-1686

[6] Miller JR, Strickler KL. Plant herbivore relationships: Finding and accepting host plants. In: Bell WJ, Cardé RT, editors. Chemical Ecology of Insects. Boston, MA: Springer; 1984. pp. 127-157. DOI: 10.1007/978-1-4899-3368-3_6

[7] Nordlund DA, Lewis WJ. Terminology of chemical releasing stimuli in intraspecific and interspecific interactions. Journal of Chemical Ecology. 1976;**2**(2):211-220. DOI: 10.1007/BF00987744

[8] Dicke M, Sabelis MW. Infochemical terminology: Based on cost-benefit analysis rather than origin of compounds? Functional Ecology. 1988;**2**(2):131-139. DOI: 10.2307/2389687

[9] Reddy GVP, Guerrero A. Interactions of insect pheromones and plant semiochemicals. Trends in Plant Science. 2004;**9**(5):253-261. DOI: 10.1016/j.tplants.2004.03.009

[10] Dressler RL. Biology of the orchid bees (Euglossini). Annual Review of Ecology and Systematics. 1982;**13**:373-394. DOI: 10.1146/annurev.es.13.110182.002105

[11] Nishida R. Sequestration of defensive substances from plants by Lepidoptera. Annual Review of Entomology. 2002;**47**:57-92. DOI: 10.1146/annurev.ent.47.091201.145121

[12] Karlson P, Butenandt A. Pheromones (Ectohormones) in insects. Annual Review of Entomology. 1959;**4**:39-58. DOI: 10.1146/annurev.en.04.010159.000351

[13] Wilson EO, Bossert WH. Chemical communication among animals. Recent Progress in Hormone Research. 1963;**19**:673-716. DOI: 10.1016/B978-0-08-012210-6.50074-0

[14] Wyatt TD. Pheromones and Animal Behaviour: Communication by Smell and Taste. Cambridge, UK: Cambridge University Press; 2003. p. 391. DOI: 10.1017/CBO9780511615061

[15] Pasteels JM, Bordereau C. Releaser pheromones in termites. In: Vander Meer RK, Breed MD,

Winston M, Espelie KE, editors. Pheromone Communication in Social Insects: Ants, Wasps, Bees, and Termites. Colorado, USA: Westview Press; 1998. pp. 193-215

[16] Winston ML. The Biology of the Honey Bee. Cambridge, Massachusetts, USA: Harvard University Press; 1987. p. 294. DOI: 10.1016/S0003-3472(88)80187-8

[17] Schmidt JO. Mass action in honey bees: Alarm, swarming and the role of releaser pheromones. In: Vander Meer RK, Breed MD, Winston M, Espelie KE, editors. Pheromone Communication in Social Insects: Ants, Wasps, Bees, and Termites. Colorado, USA: Westview Press; 1998. pp. 257-290

[18] Quintana A, Reinhard J, Faure R, Uva P, Bagneres AG, Massiot G, et al. Interspecific variation in terpenoid composition of defensive secretions of European *Reticulitermes* termites. Journal of Chemical Ecology. 2003;**29**(3):639-652. DOI: 10.1023/A:1022868603108

[19] Greenfield MD. Signalers and Receivers: Mechanisms and Evolution of Arthropod Communication. Oxford, UK: Oxford University Press; 2002. p. 432. Available from: https://www.semanticscholar.org/paper/Signalers-and-Receivers%3A-Mechanisms-and-Evolution-Greenfield/e5cde24093d07b2799f531d46a86f013cb45cb8b

[20] Vilela EF, Della Lucia MT. Insect Pheromones: Biology, Chemistry and Employment in Pest Management. 2nd ed. Ribeirão Preto, Brazil: Holos Publishing House; 2001. p. 206

[21] Poland TM, Borden JH. Attraction of a bark beetle predator, *Thanasimus undatulus* (Coleoptera: Cleridae), to pheromones of the spruce beetle and two secondary bark beetles (Coleoptera: Scolytidae). Journal of the Entomological Society of British Columbia. 1997;**94**(1):35-41. Available from: https://www.ncrs.fs.fed.us/pubs/jrnl/1997/nc_1997_Poland_001.pdf

[22] Turlings TC, Tumlinson JH, Lewis WJ. Exploitation of herbivore-induced plant odors by host-seeking parasitic wasps. Science. 1990;**250**(4985):1251-1253. DOI: 10.1126/science.250.4985.1251

[23] Evenden ML, Judd GJR, Borden JH. A synomone imparting distinct sex pheromone communication channels for *Choristoneura rosaceana* (Harris) and *Pandemis limitata* (Robinson) (Lepidoptera: Tortricidae). Chemoecology. 1999;**9**(2):73-80. DOI: 10.1007/s000490050036

[24] Reinhard J, Lacey MJ, Ibarra F, Schroeder FC, Kaib M, Lenz M. Hydroquinone: A general phago stimulating pheromone in termites. Journal of Chemical Ecology. 2002;**28**(1):1-14. DOI: 10.1023/A:1013554100310

[25] Dougherty MJ, Guerin P, Ward RD. Identification of oviposition attractants for the sandfly *Lutzomyia longipalpis* (Diptera: Psychodidae) present in volatiles of faeces from vertebrates. Physiological Entomology. 1995;**20**(1):23-32. DOI: 10.1111/j.1365-3032.1995.tb00797.x

[26] Shorey HH. Behavioural response to insect pheromones. Annual Review of Entomology. 1973;**18**:349-380. DOI: 10.1146/annurev.en.18.010173.002025

[27] Panda N, Kush GS. Host Plant Resistant to Insect. Wallingford, UK: CAB International; 1995. p. 448. Available from: https://www.cabdirect.org/cabdirect/abstract/19951111984

[28] Abd El-Ghany NM, Abd El-Aziz SE. External morphology of antennae and mouthpart Sensillae

of the granary weevil (Coleoptera: Curculionidae). Journal of Entomological Science. 2017;**52**(1): 29-38. DOI: 10.18474/JES16-19.1

[29] Abd El-Aziz SE, Abd El-Ghany NM. Impact of diatomaceous earth modifications for controlling the granary weevil, *Sitophilus granarius* (L.), (Coleoptera: Curculionidae). Journal of Agricultural Science and Technology. 2018;**20**(3):519-531

[30] Kaissling KE, Thorson J. Insect olfactory sensilla: Structural chemical and electrical aspects of the functional organization. In: Satelle DB, Hall LM, Hildebrand JG, editors. Receptors for Neurotransmitters, Hormones, and Pheromones in Insects. Amsterdam: Elsevier/North-Holland Biomedical Press; 1980. pp. 261-282. Available from: http://hdl.handle.net/11858/00-001M-0000-0029-7443-D

[31] Clyne PJ, Warr CG, Freeman MR, Lessing D, Kim JH, Carlso JR. A novel family of divergent seven-transmembrane proteins: Candidate odorant receptors in *Drosophila*. Neuron. 1999;**22**(2):327-338. DOI: 10.1016/s0896-6273(00)81093-4

[32] Vosshall LB, Amrein H, Morozov PS, Rzhetsky A, Axel R. A spatial map of olfactory receptor expression in the *Drosophila* antenna. Cell. 1999;**96**(5):725-736. DOI: 10.1016/S0092-8674(00)80582-6

[33] Pilpel Y, Lancet D. Good reception in fruitfly antennae. Nature. 1999;**398**(6725):285-287. DOI: 10.1038/18552

[34] Hansson BS. Insect Olfaction. Berlin, Germany: Springer Verlag; 1999. p. 377. DOI: 10.1007/978-3-662-07911-9

[35] Kaib M. Chemoreception. In: Dettner K, Peters W, editors. EntomologyTextbook (Chemoreception. In: Lehrbuch der Entomologie). Stuttgart, Germany: Gustav Fischer Verlag; 1999. pp. 302-320. Available from: https://epdf.pub/lehrbuch-der-entomologie.html. DOI: 10.1007/978-3-8274-2618-5

[36] Menzel R, Giurfa M. Cognitive architecture of a mini-brain: The honeybee. Trends in Cognitive Sciences. 2001;**5**(2):62-71. DOI: 10.1016/s1364-6613(00)01601-6

[37] Karban R, Baldwin IT. Induced Responses to Herbivory. Chicago, USA: University of Chicago Press; 1997. p. 319. Available from: https://www.press.uchicago.edu/ucp/books/book/chicago/I/bo3644508.html

[38] Weinzierl R, Henn T, Koehler PG, Tucker CL. Insect Attractants and Traps. USA: University of Florida IFAS Extension; 2005. p. 9. Available from: http://ufdc.ufl.edu/IR00002794/00001

[39] Cook SM, Khan ZR, Pickett JA. The use of push-pull strategies in integrated pest management. Annual Review of Entomology. 2007;**52**: 375-400. DOI: 10.1146/annurev.ento.52.110405.091407

[40] Stelinski LL. On the physiological and behavioral mechanisms of pheromone-based mating disruption. Pestycydy. 2007;**3**(4):27-32

[41] Heuskin S, Godin B, Leroy P, Capella Q, Wathelet JP, Verheggen F, et al. Fast gas chromatography characterization of purified semiochemicals from essential oils of *Matricaria chamomilla* L. (Asteraceae) and Nepetacataria L. (Lamiaceae). Journal of Chromatography A. 2009;**1216**(14):2768-2775. DOI: 10.1016/j.chroma.2008.09.109

[42] Larsson MC. Pheromones and other semiochemicals for monitoring rare and endangered species. Journal of Chemical Ecology.

2016;**42**(9):853-868. DOI: 10.1007/s10886-016-0753-4

[43] Borden JH, Pureswaran DS, Lafontaine JP. Synergistic blends of monoterpenes for aggregation pheromones of the mountain pine beetle (Coleoptera: Curculionidae). Journal of Economic Entomology. 2008;**101**(4):1266-1275

[44] Jones BC, Evenden ML. Ecological applications for pheromone trapping of *Malacosoma disstria* and *Choristoneura conflictana*. Canadian Entomologist. 2008;**140**(5):573-581. DOI: 10.4039/n08-013

[45] Carter ME, Smith MT, Turgeon JJ, Harrison RG. Analysis of genetic diversity in an invasive population of Asian long-horned beetles in Ontario, Canada. Canadian Entomologist. 2009;**141**(6):582-594. DOI: 10.4039/n09-026

[46] Finch S, Skinner G. Some factors affecting the efficiency of water-traps for capturing cabbage root flies. Annals of Applied Biology. 1974;**77**:213-226. DOI: 10.1111/j.1744-7348.1974.tb01398.x

[47] Prokopy RJ. Attraction of Rhagoletis flies (Diptera: Tephritidae) to red spheres of different sizes. Canadian Entomologist. 1977;**109**(9):593-596. DOI: 10.4039/Ent109593-4

[48] Abd El-Ghany NM, Abdel-Wahab ES, Ibrahim SS. Population fluctuation and evaluation the efficacy of pheromone-based traps with different color on tomato leafminer moth, *Tuta absoluta* (Lepidoptera: Gelechiidae) in Egypt. Research Journal of Pharmaceutical, Biological and Chemical Sciences. 2016;7(4):1533-1539. Available from: https://www.rjpbcs.com/pdf/20167(4)/[199].pdf

[49] Abdel-Razek AS, Abd El-Ghany NM, Djelouah K, Moussa A. An evaluation of some eco-friendly biopesticides against *Bemisia tabaci* on two greenhouse tomato varieties in Egypt. Journal of Plant Protection Research. 2017;**57**(1):9-17. DOI: 10.1515/jppr-2017-0002

[50] Abd El-Ghany NM, Abdel-Razek AS, Djelouah K, Moussa A. Efficacy of some eco-friendly biopesticides against *Tuta absoluta* (Meyrick). Bioscience Research. 2018;**15**(1):28-40. Available from: https://www.researchgate.net/publication/323120096

[51] Andrei VA, Russell HM, Jian JD. Visual and olfactory stimuli and fruit maturity affect trap captures of oriental fruit flies (Diptera: Tephritidae). Journal of Economic Entomology. 2000;**93**(3):644-649. DOI: 10.1603/0022-0493-93.3.644

[52] Blackmer JL, Cañas LA. Visual cues enhance the response of *Lygus Hesperus* (Heteroptera: Miridae) to volatiles from host plants. Environmental Entomology. 2005;**34**(6):1524-1533. DOI: 10.1603/0046-225x-34.6.1524

[53] Kendrick AP, Raffa KF. Sources of insect and plant volatiles attractive to cottonwood leaf beetles feeding on hybrid poplar. Journal of Chemical Ecology. 2006;**32**(12):2585-2594. DOI: 10.1007/s10886-006-9184-y

[54] Navarro-Silva MA, Marques FA, Duque JE. Review of semiochemicals that mediate the oviposition of mosquitoes: A possible sustainable tool for the control and monitoring of Culicidae. Revista Brasileira de Entomologia. 2009;**53**(1):1-6. DOI: 10.1590/s0085-56262009000100000

Chapter 3

Mandibular Structure, Gut Contents Analysis and Feeding Group of Orthopteran Species Collected from Different Habitats of Satoyama Area within Kanazawa City, Japan

Wael M. ElSayed, Shahenda Abu ElEla and Koji Nakamura

Abstract

A study was conducted on assemblies of various orthopteran species from distinct habitats in the Satoyama region, Kanazawa City, Ishikawa Prefecture, Japan, and a total of 50 distinct orthopteran species were registered. These species were represented by 10 families and were belonged to 17 subfamilies and 27 tribes. Results based on stereo-microscopic examination of the mandibular morphology and the analysis of gut contents suggested seven proposed feeding groups for these collected orthopteran species. Among the examined subfamilies, family Tettigoniidae proved to be the most diverse in mandibular structure and four feeding groups were assigned. This was followed by family Acrididae, which showed three feeding groups. Other families contained only single feeding group. It was noted that only five species, from family Acrididae, were graminivorous with their mandibles characterized by comparatively very short incisors and relatively wide molar regions. The analysis of gut contents of these five species proved to contain more than 80% monocotyledonous plant species. Predation and scavenging as feeding habits were also recorded in some orthopteran species.

Keywords: orthoptera, herbivorous, mandibular structure, gut contents analysis, feeding guild

1. Introduction

The strong relationship with diet makes mouthpart morphology an important trait for insect evolutionary biologists [1, 2] and systematists [3]. Isley was one of the first to study the structure and morphology of mouthparts in details and correlate morphological characteristics with various feeding habits [4].

In general, Isely defined three groups of mandibles according to the overall structure and distinctive diet: (i) graminivorous (grass feeding type) with grinding molar and incisors typically merged into a scythe like edge, (ii) forbivorous (forb or broadleaf plant feeding type) having a molar region composed of a depression encircled by elevated teeth and a strong interlocking incisor, (iii) herbivorous (mixed feeding type) that have features of both of the previously mentioned groups. Isely's initial results on mandible group [4] have since been shown to be prevalent in grasshoppers and other entomological taxa. Many authors have performed further thorough research in different locations, important among them.

Isley's initial results on mandible groups [4] have since been shown to be prevalent in grasshoppers and other insect taxa. Further itemized investigations have been directed by numerous authors in various regions, significant among them were Snodgrass [1], Gangwere [5, 6], Gangwere et al. [7] and Patterson in North America [8]; Liebermann; Gangwere and Ronderos in South America [9, 10]; Williams; Kaufmann; Gangwere and Morales [11–13] in Europe; Gangwere and Spiller; Gangwere et al. in the Mediterranean islands [14, 15]; Feroz and Chaudhry; Gapud; Kang et al. and Le Gall et al. in Asia [16–19]; Chapman in Africa [20].

A general scheme for explicating the diet of a given insect species could be started with a prudent inspections of their mandibular morphology [3, 8]. Specifically, the morphological characters of mandibles, incisors and molar surfaces are helpful tools for identifying particular species as either grass feeder or forb feeder [18, 20, 21]. Although most species with forb feeding mandibles feed on a mixture of grasses and forbs; determining an insect's diet should be followed by analyzing the gut contents for further confirmations [22, 23].

Although orthopteran species have often been regarded as polyphagous herbivores; most of these species, particularly grasshoppers, are regarded selective in their diet to some degree of selectivity, demonstrating particular food preferences [3, 24]. Occasionally, grasshoppers with forb-feeding mandibles may regularly feed on grasses or vice versa [20].

Nevertheless, there are some values in evaluating the structure of mouthparts relative to anticipating both the diet selectivity and preference to specific habitat, particularly for many rare or non-economic species that are unlikely to be studied in details. Information pertaining the feeding habits and mouthparts of different orthopteran species co-occurring in diverse habitats located in Satoyama area, Kanazawa City, Ishikawa Prefecture, Japan are fragmentary [25, 26] and there is a shortage of knowledge concerning the mandibular morphology of many orthopteran species inhabiting Satoyama. Thus, there is an urge to study the morphological characteristics and structural adaptations of the mandibles of orthopteran species co-occurring in Satoyama.

2. Materials and methods

2.1 Study area

Survey of orthopteran assemblies was achieved among four sampling sites located in Satoyama area in Kanazawa City, Ishikawa Prefecture, Japan. Satoyama is a region occupying ~74 ha, at an altitude of 150 m and positioned 5 km southwest from the center of Kanazawa City. It includes various habitats ranging from secondary forests occupied predominantly by *Quercus serrata* (Japanese vernacular name is known as Konara), *Q. variabilis* (Abemaki), *Phyllostachys pubescens* (Moso-chiku) and *Cryptomeria japonica* (Sugi) to *Cynodon dactylon* (Gyougi—shiba) and various artificial ecosystems, such as ponds, paddy lands and ordinary and reclaimed farmlands.

2.2 Protocols of sampling and collection

A standard entomological sweep net sampling technique was used for sampling and collecting different orthopteran species from the diverse habitats within

Satoyama area during different seasons of the year. The time for sampling was adjusted between 1000 and 1400 h of the day. Collected specimens were promptly killed in the field and a one liter-capacity containers including a cotton piece soaked with 70% ethanol was used for preserving the collected specimens. These containers were tightly secured with a rubber plug. The collected species were later identified, counted, sorted and kept in individually marked clean glass vials in the laboratory. These vials could be stored in freezer for a year with no apparent damage or impairment to the preserved specimens [22, 24, 27–29].

The collected Orthoptera were identified to species level by following the taxonomic key of Ichikawa et al. [25]. Furthermore, the collected species were also compared with identified museum specimens in Kanazawa University repository for further affirmation.

2.3 Mandibular structure

Mandibles were deliberately and precisely removed from the specimens by lifting the labrum and hauling out each mandible independently with the guide of fine forceps.

For easier manipulation for examination and photographing; the air-dried mandibles were pasted to the head of a #3 or #2 insect pin (depending on mandibular size). The mandibular morphological characters and apparent structure (for both ventral and dorsal sides) of 50 species of orthoptera from 10 families (Acrididae, Eneopteridae, Gryllidae, Mecopodidae, Phaneropteridae, Pyrgomorphidae, Mantidae, Tetrigidae, Tettigoniidae and Trigonididae) were examined under a stereo-fluorescence microscope (Nikon® SMZ800 series) equipped with digital camera for taking photos and TFT LCD Nikon® monitor for easier inspections.

Digital photographs were taken by Nikon® digital camera and these photos were montaged by using the Auto-Montage Syncroscopy scheme facilitated by the Laboratory of Biodiversity at Kanazawa University. For simplifying the categorization of the orthopteran species into proper feeding groups; we adopted the descriptions of mandible types declared by Isley [4]. The examined species were categorized into seven major groups: Forbivorous (referred subsequently as F), Herbivorous or Mixed-feeders (H), Graminivorous (G), Scavengers (S), Herbivorous with observed scavenging behavior (H_S), Forbivorous with scavenging behavior (F_s) and Predators (P). Detailed explanations of these groups are given hereafter in the methodology.

2.4 Field cages experiment

Live different orthopteran species were held and reared under natural environmental conditions in proper wooden cages and were supplied with almost all accessible plant species collected from the sampling field to minimize the hunger effect. Continuous inspections on feeding behavior were performed for continuous 3 h in three replicate field cages in each season of the year. Obtained results from field cages experiments were compared to those acquired from mandibular examination.

2.5 Gut contents analysis and feeding groups

In an attempt to glean the feeding category of each orthopteran species; gut contents analysis was carried out in concordance with mandibular structure examination.

The proportions of four main categories (monocotyledonous plant species, dicotyledonous species, orthopteran or animal parts, and scavenging observations in caged species) were calculated and considered to classify each of the orthopteran species into one of the suggested seven feeding categories:

1. Herbivorous (H): in which the number of fragments of dicotyledonous plant is almost equal to the number of fragments of monocotyledonous species.

2. Herbivorous with scavenging behavior (H_s): the same as herbivorous group with some scavenging actions were recorded in laboratory caged species.

3. Graminivorous (G): the number of fragments of monocotyledonous species is more than 75% of the gut contents.

4. Forbivorous (F): the number of fragments of dicotyledonous plant species is more than 75% of the gut contents.

5. Forbivorous with scavenging behavior (F_s): the same as forbivorous group with some scavenging actions were recorded in laboratory caged species.

6. Scavengers (S): plants species (especially roots or tubers) and dead orthopteran and/or oligochaeta parts were encountered in almost equal proportions.

7. Predators (P): all contents of the gut were insect or other orthopteran body parts with no occurrence of plant fragments.

Collected orthopteran specimens were deposited in a catalogued repository in Kanazawa University in special boxes containing small naphthalene coated sachets for further specimen protection against destructive pests.

3. Results and discussion

A total of 50 orthopteran species, belonging to 10 families representing 17 subfamilies and 27 tribes, were collected from different habitats of Satoyama. These species were sampled from different habitats including open grasslands, forest margins, ponds and paddy fields. The stereo-microscopic examination of these 50 orthopteran species revealed that these species belonged to seven major feeding groups (**Table 1**). Among the examined subfamilies, family Tettigoniidae proved to be the most diverse in mandibular type with four feeding groups could be observed. This was proceeded by family Acrididae which possessed three feeding groups. Other families were observed to contain only sole feeding group (**Table 2**).

Species from the family Acrididae, short-horned grasshoppers, and family Tettigoniidae, long-horned grasshoppers, can occur in a diverse of habitats, usually in dense vegetation like open grasslands or around paddy fields or even pond localities. Species belong to these two families were found to be quite active in both walking and flying. It is interesting to note that species with graminivorous type mandibles, were characterized by extremely slender and elongated bodies and were encountered on the edges of ponds. This was in accordance with the findings of other authors [4, 30, 31]. These species typically grasp the stems of emergent grass or grass-like vegetation such as sedges or cattails, blending in almost perfectly. On the other hand, collected species from family Oedipodinae were split into three mandibular types: graminivorous, forbivorous and herbivorous [32]. This signifies a more grass-dominated diet in their feeding behavior.

Family	Subfamily	Tribe	Acridid species	Feeding group*
Acrididae	Acridinae	Acridini	*Acrida cinerea*	G
		Parapleurini	*Stethophyma magister*	G
	Melanoplinae	Podismini	*Parapodisma Mikado*	F
	Oedipodinae	Aiolopini	*Aiolopus thalassinus tumulus*	H
		Oedopodini	*Sphingonotus japonicus*	F
		Locustini	*Oedaleus infernalis*	G
		Trilophidiini	*Trilophidia annulata*	G
	Oxyinae	Oxyini	*Oxya yezoensis*	G
Eneopteridae	Oecanthinae	Oecanthini	*Oecanthus simulator ichikawa*	F
Gryllidae	Gryllinae	Gryllini	*Acheta domesticus*	S
			Loxoblemmus equestris	S
			Loxoblemmus sylvestris	S
			Loxoblemmus tsushimensis ichikawa	S
			Stethophyma magister	S
			Teleogryllus occipitalis	S
			Teleogryllus emma	S
			Velarifictorus asperses	S
			Velarifictorus Mikado	S
			Velarifictorus ornatus	S
		Modicogryllini	*Modicogryllus siamensis*	S
	Sclerogryllinae	Sclerogryllini	*Sclerogryllus punctatus*	S
Mecopodidae	Mecopodinae	Mecopodini	*Mecopoda niponensis*	F_s
Phaneropteridae	Phaneropterinae	Ducetini	*Ducetia japonica*	F_s
		Phaneropterini	*Phaneroptera falcate*	F_s
			Phaneroptera nigroantennata	F_s
Pyrgomorphidae	Pyrgomorphinae	Atractomorphini	*Atractomorpha lata*	F
Mantidae	Mantinae	Mantini	*Tenodera angustipennis*	P
			Tenodera aridifolia	P
Tetrigidae	Scelimeninae	Criotettigini	*Criotettix japonicas*	F
	Tetrigidae	Tetrigini	*Euparatettix tricarinatus*	F
			Tetrix japonica	F
			Tetrix macilenta	F
			Tetrix minor ichikawa	F
			Tetrix nikkoensis	F
			Tetrix silvicultrix ichikawa	F

Family	Subfamily	Tribe	Acridid species	Feeding group*
Tettigoniidae	Conocephalinae	Conocephalini	*Conocephalus japonica*	F_S
			Conocephalus melaenus	F
		Copiphorini	*Euconocephalus varius*	F
			Ruspolia dubia	F_S
	Tettigoniinae	Decticini	*Chizuella bonneti*	F_S
			Eobiana gradiella ishikawa	H
			Eobiana engelhardti subtropica	F_S
		Gampsocleidini	*Gampsocleis Mikado*	H_S
		Hexacentrinae	*Hexacentrus japonicas*	H_S
	Tettigoniinae	Tettigoniini	*Tettigonia orientalis*	F
			Tettigonia sp. 6**	F
			Tettigonia sp. 8	F
Trigonididae	Nemobiinae	Pteronemobiini	*Dianemobius furumagiensis*	S
			Pteronemobius fascipes	S
	Trigonidinae	Trigonidini	*Trigonidium pallipes*	S

**F, Forbivorous (Forb-feeder); H, Herbivorous (Mixed-feeder); G, Graminivorous (Grass-feeder); S, Scavengers; H_S, Herbivorous with observed scavenging behavior; F_S, Forbivorous with observed scavenging behavior; P, Predator; **Species 6 and 8 according to Ichikawa et al. [25].*

Table 1.
Check-list of orthopteran species inhabiting different habitats of Satoyama area with their family, subfamily, tribe and feeding group.

However, these species are much more divergent in their feeding category in which some may be completely graminivorous or forbivorous. Most of the species were found on the ground in open areas or on bare soil, rarely on plant species or grasses. As a general finding, the greatest diversity in mandibular structure was observed in subfamily Oedipodinae in comparison with other orthopteran subfamilies [32]. Equitably even distribution of the three mouthpart types in this group were also recorded [4, 6, 18].

Mantidae, on the other hand, where observed to be represented by only two *Tenodera* species (*Tenodera angustipennis* and *T. aridifolia*). Concerning their feeding habit, these two species were completely predacious. Their mandibles were characterized by sharp incisor points used to capture and pierce the captured prey, and relatively long terebral ridge used to kill and slice prey into small pieces. Results from the analysis of gut contents of these mantid species revealed fragments of chitinous arthropod exoskeleton and other body parts including parts from antennae, wings or even legs affirming their zoophagous feeding behavior.

It was interesting to observe that tenth of the collected orthopteran species, five species out of the 50 species, were graminivorous, all were from the family Acrididae. These five species were characterized by very short incisors and relatively wide molar region in their mandibles. The molar area of some individuals of *Oxya yezoensis* (as one of these five species) showed a severe erosion in the molar region. It has to be mentioned that, feeding on grasses could be one avenue by which some orthopteran species may avoid toxic chemicals [22, 23, 33].

Family	Number			
	Subfamilies	**Tribes**	**Observed species**	**Feeding group**
Acrididae	4	8	8	3
Eneopteridae	1	1	1	1
Gryllidae	2	3	12	1
Mecopodidae	1	1	1	1
Phaneropteridae	1	2	3	1
Pyrgomorphidae	1	1	1	1
Mantidae	1	1	2	1
Tetrigidae	2	2	7	1
Tettigoniidae	2	6	12	4
Trigonididae	2	2	3	1
Total	17	27	50	7

Table 2.
Number of families, subfamilies, tribes, species and feeding group of orthopteran species co-occurring in different habitats of Satoyama area.

In this process, little or no energy, or other resources, would need to be spent on the detoxification process [34]. Contents of the gut of these five graminivorous species contained silica particles in relatively minor amounts. It is assumed that these silica particles could be ingested accidentally during feeding regime and thus accelerating the erosion of the molar area especially in old individuals due to severe and continuous friction [22, 23].

The analysis of gut contents of these five species revealed that the contents contained more than 80% monocotyledonous plant species. Controversially, *Acrida cinerea* as a graminivorous species, less than 12% of dicotyledons plant species were also encountered in their guts. Some authors in their field and laboratory works on a related acridid, *Acrida pellucida*, observed that this species may select non-graminous plants (dicotyledons) for enhancing the reproductive potential since these dicotyledon species showed a pronounced effects on both fecundity and developmental rates in laboratory rearing and food-choice tests [19, 22, 23, 35–37]. It could be assumed that the acridid, *Acrida cinerea*, may exploit some dicotyledonous plant species for augmenting specified biological and physiological processes. In this study, the acridid species belonging to the subfamily Acridinae are typically considered to be grass-feeders, displaying the classic graminivorous type mandibles [4, 20, 22].

Family Gryllidae was typically represented by 12 gryllidae species. These species showed mandibles with sharp incisors and comparatively long knife-shape terebral ridge. These mandibular modifications could delineate a predacious feeding habit. However, the gut content analysis revealed that parts from plant roots, tubers or even debris (38%) and subterranean arthropod species including amphipod and isopod species (62%) were collected from their guts. Consequently, the feeding group of these 12 gryllidae species could be confined to the scavenging habit.

Examination of Tetrigidae, represented by seven species, revealed that these species were mainly forbivorous (F_m). Their mandibles were characterized by pointed and sharp incisor points while their molar region was relatively small. The contents of the gut of these species contained dicotyledonous plan species without any presence of monocotyledonous ones.

Due to merely one representative species from the three subfamilies Eneopteridae, Mecopodidae and Pyrgomorphidae; determination of the mandibular structure of these families was relatively limited (**Table 1**). However, the supposed major mandible type and in turn the feeding group was mostly confined to the forbivorous type (F) where more dicotyledonous plants with nearly 79% dominance were consumed in much greater amount than monocotyledonous species (21%) as emphasized by gut contents analysis.

The determination of the mandibular structure of three subfamilies, Eneopteridae, Mecopodidae and Pyrgomorphidae, was comparatively restricted due to the fact that only one representative species from each of the three subfamilies could be collected (**Table 1**). However, the main mandibular type and in turn the feeding category was mostly restricted to the forbivorous group where more dicotyledonous species (79%) were devoured in relatively greater quantities than monocotyledonous ones (21%) as verified by the assessment of the analysis of gut contents.

At family level, it is noted that family Tettigoniidae with 12 species, was the most diverse family in both mandibular type and feeding group. It was obvious that four different feeding groups could be detected in Tettigoniidae. This was proceeded by family Acrididae in term of feeding group. Acrididae which harbored eight species had displayed three distinctive feeding groups as indicated in **Table 2**. Diversely, other families possessed only a single feeding group irrespective to the number of species (**Table 2**). Moreover, results perceived from **Table 2** showed that both family Gryllidae and family Tetrigidae (12 and 7 species, respectively) retained only one type of mandible and a single feeding group.

In all cases, a range of food of plant and/or animal origin was used in their diet, even though some were used infrequently. Thus these orthopteran species inhabiting different habitats in Satoyama area could be considered polyphagous species.

Cates [38] depicted the degree of diet specialization into the following three criteria: (1) monophagy: where one or more species within a genus; (2) oligophagy: two or more tightly associated genera; and (3) polyphagy: two or more plant families. In fact, none of the orthopteran species regarded in this research can be considered as either monophagic or oligophagic species. A variety of plant and/or animal foods were consumed in their diet in all instances, although some were rarely devoured. Thus, these species of orthopteran co-occurring in the diverse habitats of Satoyama area could be regarded as polyphagic species.

4. Conclusion

In conclusion, it is far from accuracy to roughly connect between mouthparts morphology and diet. Some authors like Mulkern was persuaded that only the grossest associations between mandibular structure and their diet regime (i.e., graminivorous, forbivorous, and herbivorous) could be made [3]. Some orthopteran species, especially grasshoppers with forb-feeding adapted mandibles, occasionally feed on grasses on a regular basis or vice versa [20, 22]. Nevertheless, the evaluation of mouthpart structure and morphology as a predictive avenue in determining diet and habitat preference in orthopteran species has some importance, particularly for the rare or non-economic species that are unlikely to be studied in details. Thus, the analysis of gut contents in parallel with laboratory examinations and precise observations on feeding behavior could be used as confirmation cues for the discovery of the mandibular structural adaptations. This would solve some hidden aspects that could not be deduced from the morphological characters of the mandibles if they were adopted alone.

Acknowledgements

The authors wish to acknowledge the staff at the Laboratory of Ecology, Graduate School of Natural Science and Technology, Kanazawa University for the laboratory facilities provided. Also, sincere gratitude is extended to Prof. Dr. Nakamura Koji (Kanazawa University, Japan) for his keen hospitality and encourages for completing the research and writing the manuscript.

Conflict of interest

The authors declare that there are no conflicts of interest associated with this article.

Author details

Wael M. ElSayed[1], Shahenda Abu ElEla[1,2] and Koji Nakamura[1,3]*

1 Department of Entomology, Faculty of Science, Cairo University, Giza, Egypt

2 Graduate School of Natural Science and Technology, Kanazawa University, Kakuma Campus, Kanazawa, Japan

3 Division of Biodiversity, Institute of Nature and Environmental Technology, Kanazawa University, Kakuma Campus, Kanazawa, Japan

*Address all correspondence to: koji@kenroku.kanazawa-u.ac.jp

References

[1] Snodgrass RE. Morphology and evolution of the insect head and its appendages. Smithsonian Miscellaneous Collection. 1928;**81**:1-158

[2] Brues CT. Food, drink, and evolution. Science. 1939;**90**:145-149

[3] Mulkern GB. Food selection by grasshoppers. Annual Review of Entomology. 1967;**12**:59-78

[4] Isley FB. Correlation between mandibular morphology and food specificity in grasshoppers. Annals of the Entomological Society of America. 1944;**37**:47-67

[5] Gangwere SK. Food selection in the Oediponidae grasshopper *Arphia sulphurea*. American Midland Naturalist. 1965;**74**:67-75

[6] Gangwere SK. Relationships between mandibles, feeding behavior and damage inflicted on plants by the feeding of certain acridids (Orthoptera). Michigan Entomology. 1966;**1**:13-16

[7] Gangwere SK, Evans FC, Nelson ML. The food habits and biology of Acrididae in an old-field community in Southeastern Michigan. Great Lakes Entomology. 1976;**9**:83-123

[8] Patterson BD. Correlation between mandibular morphology and specific diet of some desert grassland Acrididae (Orthoptera). American Midland Naturalist. 1984;**111**:296-303

[9] Liebermann J. The mandibles of grasshoppers of the subfamily Chilacridinae. Revista de Investigaciones Agropecuarias. 1968;(5):53-62. (In Spanish)

[10] Gangwere SK, Ronderos RA. A synopsis of food selection in argentine Acridoidea. Acrida. 1975;**4**:173-194

[11] Williams LH. The feeding habits and food preferences of Acridinae and the factors which determine them. Transaction of the Royal Entomological Society of London. 1954;**105**:423-454

[12] Kaufmann T. Biological studies on some Bavarian Acridoidea (Orthoptera), with special reference to their feeding habits. Annals of the Entomological Society of America. 1965;**58**:791-801

[13] Gangwere SK, Morales A. Food selection and feeding behaviourin Iberian Orthopteroidea. Anales del Instituto Nacionalde Investigaciones Agrarias. Serie proteccion vegetal. 1973;**3**:251-337

[14] Gangwere SK, Spiller DO. Food selection and feeding behavior in selected Orthoptera sen. lat. of the Balearic Islands, Spain. Journal of Orthopteran Research. 1995;**4**:147-160

[15] Gangwere SK, Mc kinney JC, Ernemann MA, Bland RG. Food selection and feeding behavior in selected Acridoidea (Insecta: Orthoptera) of the Canary Islands, Spain. Journal of Orthopteran Research. 1998;**7**:1-21

[16] Feroz M, Chaudhry MA. Studies on mandibles of some grasshoppers of Lahore. Biologia (Lahore). 1975;**21**: 211-225

[17] Gapud VP. The external morphology of the head and mouthparts of some Philippine Orthoptera. Philippine Entomology. 1968;**1**:11-32

[18] Kang L, Gan Y, Li SL. The structural adaptation of mandibles and food specificity in grasshoppers on Inner Mongolian grasslands. Journal of Orthopteran Research. 1999;**8**:257-269

[19] Le Gall P, Djihou Z, Tchenga G, Lomer CJ. Diet of *Zonocerus variegatus*

(Linné, 1758) (Orthoptera: Acrididae) in cassava field in Bénin. Journal of Applied Entomology. 2003;**127**(7):435-440

[20] Chapman RF. The structure and wear of the mandibles in some African grasshoppers. Proceedings of the Zoological Society of London. 1964;**142**:107-121

[21] Bernays EA, Barbehenn R. Nutritional ecology of grass foliage-chewing insects. In: Slansky F, Rodriguez JG, editors. Nutritional Ecology of Insects, Mites, Spiders, and Related Invertebrates. New York: John Wiley & Sons; 1987. pp. 147-175

[22] ElSayed WM. Grasshoppers (Orthoptera: Acrididae) communities in Abu-Raûwash district (Giza, Egypt) [MSc. Thesis]. Cairo University; 2005. pp. 169

[23] ElShazly M, ElSayed WM. The structure and host plant selection of an acridid community on the edge between a desert- and an agro-ecosystem in Egypt. Environmental Informatics Archives. 2006;**4**:1-25

[24] Uvarov B. Grasshoppers and Locusts. A Handbook of General Acridology. Vol. 2. Cambridge, United Kingdom, London: Centre for Overseas Pest Research; 1977. pp. 613

[25] Ichikawa A, Kano Y, Kawai Y, Kawai M, Tominago O, Murai T. Orthoptera of the Japanese Archipelago in Color. 2nd ed. 2006. pp. 687. (In Japanese)

[26] ElSayed WM. Diversity and structure of ground beetles (Coleoptera: Carabidae) assemblages in Satoyama. [Ph.D. thesis]. Japan: Kanazawa University; 2010. p. 123

[27] Mulkern GB, Anderson JF. A technique for studying to the food habits and preferences of grasshoppers. Journal of Economic Entomology. 1959;**52**:342

[28] Brusven MA, Mulkern GB. The Use of Epidermal Characteristics for the Identification of Plants Recovered in Fragmentary Condition from the Crops of Grasshoppers. Vol. 3. North Dakota, USA: North Dakota Agricultural Experimental Station Research Reports; 1960. p. 11

[29] Ohabuike JE. Grass availability and food preference of the African migratory locust, Locusta migratoria migratoriodes (R. and F.). Zeitschrift für Angewandte Entomologie. 1979;**88**:354-363

[30] Squitier JM, Capinera JL. Habitat associations of Florida grasshoppers (Orthoptera: Acrididae). Entomologist. Florida. 2002;**85**:235-244

[31] Smith TR, Capinera JL. Host preferences and habitat associations of some Florida grasshoppers (Orthoptera: Acrididae). Environmental Entomology. 2005;**34**:210-224

[32] Capinera JL. Relationships between insect pests and weeds: An evolutionary perspective. Weed Science. 2005;**53**:892-901

[33] Bernays EA, Chapman RF. Plant chemistry and acridoid feeding behavior. In: Harborne HB, editor. Biochemical Aspects of Plant and Animal Coevolution: Annual Proceedings of the Phytochemical Society of Europe. No.15. New York: Academic Press; 1978. pp. 99-141

[34] Joern A. Host plant utilization by grasshoppers (Orthoptera: Acrididae) from a sandhills prairie. Journal of Range Management. 1983;**36**:793-797

[35] Le Gall P, Mingouolo E, Bani G. Diets of *Zonocerus variegates* (L.) (Orthoptera: Acrididae) in cassava fields in Congo. Journal of Applied Entomology. 1998;**122**:9-13

[36] Hafez M, Ibrahim MM. Ecological and biological studies on *Acrida pellucida* Klug in Egypt. Bulletin of the Society of Entomology of Egypt XIII. 1958a:163-183

[37] Hafez M, Ibrahim MM. Studies on the egg and nymphal stages of *Acrida pellucida* Klug in Egypt. Bulletin of thc Society of Entomology of Egypt XIII. 1958b:183-198

[38] Cates RG. Feeding pattern of monophagous, oligophagous and polyphagous herbivores: The effect of resource abundance and plant chemistry. Oecologia. 1980;**46**:22-31

Chapter 4

Practical Approaches to Pest Control: The Use of Natural Compounds

Sladjan Stankovic, Miroslav Kostic,
Igor Kostic and Slobodan Krnjajic

Abstract

Food production is challenged by different factors: climate changes, market competitiveness, food safety, public demands, environmental challenges, new and invasive pests, etc. Intensive food production must be protected against pests, which is nowadays impossible with traditional techniques. The use of eco-friendly biopesticides based on essential oils (EOs), plant extracts (PE), and inert dusts appears to be a complementary or alternative methodology to the conventional chemically synthesized insecticides. The use of such biopesticides reduces the adverse pesticide effects on human health and environment. Biopesticides can exhibit toxic, repellent, and antifeeding effects. Development of bio-insecticides tackles the problem of food safety and residues in fresh food. Innovation within this approach is the combination of several types of active ingredients with complementary effects. Essential oils are well-known compounds with insecticide or repellent activities. New approaches, tools, and products for ecological pest management may substantially decrease pesticide use, especially in fruit and vegetable production. A win-win strategy is to find an appropriate nature-based compound having impact on pests, together with pesticide use, when unavoidable. Toxic or repellent activity could be used for pest control in the field conditions, as well as attractiveness of some compounds for mass trapping, before pests cause significant economic damage.

Keywords: insect, pests, essential oil, nature-based compound, metabolites, defense, antifeeding

1. Introduction

The current agricultural production, especially food production (whole production-market chain) in the fruit and vegetable sector, is challenged by climate changes, worldwide market competitiveness, food safety, environmental and public demands, new and invasive pests and diseases, etc. New invasive and destructive pests that recently appeared, especially in fruit and vegetable production, limited the use of chemical control agents because of their high persistence in the fresh food chain. For humans, fruits and vegetables are a rich source of vitamins, minerals, fibers, acids, sugars and secondary metabolites in biologically functional forms. Generally, a higher fruit and vegetable consumption is important in improving human's health. Additionally challenged, by newer standards and climate changes, intensive food production is unthinkable without protection from pests and diseases, which is nowadays impossible using only commonly used plant protection techniques. Different approaches such as better hygiene, standards in production (e.g. GlobalG.A.P.), agro- and pomotechnical measures, prophylactic measures, beneficial insects, mechanical intervention, biocontrol products and less sensitive varieties have been developed. However, a wide use of pesticides is still necessary, but none of the pesticide control techniques, during the long-lasting history, developed against important economic pests has provided long-term protection against pest-resistant species [1, 2]. Also, it may result in higher residues on food and food products than the allowed maximum residue level (MRL) when produced under good agricultural practices (GAP), legally determined by regulations (e.g. EU regulation, WTO, CEFTA, etc.). Multiple pesticide residues were found in 48% of the analyzed apples, 55% of the peaches and 56% of the cherries in 2015 [3]. Additionally, pesticides have an impact on the environment. In several European countries, groundwater pesticide concentrations exceed the European quality standards. Increasing customers and consumers and society's concern about the effects of pesticide utilization on human health and the environment have led to continuous changes in exploring techniques for pest and plant disease management. Even though significant improvements have been made, there is a need for alternative methodologies to ensure a lower utilization of pesticides that have less impact on the environment and guarantee that fruits are practically free from pesticide residues.

The use of eco-friendly biopesticides based on essential oils (EOs), plant extracts (PE) and inert dusts appears to be a complementary or alternative methodology to the chemically synthesized insecticides. Within plant protection practices, modern environmental requirements impose the need for expanding the biological control measures. Investigations of biological activity of plant derivatives lead to this goal, and some researchers have demonstrated certain promising natural substances that can be used for this purpose [4–7]. Natural semiochemicals with low toxic potential which would not cause ecosystem disturbance due to the high mortality of the target insect population could become the predominant method of pest control in the future [8], relying on naturally acquired plant defense mechanisms. Antifeedant activities of essential oils or extracts of different plant species seem to interfere with insect chemoreceptors. Plants produce alkaloids, steroids, flavonoids, terpenoids and saponins that possess high antifeedant activities against different insects; therefore, these compounds could be used in certain formulations and products that would be suitable in integrated insect management programmes. Generally EOs and their components have been considered safer than other plant-derived chemicals like rotenone and pyrethrum, as well as the use of several inert dusts for pest and plant disease control [9–11]. Novel strategies are important and necessary, having in mind the challenges arising due to climate change (increased areas of pest species, number of generations, etc.), public demands and standards in production practice.

2. Plant-pest interactions

2.1 Defensive mechanisms of plants under insect infestations

In all natural ecosystems, plants are exposed to stressful situations caused by biotic and abiotic factors that are largely responsible for significantly reducing crop productivity. For these reasons, plants produce secondary metabolites that protect them in adverse conditions [12]. When it comes to biotic stress, there are three basic strategies that plants use to defend their enemies: [1] direct defense, [2] indirect defence and [3] tolerance [13]. These strategies are similar to those described by Berryman [14] who stated that plants either may tolerate attack or will use defence mechanisms. Which plant defence strategy will be used depends on the insect species that is causing the damage [15]. During the co-evolution of plants and insects, plants have developed certain responses to attacks of herbivores: changes in the chemical composition of their leaves, as well as their different morphological and physiological properties [16]. Considering the abiotic stresses, for example, the lack of water can significantly affect the choice of the plant defence mechanism. Lack of water in a negative sense causes physiological and morphological changes on plants [17]. The represented defence mechanisms in plants are directly related to the origin and intensity of stress, and it can be classified as indirect and direct defence mechanisms. As stress increases, the number of possible defence scenarios is decreasing.

Indirect defence mechanisms include all plant features that increase the attraction of pest natural enemies [13] or prevent pest oviposition [18, 19]. In contrast, direct defence mechanisms are morphological (e.g. thorns, hairs) or chemical in nature (primary and secondary metabolites), or as their combination, the leaves of some plant species have hairs that directly adversely affect herbivores and, in addition, glands that secrete secondary metabolites [20] and often have a toxic effect (e.g. alkaloids, terpenoids, phenols) and may also inhibit digestive enzymes [21] forcing them to detoxify, causing poorer growth and development of herbivores. If the level of biotic stress is of lower intensity, tolerance is represented. Tolerance is considered when a plant may lose tissue by the herbivore while continuing its further development [22].

The defence mechanisms of direct and indirect defenses can be further divided into passive or constitutive and dynamic or induced defence described in the following paragraphs.

2.1.1 Constitutive defence

Constitutive defence is a passive type of defence of a plant against herbivores and other pathogens and is recognizable by the use of accumulated secondary metabolites under favorable conditions for defensive purposes, caused by the resulting stress [16, 17]. It is a characteristic of perennial plants and is effective in fighting generalists such as the gypsy moth—*Lymantria dispar* L. (Lepidoptera:Erebidae). This type of defence is based on carbon and is present in plants growing under conditions that cause chronic excess of carbon, which provokes accumulation of carbon-based allelochemicals: lignin, tannins and other phenolic compounds, terpenes and resins. These herbal compounds that have negative effects on the growth, development or survival of another organism are considered as toxins. Plants that endure stressful situations by constitutive chemical defence must at the same time be able to sustainably synthesize and accumulate toxic substances without negative consequences on their physiology [23].

However, insects and other plant-borne pathogens have developed various mechanisms to respond to plant toxins [23] and often use them to identify plants as hosts for nutrition and oviposition [24]. Hilker and Meiners [25] consider that the presence of a particular insect species, which has developed adaptability to biochemical mechanisms to the toxic effects of plant secondary metabolites, enhances plant defence in the event of a subsequent herbivore attack. Nevertheless, constituent secondary metabolites having antifeeding action protect plants from most unadapt insects [26] and at high concentrations adversely affect specialized insects [27].

2.1.2 Induced defence

Induced defence in plants is based on their secondary metabolites (terpenes, phenols) and physical structures (cell lignification) as well as a reduction in the production of essential substances to attract herbivores in response to their attack [14]. The type of plant response depends on the balance between primary and secondary metabolites [28]. If the current reserves are reduced by stressful conditions (drought, nutrient deficiency), the presence of herbivore populations is more pronounced. Increased plant resistance reduces the presence and the harmful effects of insects. The minimal length of latency for a plant depends on the rate of decline of plant resistance (e.g. time needed for the plant to recover from defoliation) [29]. The response of plants to the harmful effects of insects is measurable over time (evolutionary time), ranging from a few minutes to a longer period [28].

Additional research has been focused on increased concentrations of secondary metabolites, induced by the attack of insects or other pathogenic organisms. Terpenoids are considered to be the most abundant and diverse metabolic class of plant bioactive products (more than 40,000 structures). They have antifeedant, repellent and toxic effects and can act as regulators of insect development [30]. Bioactive natural products such as alkaloids possess well-known metabolic effects on mammals (e.g. caffeine, nicotine, morphine, strychnine and cocaine) and have probably evolved as a defence against herbivore insects [31]. It is known that the feeding of autumnal moth, *Epirrita autumnata* (Borkhausen) (Lepidoptera:Geometridae), with birch leaves increases the content of phenolic compounds [32]. Gypsy moth (*L. dispar*) feeding increases the content of tannins in oak leaves [33], while after the attack of bark beetles, terpenes and phenolics levels rise in the phloem of attacked trees [34]. Defensive proteins that act on insect digestive enzymes have also been identified in plants. For example, protease inhibitors [21] play a special protective role against insects and microorganisms, in addition to their primary role in the regulation and control of endogenous protease activity, and serve as reserve proteins [35]. The synthesis of protease inhibitors is a part of the induced defence of plants from insect attack. Thanks to the advances in genetic engineering, there is possibility to grow plants with increased levels of protease inhibitors with herbivore defence mechanisms.

2.2 The role of secondary metabolites in insect-plant interactions

Secondary metabolites are organic compounds including terpenes, phenols, alkaloids, proteins and enzymes. They are not directly involved in the development or reproduction of plants (as primary metabolites), but they are often represented in plant defence mechanisms. Usually found in only one plant species or genus, with limited distribution, their production in plants impairs plant growth and reproduction [36]. These compounds are considered as waste products of metabolism without essential function in plant survival [37].

Plants produce different chemical compounds that can be toxic or indigestive for animals [38]. Plant chemical defence is classified into two categories:

1. Quantitative defence, with massive production of indigestible substances; and

2. Qualitative defence, with limited production of toxic substances [39].

By the theory of apparency, plants with their organs are classified into apparent or unapparent [39]. The theory on the balance of growth and differentiation (plant's "dilemma" for the determination between cell growth or division and differentiation) that creates specialized organs and compounds for defence has also been proved [38].

The presence and availability of nutrients in soil significantly contribute to the level of constituents and induced allelochemicals in plants [40, 41]. There are numerous examples for such actions [42]. Nitrogen fertilization affects the increase in induced poplar resistance after continuous feeding of gypsy moth caterpillars for only 72 h. The composition and concentration of secondary metabolites indicate the interspecies variation is not the case with the primary metabolites. Significant variation was observed between genotypes within the same species, different ages and different branches of one tree and between leaves of different ages on one branch.

2.2.1 Terpenes

Terpenes are the largest class of secondary metabolites (over 22,000 compounds described) and occur in all plants and are classified by the number of isoprene units: monoterpenoids (two units), isoprene sesquiterpenoids (three units), diterpenoids (four units) and triterpenoids (six units). Isoprene (C_5H_8) is the simplest terpenoid to protect cell membranes from damage under adverse conditions (high temperatures). The primary components of essential oils are monoterpenoids and sesquiterpenoids. They are volatile, and their aromas are characteristic of certain plants. They are toxic to insects and pathogens. Monoterpenoids can be used as insecticides, for example, pyrethrins (a compound from Chrysanthemum) acts as a neurotoxin to insects. Synthetic analogues of pyrethrin are pyrethroids, a chemical group of pesticides with a large number of commercial insecticides. Alpha- and beta-pinenes are known for repellent action. They are found in pine resin and are known as potent repellents. Monoterpenoids can also be used as spices and perfumes while being relatively harmless to humans. Diterpenoids may have antifungal and antibacterial properties such as gossypol, which is a component of cotton. Triterpenoids are similar in their molecular structure to plant and animal sterols and steroid hormones, which are imitations of insect-coated hormones. For example, azadirachtin is a limonoid isolated from Indian wood (*Azadirachta indica*) that has antifeedant activity and causes sterility. Limonoids also include citronella essential oil isolated from *Cymbopogon citratus* and in the United States is popular as a mosquito repellent for its low toxicity [43]. In addition to defence against the harmful insects and microorganisms, they have a role as a signal in attracting pollinators [44].

2.2.2 Phenols

Phenols are also a large class of plant secondary metabolites and comprise a wide range of compounds (flavonoids, anthocyanins, phytoalexins, tannins, lignins, furanocoumarins). They have different effects on harmful organisms. Tannins have a toxic effect on insects by binding to proteins and salivary digestive enzymes, including trypsin, leading to protein inactivation. By ingesting a large amount of tannins, herbivorous insects do not gain weight and finally die. Lignins are entrenched in the cell walls of plants and provide an excellent physical barrier against pathogens. Furanocoumarins are produced by a wide variety of plants in response to pathogens and are activated by UV light; they are toxic to vertebrates and invertebrates due to integration into DNA and affect at the cellular level [43].

2.2.3 Alkaloids

Alkaloids are a large class of bitter-tasting nitrogen compounds and are found in many vascular plants (caffeine, cocaine, morphine, nicotine). They are derived from the amino acids aspartate, lysine, tyrosine and tryptophan. They have powerful effects on the physiological processes of animals. Caffeine is toxic to insects and fungi and also inhibits seed germination in the vicinity of other growing plants (allelopathy). Nicotine is produced at the root of the tobacco plant and is transported in leaves where it is stored in vacuoles and in the presence of herbivores is released and has toxic effects. Plants that produce cyanogenic glycosides also produce enzymes that convert these compounds into the hydrogen cyanide, including glycosides that are stored in separate cells, and toxic cyanotic hydrogen is secreted by these tissues [43].

2.2.4 Proteins

In contrast to the simple chemicals such as the terpenoids, alkaloids and phenols, proteins require a large expenditure of energy from plants and are formed in significant amounts after the attack of pathogens. Once activated, the defence proteins and enzymes effectively inhibit fungi, bacteria, nematodes and herbivorous insects. Defence against herbivores is obtained by forming an enzyme complex which leads to enzyme inhibition. They include defensins, amylase inhibitors, lecithins and proteinase inhibitors. Defensins have broad antimicrobial activity. First isolated from barley endosperm (*Hordeum vulgare* L., Poales:Poaceae) and wheat (*Triticum aestivum* L., Poales:Poaceae), they are widely distributed and found in most plants. They are most prevalent in seeds but can be found in almost all plant tissues. In addition to inhibiting the growth and development of many fungi and bacteria, they inhibit the digestive proteins of herbivores and impair the cellular balance of ions. Proteins are inhibitors of digestive enzymes and block the normal process of digestion and absorption of nutrients in vertebrates and invertebrates of herbivores. Alpha-amylase interferes with starch digestion, lecithin has a wide range of functions including impaired digestion in insects and blood cell disintegration in vertebrates, and ricin (toxin) produced in castor (*Ricinus communis* L., Malpighiales:Euphorbiaceae) is a highly potent toxin and inhibits protein synthesis. Plants in response to the attack of herbivores produce proteases that inhibit digestive enzymes including trypsin and chymotrypsin and are widespread in nature [43].

2.2.5 Enzymes

A special group of proteins, enzymes, are produced in plants in response to the presence of pathogenic organisms and often accumulate in extracellular spaces where they degrade the cell walls of pathogenic fungi. Chitinases are enzymes that catalyze the degradation of chitin, a cellulose-like polymer present in the cell walls of fungi. Glucanases are enzymes that degrade glyosidic bonds, a class of cellulose-like polymers present in the cell walls of many oomycetes, while lysozyme is a hydrolytic enzyme capable of degrading bacterial cell walls [43]. Chitinase and glucanase enzyme activity lyses pathogen cells [45].

2.2.6 The effects of secondary plant metabolites on harmful insects: state of the art

There is strong public pressure for the production of health food, i.e. food without pesticide residues. For these reasons, extensive testing is being carried out such as the use of secondary metabolites as an alternative to pesticides, the creation of resistant varieties, the application of nanoparticles, the joint cultivation of cultivated plants with plants that would be confusing on harmful insect and other research.

Plants have created many strategies during co-evolution with insects for effective protection. The most important defence mechanism in plants is the synthesis of biologically active compounds, the so-called secondary metabolites, which can act directly as insecticides or affect indirectly the behaviour of insects—these are called allelochemicals. Allelochemicals are divided into four subgroups, allomones, keiromones, synomones and apneumones, and can be used in plant protection.

Metabolites from allomone subgroup represent a respectable group with the currently highest potential [46]. However, it is known that plant secondary metabolites (essential oils, alkaloids, saponins, glucosides, tannins, flavonoids, organic acids) are involved in the defence of harmful insects [4, 6, 7, 47, 48] leading to attempts for field application (spraying) of plant extracts. In recent decades, there are increased evidences of the diverse ecological, physiological and biochemical role of these compounds [37, 49, 50]. The antifeeding properties of plant sprays against harmful insects are thought to have no negative effects on predators or pollinators [51], thus providing an ideal opportunity for pest control [52]. Numerous secondary metabolites, plant extracts and essential oils have insecticidal properties [53, 54]. These substances have oral, contact or inhalation toxic effect to insects, together with antifeeding and repellent effects, which cause a decrease in reproductive potential and change in normal behaviour [55]. Plants produce a wide range of chemicals in various parts above and below ground that are used to defend against stress caused by biotic and abiotic factors but also for communication with other plants and organisms. On the other side, insects have developed strategies to avoid these chemicals [56] or effective detoxification systems specific to individual insect taxa [57], which can be very different between species feeding on the same plant [27, 58].

The insect's orienting abilities include receiving information about the spatial relationships of an organism, processing them and transmitting this information to effectors that can change the relationships. This can be redefined as the relationship between the input and output state of the system (insect/plant ratio); therefore, the chemosensory system allows insects to maintain a constant course, find a host or turn to a sexual partner [59]. Insects often use more than one substance to detect differences between host plants, and the use of secondary metabolites for these purposes is a consequence of evolution.

Dethier et al. [60] described the reactions of insects to chemical compounds:

1. Attractant: A chemical that causes the insect to orientate towards the source.

2. Repellent: A chemical that causes the insects to move away from the source.

3. Arrestant: A chemical that causes confusing action and slows the movement of an insect towards the source.

4. Feeding or ovipositional stimulant: A chemical that causes nutrition and egg laying (oviposition).

5. Deterrent: A chemical that causes an inhibition of nutrition and prevents egg laying (oviposition), and in that area the insect would otherwise feed and lay eggs.

This terminology is generally accepted in describing and considering the reaction of insects to chemical compounds that have been applied in the plant or targeted for protection against herbivores.

An essential biological characteristic of herbivores is nutrition, that is, whether they feed on a single plant species (monophagous), several plant species in one family (oligophagous) and various plant species (polyphagous) whose diet, oviposition and overall biological cycle unfold smoothly across different plant species of different families. In recent decades, extensive research has been done on the impact of secondary plant metabolites on harmful insects, regardless of which group they are classified in according to the nutrition classification.

Effects on stored product pests were widely investigated. Bioactive substances from *Myristica fragrans* Houtt. (Magnoliales:Myristicaceae) oil have been found to have repellent and antifeeding (contact and fumigant) activity and significantly affect offspring reduction in *Sitophilus zeamais* (Motschulsky) (Coleoptera:Curculionidae) and *Tribolium castaneum* (Herbst) (Coleoptera:Tenebrionidae) species [61]. *Elettaria cardamomum* L. (Zingiberales:Zingiberaceae) seed oil possesses contact and fumigant toxicity and antifeeding activity against *S. zeamais* and *T. castaneum* [62]. This essential oil causes reduction in the number of egg laying and egg hatching of *T. castaneum*. Extracts obtained from seeds of the *Basella alba* plant and leaves of *Operculina turpethum* and *Calotropis gigantea* act as inhibitors of *S. zeamais* development [63]. Essential oils obtained from the leaves of *Eucalyptus dunnii*, *E. saligna*, *E. benthamii*, *E. globulus* and *E. viminalis* (Myrtaceae) showed a pronounced insecticidal and repellent effect on *S. zeamais* [64, 65]. Somewhat weaker but also a very toxic and repellent effect on *S. zeamais* and *T. castaneum* showed the essential oil obtained from the leaves of *Cupressus sempervirens*, as well as cymene, the dominant component of the essential oils of *E. saligna* and *C. sempervirens* [65]. Both cinnamon extracts (*Cinnamomum zeylanicum*) and essential oils of the plants *Etlingera elatior*, *E. pyramidosphaera* and *Zingiber officinale* show strong repellent activity towards *S. zeamais*, while the moderate repellent activity is shown by the extracts of *Curcuma longa* and *Piper nigrum* [66]. Essential oils of *Ocimum basilicum* L. and *Salvia officinalis* L. caused significant mortality and repellent and anti-reproductive effect [67]. Examination of five ethanol extracts of medicinal aromatic plants for bean protection from weevil *Acanthoscelides obtectus* Say on repellent and toxic action as well as reducing F1 offspring showed a significant insecticidal activity of concentrated extracts of *Urtica dioica* L. and *Taraxacum officinale* L, while *Achillea millefolium* L. extract had repellent effect and caused a decrease in F1 offspring [68]. Similar tests on *A. obtectus* with the essential oils of *Thymus vulgaris* L., *Rosmarinus officinalis* L. and *Ocimum basilicum* L. and their dominant components (thymol, alpha-pinene, 1,8-cineol and linalool) showed that *T. vulgaris* EO and thymol have promising efficiencies and can be used as alternatives to synthetic pesticides [69].

Colorado potato beetle (CPB) (*Leptinotarsa decemlineata* Say, Coleoptera: Chrysomelidae) is an oligophagous pest. The major components in the EOs of potato leaves responsible for the attractive action on potato sprouts have been identified and are referred as "volatile green leaves." Basically, they are represented by a chain of saturated and unsaturated aldehydes and alcohols, formed by the oxidative degradation of plant lipids. The relative proportions of these end products (mainly alcohols and aldehydes) vary among different plant species within the same genus, as well as seasonally within one species, due to the aging and injury of the plants, all of which affect the degree of attraction of the CPB. It is reported what are the volatile components that attract potato gold: trans-2-hexen-1-ol, hexanol-1, cis-3-hexen-1-ol, trans-2-hexenal, and linalool in the following ratios (expressed as a percentage): 100: 17: 7: 7: 4 [70]. Host attractiveness to insects related to secondary metabolites, based on the molecular interaction of CPB with plant species of the family Solanaceae, was investigated by Lawrence et al. [71].

The neem extract (i.e. azadirachtin) prepared against the third-stage larvae of *L. decemlineata* has significant antifeedant effect and low toxicity and can be used to control oligophagous herbivores [4]. In biological studies of residual toxicity and antifeedant action of ethanolic derivatives of sage, *Salvia officinalis* L. (Lamiaceae) (essential oil, five fractions of the same oil F1–F5 and camphor), low toxicity was observed on the second-stage larvae and CPB adulthood, not affecting embryonic development, and the antifeedant activity on the larvae in the first 96 h was very significant for the subsequent activity declined [5]. The possibility of disturbing the attractive properties of the potato leaf on the female potato pollen in the olfactometer was investigated by applying an ethanolic solution of sage oil and five fractions (F1–F5) of this oil. The most pronounced impediments to the recognition of potato leaf are from the sage essential oil and the least expressed by fraction one (F1) [72]. Extracts of five plant species collected in Turkey (*Arctium lappa* L., *Bifora radians* M.Bieb., *Humulus lupulus* L. or *Xanthium strumarium* L. and *Verbascum songaricum* (Schrenk)) were used to investigate the antifeedant effect on *L. decemlineata* larvae. In the first 15 min, the interaction between the larvae and the leaf mass of the potatoes was significantly affected, and during the first 24 h, nutrition was reduced. Gökçe et al. [73, 74] observed that the toxic effect on CPB was obtained by the extracts of the dried rhizome of *Veratrum album* ($CHCl_3$, acetone and NH_4OH / benzene) and the compounds oxyresveratrol, b-sitosterol-3-O-b-D-glucopyranoside and jervine have the potential to be used as natural insecticides. Biological effects of 24 terpenes, commonly found in aromatic plants in the Mediterranean region, have been investigated to determine their antifeedant effect n and CPB as well as allelopathic impact. Terpene (−) α-bisabolol possesses high antifeeding and low phytotoxic activity [44].

Gypsy moth is a polyphagous insect and belongs to the group of the most harmful butterflies. The caterpillar feeds on the leaves of almost all types of hardwoods, conifers and the green mass of many agricultural, fruit and vegetable crops. Protection against the damaging effect of gypsy moth must involve knowledge that secondary metabolites are involved in the defence of insect plants [4, 6, 8, 47]. Other EOs and their components have antifeeding activity against caterpillars: Kostic et al. [6] found that *Ocimum basilicum* EO and its dominant component linalool cause antifeedant activity against second-stage larvae, and Popovic et al. [8] found that fractions of *O. basilicum* EO also act as antifeedant on gypsy moth caterpillars of the second-instar (L2) as well as EOs of *Athamanta haynaldii* and *Myristica fragrans* [7]. Also, neem (0.09% azadirachtin, safer), shows good antifeedant activity against L2 and low digestive toxicity [4], which were confirmed in other investigations [6–8].

Pavela [46] found that *Foeniculum vulgare* EO has a very pronounced digestive toxicity to fourth-instar (L4) caterpillars of *Spodoptera littoralis*. Singh et al. [75] found that trans-anethole exhibited moderate digestive toxicity to first-instar (L1) *Chilo partellus* (Swinhoe) (Lepidoptera:Crambidae) caterpillars, whereas it showed significantly lower toxicity to second-instar caterpillars.

There are numerous positive properties that herbal extracts and EOs have compared to those of the conventional insecticides such as the absence of adverse environmental effects, the disturbance of biocenosis, the absence of nonspecific effects on predators and parasitoids, the minimal toxicity to mammals, the ease of detection and finally the inability to develop resistance. Some disadvantages must be overcome in order to make their application as efficient and easy as possible. The problems encountered in dealing with EOs are their high volatility, incoherence, inadequate formulation, limited shelf life and action on a very limited number of pests [76, 77].

When insects develop resistance to certain plant secondary metabolites, they also develop resistance to the associative molecules of these metabolites generating synergistic effects. For example, in oak leaves, the tannin-binding protein forms complexes with tannins, difficult to digest. Fenny [39] concludes that tannins, as part of a wide range of defence mechanisms, have repellent, antibiotic and growth-inhibiting properties, via their effect on protein availability. However, for gypsy moth, tannic acid is an attractant, and the alkaline pH value of the digestive tract prevents the formation of tannin protein complexes. Insects often use more than one substance to detect differences between host plants, and the use of secondary matter for these purposes is a consequence of evolution. In recent decades, there has been increasing evidence of the diverse ecological, physiological and biochemical role of these compounds [49, 50].

2.3 Inorganic compounds

One of the alternative methods of crop protection and protection of stored agricultural products in warehouses has been the use of various inorganic dusts in recent years.

So far, diatomaceous earth (DE) preparations have been mostly registered and applied in agricultural practice. The diatomaceous earth was created by the fossilization of tiny aquatic algae (microscopic algae) by organisms called diatoms. The main constituent of their skeleton is called silica, which in contact with water and oxygen forms silicon dioxide. The compositions on the basis of DE consist mainly of an amorphous form of silicon dioxide (amorphous silicon dioxide) and a smaller part of the crystalline silicon dioxide (crystalline silicon dioxide). The first registered composition on the basis of DE was registered in 1960 in the United States for control of insects and mites. To date, over 150 preparations for various uses have been registered. They are used to counter bedbugs, cockroaches, crickets, fleas, ticks, spiders and many other pests. They have also found application in the protection of stored products, except in conventional agricultural production and in IPM and organic production [78].

In addition to DE, many other inorganic powders such as silicophosphate, rock phosphate, sand, kaolinite, clay, zinc oxide, titanium dioxide, vermiculite dust, zeolite, alumina, etc. have also been studied [9, 10, 79–82]. In addition to natural dusts, the possibility of obtaining and applying nano-dusts has been increasingly studied in recent years. The application of modern nano-methods yields nanopowders of improved properties (**Figure 1**) and efficiency and is more environmentally friendly (less toxic to mammals and plants, durability, eco-friendly, less harmful to the environment than the conventional) [9, 79, 80, 83, 84].

The mechanism of action of native and nano-dusts is not fully understood. Some authors believe that the particles of these preparations bind to the exoskeleton and

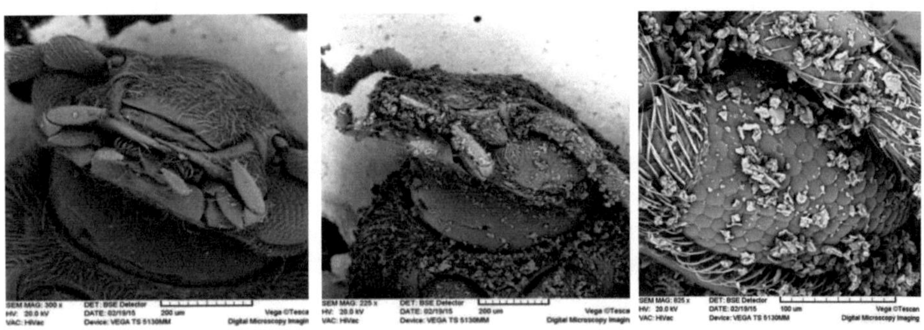

Figure 1.
SEM images of untreated (left) insect treated with Al_2O_3 (middle) and enlarged image of insect treated with Al_2O_3 (right) [9].

that they adsorb lipids from the cuticle and cause dehydration of the insect [80]. In contrast, other authors believe that dust particles can physically damage the cuticle and lead to dehydration, that they can ingest damage to the intestinal tract of insects, and that they can block the trachea and thus the insect's breathing [85, 86], like abrasion of the cuticle, absorption of the cuticular waxes from the epicuticle surface, damaging of the digestive tract, blocking of spiracles and tracheae, surface enlargement combined with dehydration and repellence caused by the physical presence of the dust. It is assumed that such chemically inert compounds attached to an exoskeleton are able to adsorb cuticular lipids, thus causing rapid dehydration of insects.

Mineral elements (macronutrients and trace elements) play an important nutritional role in plants and are necessary for the normal course of many cellular processes such as primary and secondary metabolism, defence, gene regulation, hormone perception, energy metabolism, reproduction and signal transduction [87]. A series of functions performed in plants can be affected by the increase in their resistance and protection against harmful organisms. According to Reynolds et al. [88], silicon (Si), which has been found to play a significant role in overcoming the various biotic and abiotic stress factors to plants, may have an indirect and direct effect on enhancing the defence capabilities of plants against harmful insects as part of the mechanisms of physical and induced chemical defence of plants. The physical defence mechanisms involving Si are mainly related to the deposition of Si, mainly in the form of opalina phytoliths, in the cell walls, especially in the epidermal cells of the plants, thereby increasing their firmness and abrasiveness, which in insects can lead to difficult nutrition and damage to the oral apparatus. Also, such a plant food for the insects is reduced digestibility which negatively affects the parameter growth and feeding insects and which is reflected in their reduced growth, length of life and fertility. The presence of Si in the plant may also initiate or accelerate a number of different chemical defence mechanisms that protect the plant from harmful insects. Si can cause a significant increase in defence enzymes such as peroxidase, phenylalanine ammonia-lyase (PAL) and polyphenol oxidase involved in the processes of lignification and synthesis of suberin (peroxidase), increased production of phenolic compounds (PAL) and oxidation of phenolic compounds (polyolase) which increases the hardness of plant tissue and the production of compounds that have detergent and toxic properties while reducing the nutritional quality of food and the digestion of proteins. Also, silicon exerts a positive effect on the biosynthesis of volatile compounds such as jasmonic acid and salicylic acid, in which herbivore-invaded plants emit to attract the natural enemies (predators and parasitoids) of the insects that attack them. Silicon definitely may be considered as an environmentally friendly option in the concept of sustainable agriculture.

3. Conclusion

Intensive food production must be protected against pests and diseases, which is nowadays impossible with single and traditional techniques. However, a wide use of pesticides is still necessary, which may result in higher residues on food and food products than the allowed maximum residue level (MRL). The use of eco-friendly biopesticides based on essential oils (EOs), plant extracts (PE) and inert dusts appears to be a complementary or alternative method to chemically synthesized insecticides. The use of biopesticides may reduce the adverse effects of chemical pesticides on human health and environment. Biopesticides can exhibit toxic, repellent and antifeedant effects on different insect species. Investigations for developing a new bio-insecticide tackle the problem of food safety and residues in fresh food. Innovation within this approach is the combination of several types of active ingredients with complementary effects. Essential oils are well-known for their insecticide or repellent activity. But so far their use in practice is limited due to their high volatility and short period of action. This problem could be solved by their encapsulation with natural coating materials. Regarding such formulation, their volatility should be prolonged, and EOs will have a chance to provide satisfactory efficacy against pests. New approaches, tools and products for ecologically improved pest management may substantially decrease pesticide use against pests, especially in the fruit and vegetable sector. A win-win strategy is to find an appropriate nature-based compound which will have a wide spectrum of impacts on pest populations. Toxic or repellent activity could be used to control their presence in the field conditions, combined with the use of attractants of some compounds for pest mass trapping, followed by pesticide use when unavoidable.

Acknowledgements

The paper is the result of research under a contract for the implementation and financing of scientific research in 2020 between the Institute for Science Application in Agriculture, Belgrade and the Ministry of Education, Science and Technological Development of Republic of Serbia, record number of contract: 451-03-68/2020-14/200045.

The paper is also the result of research under a contract for the implementation and financing of scientific research in 2020 between the Institute for Medicinal Plant Research "Dr Josif Pancic", Belgrade and the Ministry of Education, Science and Technological Development of Republic of Serbia (record number of contract: 451-03-68/2020-14/200003) and contract for the implementation and financing of scientific research in 2020 between Institute for Multidisciplinary Research, University of Belgrade and the Ministry of Education, Science and Technological Development of Republic of Serbia (record number of contract: 451-03-68/2020-14/200053).

Author details

Sladjan Stankovic[1*], Miroslav Kostic[2], Igor Kostic[3] and Slobodan Krnjajic[3]

1 Institute for Science Application in Agriculture, Belgrade, Serbia

2 Institute for Medicinal Plant Research "Dr Josif Pancic", Belgrade, Serbia

3 Institute for Multidisciplinary Research, University of Belgrade, Belgrade, Serbia

*Address all correspondence to: sstankovic@ipn.bg.ac.rs

References

[1] Stankovic S, Kostic M. Role of Carboxylesterases (ALiE) Regarding Resistance to Insecticides: Case Study of Colorado Potato Beetle (*Leptinotarsa decemlineata* Say). In: Shields V, editor. Insect Physiology and Ecology. Rijeka, Croatia: InTech; 2017. pp. 159-178. DOI: 10.5772/66254

[2] Kostic M, Stankovic S, Kuzevski J. Role of AChE in Colorado potato beetle (*Leptinotarsa decemlineata* Say) resistance to carbamates and organophosphates. In: Trdan S, editor. Insecticides Resistance. Rijeka, Croatia: InTech; 2015. pp. 19-40. DOI: 10.5772/61460

[3] EFSA (European Food Safety Authority). The 2015 European Union report on pesticide residues in food. EFSA Journal. 2017;**15**(4):4791. DOI: 10.2903/j.efsa.2017.4791

[4] Zabel A, Manojlović B, Rajković S, Stanković S, Kostić M. Effect of Neem extract on *Lymantria dispar* L. (Lepidoptera: Lymantriidae) and *Leptinotarsa decemlineata* Say. (Coleoptera: Chrysomelidae). Journal of Pesticide Science. 2002;**75**:19-26

[5] Kostić M, Dražić S, Popović Z, Stanković S, Sivčev I, Živanović T. Developmental and feeding alternations in *Leptinotarsa decemlineata* Say. (Coleoptera: Chrysomelidae) caused by *Salvia officinalis* L. (Lamiaceae) essential oil. Biotechnology and Biotechnological Equipment. 2007;**21**(4):426-430

[6] Kostić M, Popović Z, Brkić D, Milanović S, Sivčev I, Stanković S. Larvicidal and anti feedant activity of some plant-derived compounds to *Lymantria dispar* L. (Lepidoptera: Lymantriidae). Bioresource Technology. 2008;**99**(2008):7897-7901

[7] Kostić I, Petrović O, Milanović S, Popović Z, Stanković S, Todorović G, et al. Biological activity of essential oils of *Athamanta haynaldii* and *Myristica fragrans* to gypsy moth larvae. Industrial Crops and Products. 2012;**2012**(41):17-20

[8] Popovic Z, Kostić M, Stanković S, Milanović S, Sivčev I, Kostić I, et al. Ecologically acceptable usage of derivatives of essential oil of sweet basil, *Ocimum basilicum*, as antifeedants against larvae of the gypsy moth, *Lymantria dispar*. Journal of Insect Science. 2013;**13**(1):161. DOI: 10.1673/031.013.16101

[9] Krnjajić S, Radojković A, Radović M, Kostić I, Kostić M, Mitrović J, et al. Insecticide effects of non-toxic inorganic powders against bean weevil (*Acanthoscelides obtectus*). In: XXIII International Conference Ecological Truth; June 17-20, 2015. Kopaonik, Serbia; 2015. pp. 480-486

[10] Lazarević J, Radojković A, Kostić I, Krnjajić S, Mitrović J, Kostić M, et al. Insecticidal impact of alumina powders against *Acanthoscelides obtectus* (Say). Journal of Stored Products Research. 2018;**77**:45-54

[11] Popović T, Milićević Z, Oro V, Kostić I, Radović V, Jelušić A, et al. A preliminary study of antibacterial activity of thirty essential oils against several important plant pathogenic bacteria. Pesticidi i Fitomedicina. 2018;**33**(3-4):185-195

[12] Mazid M, Khan TA, Mohammad F. Role of secondary metabolites in defense mechanisms of plants. Biology and Medicine. 2011;**3**((2)Special Issue): 232-249

[13] Karban R, Baldwin IT. Induced Responses to Herbivory. Chicago, IL. The University of Chicago Press; 1997. p. 319

[14] Berryman AA. Towards a unified theory of plant defense. In: Mattson WJ, Levieux J, Dagan BC, editors. Mechanisms of Woody Plant Defenses against Insects. New York: Springer-Verlag; 1988. pp. 39-57

[15] Mattson WJ, Lowrence RK, Haack RA, Herms DA, Charles PJ. Defensive strategies of woody plants against different insect-feeding guilds in relation to plant ecological strategies and intimacy of association with insects. In: Mattson WJ, Levieux J, Dagan BC, editors. Mechanisms of Woody Plant Defenses against Insects. New York: Springer-Verlag; 1988. pp. 3-39

[16] Baldwin IT, Preston CA. The eco-physiological complexity of plant responses to insect herbivores. Planta. 1999;**208**:137-145

[17] Hsiao TC. Plant responses to water stress. Annual Review of Plant Physiology. 1973;**24**:519-570

[18] Kessler A, Baldwin IT. Defensive function of herbivore-induced plant volatile emissions in nature. Science. 2001;**291**:2141-2144

[19] De Moraes CM, Mescher MC, Tumlinson JH. Caterpillar-induced nocturnal plant volatiles repel nonspecific females. Nature. 2001;**410**:577-580

[20] Duffey SS. Plant glandular trichomes: Their partial role in defense against insects. In: Juniper BE, Southwood TRE, editors. Insect and the Plant Surface. London: Arnold; 1986. pp. 151-172

[21] Koiwa H, Bressan RA, Hasegawa PM. Regulation of protease inhibitors and plant defense. Trends in Plant Science. 1997;**2**:379-384

[22] Stowe KA, Marquis RJ, Hochwender CG, Simms EL. The evolutionary ecology of tolerance to consumer damage. Annual Review of Ecology and Systematics. 2000;**31**:565-595

[23] Wittstock U, Gershenzon J. Constitutive plant toxins and their role in defense against herbivores and pathogens. Current Opinion in Plant Biology. 2002;**5**(4):300-307

[24] Roininen H, Price PW, Julkunen-Tiitto R, Tahvanainen J, Ikonen A. Oviposition stimulant for a gall-inducing sawfly, *Euura lasiolepis*, on willow is a phenolic glucoside. Journal of Chemical Ecology. 1999;**25**:943-953

[25] Hilker M, Meiners T. Early herbivore alert: Insect eggs induce plant defense. Journal of Chemical Ecology. 2006;**32**:1379-1397

[26] Bernays EA, Chapman RF. Deterrent chemicals as a basis of oligophagy in *Locusta migratoria* (L.). Ecological Entomology. 1976;**2**:1-18

[27] Zangerl AR, Berenbaum MR. Plant chemistry, insect adaptations to plant chemistry and host plant utilisation patterns. Ecology. 1993;**74**:47-54

[28] Karban R, Myers JH. Induced plant responses to herbivory. Annual Review of Ecology and Systematics. 1989;**20**:331-348

[29] Tuomi J, Niemela P, Chapin FS, Brayant JP, Siren S. Defensive response of trees in relation with their carbon/nutrient balance. In: Mattson WJ, Levieux J, Dagan BC, editors. Mechanisms of Woody Plant Defenses Against Insects. New York: Springer-Verlag; 1988. pp. 57-73

[30] Aharoni A, Jongsma MA, Bouwmeester HJ. Volatile science? Metabolic engineering of terpenoids in plants. Trends in Plant Science. 2005;**10**:594-602

[31] Howe GA, Jander G. Plant immunity to insect herbivores. Annual Review of Plant Biology. 2008;**59**:41-66

[32] Kaitaniemi P, Ruohomäki K, Ossipov VM, Haukioja E, Pihlaja K. Delayed induced changes in the biochemical composition of host plant leaves during an insect outbreak. Oecologia. 1998;**116**:182-190

[33] Rossiter MC, Shultz JC, Baldwin IT. Relationships among defoliation, red oak phenolics and gypsy moth growth and reproduction. Ecology. 1988;**69**:267-277

[34] Raffa KF. Induced defensive reactions in conifer-bark beetle systems. In: Tallamy, Raupp, editors. Phytochemical Induction by Herbivores. New York: Wiley-Interscience Publication; 1991. pp. 245-276

[35] Ninković S, Miljuš-Đukić J, Lazarević J. Proteinase inhibitors and the possibility of their application in the control of harmful insects (in Serbian). Ekologija. 2001;**36**(1):1-38

[36] Siemens DH, Garner SH, Mitchell-Olds T, Callaway RM. Cost of defense in the context of plant competition: *Brassica rapa* may grow and defend. Ecology. 2002;**83**(2):505-517

[37] Seigler D, Price PW. Secondary compounds in plants: Primary functions. The American Naturalist. 1976;**110**:101-105

[38] Herms DA, Mattson WJ. The dilemma of plants: To grow or defend. Quarterly Review of Biology. 1992;**67**:283-335

[39] Feeny PP. Plant apparency and chemical defense. Recent Advances in Phytochemistry. 1976;**10**:1-40

[40] Coviella CE, Stipanovic RD, Trumble JT. Plant allocation to defensive compounds: Interactions between elevated CO_2 and nitrogen in transgenic cotton plants. Journal of Experimental Botany. 2002;**53**:323-331

[41] Wall R, Appel H, Cipollini M, Schultz J. Fertility, root reserves and the cost of inducible defenses in the perennial plant *Solanum carolinense*. Journal of Chemical Ecology. 2005;**31**:2263-2288

[42] Schowalter TD. Insect Ecology: An Ecosystem Approach: Fourth Edition. Academic Press; 2016. Available from: https://www.elsevier.com/books/insect-ecology/schowalter/978-0-12-803033-2

[43] Freeman BC, Beattie GA. An overview of plant defenses against pathogens and herbivores. Plant Pathology and Microbiology Publications; 2008. Available from: https://www.apsnet.org/edcenter/disimpactmngmnt/topc/Pages/OverviewOfPlantDiseases.aspx

[44] de Elguea-Culebras GO, Sanchez-Vioque R, Berruga MI, Herraiz-Penalver D, Santana-Meridas O. Antifeedant effects of common terpenes from mediterranean aromatic plants on *Leptinotarsa decemlineata*. Journal of Soil Science and Plant Nutrition. 2017;**17**(2):475-485

[45] Duijff BJ, Kogel PA, Bakker HM, Schippers B. Influence of pseudobactin 358 on the iron nutrition of barley. Soil Biology and Biochemistry. 1994;**26**:1681-1688

[46] Pavela R. Natural products as allelochemicals in pest management. In: Dubey NK, editor. Natural Products in Plant Pest Management. CAB International; 2001. Available from: www.cabi.org

[47] Isman MB, Machial CM. Pesticides based on plant essential oils: From traditional practice to commercialization. In: Rai M, Carpinella MC, editor. Advances in Phytomedicine. Naturally Occurring Bioactive Compounds. Vol. 3. Amsterdam, the Netherlands; Elsevier: 2006. pp. 29-44

[48] Visser JH, Piron PGM. An open Y-track olfactometer for recording aphid behavioural responses to plant odours. In: Proceedings of the section Experimental and Applied Entomology of the Netherlands Entomological Society, Vol. 9. New Amsterdam; 1998. pp. 41-46

[49] Paschold A, Halitschke R, Baldwin IT. Co(i)-ordinating defenses: NaCOI1 mediates herbivore-induced resistance in *Nicotiana attenuata* and reveals the role of herbivore movement in avoiding defenses. The Plant Journal. 2007;**51**:79-91

[50] Endara MJ, Coley PD. The resource availability hypothesis revisited: A meta-analysis. Functional Ecology. 2011;**25**:389-398

[51] Munakata K. Insect antifeedants in plants. In: Wood DL, Silverstein RM, Nakajima M, editors. Control of Insect Behavior by Natural Products. New York: Academic Press; 1970. pp. 179-187

[52] Renwick JA. Nonpreference mechanisms: Plant characteristics influencing insect behavior. In: Hedin PA, editor. Plant Resistance to Insects. ACS Symposium Series No. 208. Washington D.C.:ACS; 1983. pp. 199-213

[53] Ignatowicz S, Wesolowska B. Insecticidal and deterrent properties of extracts from herbaceous plants. Ochrona Roślin. 1994;**38**:14-15

[54] Shaaya E, Kostjukovski M, Eilberg J, Sukprakarn C. Plant oils as fumigants and contact insecticides for the control of stored-product insects. Journal of Stored Products Research. 1997;**33**(1):7-15

[55] Weaver DK, Subramanyam B. Botanicals. In: Subramanyam B, Hagstrum DW, editors. Alternatives to Pesticides in Stored-Product IPM. Dordrecht: Kluwer Academics Publishers; 2000. pp. 303-320

[56] Dussourd DE. Foraging with finesse: Caterpillar adaptations for circumventing plant defenses. In: Stamp NE, Casey TM, editors. Caterpillars. Ecological and Evolutionary Constraints on Foraging. New York: Chapman and Hall; 1993. pp. 92-131

[57] Brattsten LB. Metabolic defenses against plant allelochemicals. In: Rosenthal GA, Berenbaum MR, editors. Herbivores: Their Interactions Secondary Plant Metabolites. 2nd ed. Vol. 2. San Diego: Academic Press; 1992. pp. 175-242

[58] Berenbaum M, Zangerl AR. Facing the future of plant-insect interaction research: Le retour à la "raison d'être". Plant Physiology. 2008;**146**:804-811

[59] Wood DL, Silverstein RM, Mittelstaedt MH. Control systems of orientation in insects. Annual Review of Entomology. 1962;**7**:177-198

[60] Dethier VG, Browne BL, Smith CN. The designation of chemicals in terms of the responses they elicit from insects. Journal of Economic Entomology. 1960;**53**:1,134-1,136

[61] Huang Y, Tan J, Kini RM, Ho SH. Toxic and antifeedant action of nutmeg oil against *Tribolium castaneum* (Herbst) and *Sitophilus zeamais* Motsch. Journal of Stored Products Research. 1997;**33**:289-298

[62] Huang Y, Lam SL, Ho SH. Bioactivities of essential oils from *Elettaria cardamomum* (L.) to *Sitophilus zeamais* Motschulsky and *Tribolium castaneum* (Herbst). Journal of Stored Products Research. 2000;**36**:107-117

[63] Haque MA, Nakakita H, Ikenaga H, Sota N. Development-inhibiting activity of some tropical plants against *Sitophilus zeamais* Motschulsky (Coleoptera:

Curculionidae). Journal of Stored Products Research. 2000;**36**(3):281-287

[64] Mossi JA, Astolfi V, Kubiak G, Lerin L, Zanella C, Toniazzo G, et al. Insecticidal and repellency activity of essential oil of *Eucalyptus* sp. against *Sitophilus zeamais* Motschulsky (Coleoptera, Curculionidae). Journal of the Science of Food and Agriculture. 2011;**91**:273-277

[65] Tapondjou AL, Adler C, Fontem DA, Bouda H, Reichmuth C. Bioactivities of cymol and essential oils of *Cupressus sempervirens* and *Eucalyptus saligna* against *Sitophilus zeamais* Motschulsky and *Tribolium confusum* du Val. Journal of Stored Products Research. 2005;**41**:91-102

[66] Takahiro I, Matsuzawa H, Santhanaraju VC. Repellent activity of common spices against the rice weevil, *Sitophilus zeamais* Motsch (Coleoptera, Curculionidae). Journal of Tropical Biology and Conservation. 2010;**7**:75-80

[67] Popović Z, Kostić M, Popović S, Skorić S. Bioactivities of essential oils from basil and sage to *Sitophilus oryzae* L. Biotechnology and Biotechnological Equipment. 2006;**20**(1):36-40

[68] Jovanović Z, Kostić M, Popović Z. Grain-protective properties of herbal extracts against the bean weevil *Acanthoscelides obtectus* Say. Industrial Crops and Products. 2007;**26**:100-104

[69] Jevremović S, Lazarević J, Kostić M, Krnjajić S, Ugrenović V, Radonjić A, et al. Contact application of Lamiaceae botanicals reduces bean weevil infestation in stored beans. Archives of Biological Sciences. 2019;**71**(4):665-676

[70] Visser JH, Van Straten S, Maarse H. Isolation and identification of volatiles in the foliage of potato, *Solanum tuberosum*, a host plant of the Colorado beetle, *Leptinotarsa decemlineata*. Journal of Chemical Ecology. 1979;**5**:13-25

[71] Lawrence S, Novak N, Ju C, Cookeb J. Examining the molecular interaction between potato (*Solanum tuberosum*) and Colorado potato beetle *Leptinotarsa decemlineata*. Botany. 2008;**86**(9):1080-1091

[72] Kostić M, Kostić I, Marković T, Jevđović R, Stanković S, Todorović G, Nedić N. Disruption of attractant properties of potato foliage on *Leptinotarsa decemlineata* Say by the use of *Salvia officinalis* L. essential oil. In: Proceedings of the 7th Conference on Medicinal and Aromatic Plants of Southeast European Countries, Serbia; 2012. pp. 351-356

[73] Gökçe A, Isaacs R, Whalon ME. Behavioral response of Colorado potato beetle larvae to antifeedant plant extracts. Pest Management Science. 2006;**62**:1052-1057

[74] Gökçe A, Whalon ME, Çam ME, Yanar Y, Demirtaş İ, Gören N. Plant extract contact toxicities to various developmental stages of Colorado potato beetles (Coleoptera: Chrysomelidae). Annals of Applied Biology. 2006;**149**:197-202

[75] Singh R, Koul O, Jai RP, Jindal J. Evaluation of dietary toxicity of some essential oil allelochemicals for the management of *Chilo partellus* (Swinhoe). Journal of Plant Protection Research. 2010;**50**(3):293-201

[76] Copping LG, Menn JJ. Biopesticides: A review of their action, applications and efficacy. Pest Management Science. 2000;**56**:651-676

[77] Copping LG, Duke SO. Natural products that have been used commercially as crop protection agents. Pest Management Science. 2007;**63**:524-554

[78] Bunch TR, Bond C, Buhl K, Stone D. Diatomaceous Earth General Fact Sheet. National Pesticide

Information Center, Oregon State University Extension Services; 2013. Available from: http://npic.orst.edu/factsheets/degen.html; http://npic.orst.edu/

[79] Stadler T, Butcler M, Weaver DK. Novel use of nanostructured alumina as an insecticide. Pest Management Science. 2009:577-579

[80] Stadler T, Butcler M, Weaver OK, Sofie S. Comparative toxicity of nanostructured alumina and a commercial inert dust for *Sitophilus oryzae* (L.) and *Rhyzopertha dominica* (F.) at varying ambient humidity levels. Journal of Stored Products Research. 2012;**81-90**:445-453

[81] Korunić Z. Diatomaceous earths—Natural insecticides. Pesticidi i Fitomedicina. 2013;**28**(2):77-95

[82] Gvozdenac S, Tanaskovic S, Krnjajic S, Prvulovic D, Ovuka J, Sedlar A. Effects of different inert dusts on *Sitophilus oryzae* and *Plodia interpuncella* during contact exposure. In: International Working Conference on Stored Product Protection (IWCSPP); 7-11 October 2018. Berlin, Germany; 2018. pp. 829-834

[83] Gogos A, Knauer K, Bucheli TD. Nanomaterials in plant protection and fertilization: Current state, foreseen applications, and research priorities. Journal of Agricultural and Food Chemistry. 2012;**60**:9781-9792

[84] Kostić I, Krnjajić S, Radojković A, Radović M, Savić MS, Kostić M, et al. Alumina powders as novel non-toxic insecticide against bean weevil (*Acanthoscelides obtectus* Say). In: Sixth International Scientific Agricultural Symposium "Agrosym 2015", Jahorina, Republic of Srpska, Bosnia, Book of Proceedings; 2015. pp. 943-948

[85] Korunic Z. Rapid assessment of the insecticidal value of diatomaceous earths without conducting bioassays. Journal of Stored Products Research. 1997;**33**(3):219-229

[86] Shah MA, Khan AA. Use of diatomaceous earth for the management of stored-product pests. International Journal of Pest Management. 2014;**60**:100-113

[87] Hänsch R, Mendel RR. Physiological functions of mineral micronutrients (Cu, Zn, Mn, Fe, Ni, Mo, B, Cl). Current Opinion in Plant Biology. 2009;**12**(3):259-266. DOI: 10.1016/j.pbi.2009.05.006

[88] Reynolds O, Keeping M, Meyer J. Silicon-augmented resistance of plants to herbivorous insects: A review. Annals of Applied Biology. 2009;**155**:171-186

Chapter 5

Current Advances in Mass Spectrometry Imaging for Insect Physiology and Metabolism

Fei-Ying Yang, Wei-Yi He and Min-Sheng You

Abstract

Research regarding the distribution of metabolites is a vital aspect of insect molecular biology. However, current approaches (e.g., liquid chromatography-mass spectrometry or immunofluorescence) have cons like requirement of massive tissues, low efficiency, and complicated operating processes. As an emerging technology, mass spectrometry imaging (MSI) can visualize the spatiotemporal distribution of molecules in biological samples without labeling. In this chapter, we retrospect the major types of *in situ* measurement by MSI, and the application of MSI for investigating insect endogenous and exogenous metabolites and monitoring the dynamic changes of metabolites involved with the interactions between insects and plants. Future studies that combine MSI with other genetic tools can facilitate to better explore the underlying mechanisms concerning insect physiology and metabolism.

Keywords: spatial metabolomics, *in situ* characterization, endogenous metabolites, exogenous metabolites, plant-insect interaction

1. Introduction

Insect molecular biology studies the molecular basis of biological processes in insects, including molecular synthesis, modification, mechanisms, and interactions [1]. Metabolites play key roles among all these aspects of insect molecular biology. Therefore, understanding the distribution of metabolites contributes to revealing the mechanisms of insect biology, including ontogeny, metabolism, and physiology. Research methodologies such as liquid chromatography-mass spectrometry (LC-MS) and immunofluorescence are generally used in visualizing the distribution of metabolites. However, all of them have their shortcomings. LC-MS or gas chromatography-mass spectrometry (GC-MS) usually uses the homogenate of a certain weight of specific tissue(s) or organ(s), resulting in losing *in situ* spatiotemporal information. Insect body sizes are mostly small, let alone certain tissues; so tissue-specific researches, in most case, consume a large number of insect individuals [2–4]. On the other hand, ordinary *in situ* characterization technologies such as immunofluorescence assay and fluorescence *in situ* hybridization (FISH) require labeling at specific biomolecules [5–7]. Hence, operating processes such as synthesizing probes and antibodies are usually time-consuming, inefficient, and limited to only one molecule.

As a new molecular visualization technology, mass spectrometry imaging (MSI) has drawn more and more attention in recent years. MSI can visualize the spatial distribution of molecules in specific samples without any labeling and enable simultaneous evaluation and identification of hundreds of molecules *in situ*. In comparison with LC-MS and GC-MS, MSI requires only one sample for biomolecular localization, which makes it a powerful tool to visualize the changes in organism physiology and biochemistry. The basic principle of MSI is to scan target samples such as tissue slice for desorption and ionization of molecules or ions on the surface of samples by a laser or a high-energy ion beam [8]. Mass analyzer obtains mass-to-charge ratio (m/z) and ion intensity of the molecules or ions from pixels. Mass peaks are obtained from the database of imaging software such as FlexImaging and used to visualize one-dimensional linear profiling, two-dimensional spatial distribution of molecules, or three-dimensional anatomic structure [8]. MSI has been widely applied in life sciences, such as histology [9, 10]; pathology [11, 12]; pharmacology [13, 14]; food science [15]; botany [16–19]; and microbiology [20, 21].

In this chapter, we introduce the major types of *in situ* measurement by MSI and present an example of matrix-assisted laser desorption ionization (MALDI) to elucidate the operating processes. We also discuss the advances of MSI in insect physiology and biochemistry to better promote the research in entomology.

2. Mass spectrometry imaging method

Among all the MSI technologies, we can divide them into two major groups, vacuum ionization mass spectrometry imaging system and ambient ionization mass spectrometry imaging system, based on whether the environment of the instruments is a vacuum [8]. Based on desorption or ionization ion, vacuum ionization mass spectrometry imaging system can be further divided into different categories, namely MALDI [22], secondary ion mass spectrometry (SIMS) [23], surface-assisted laser desorption ionization (SALDI) [24], and laser desorption ionization (LDI) [25]; ambient ionization mass spectrometry imaging system can be further divided into different categories, namely desorption electrospray ionization (DESI) [26], laser ablation electrospray ionization (LAESI) [27], laser electrospray mass spectrometry (LEMS) [28], electrospray laser desorption ionization (ELDI) [28], atmospheric pressure matrix-assisted laser desorption ionization (AP-SMALDI) [29], and air flow-assisted ionization (AFAI) [30]. Among all these above-mentioned technologies, MALDI-MSI is the most popular technology in life science research because it not only can be applied to a wide range from inorganic ion, small molecules to proteins but also has the characteristics of high accuracy and sensitivity [31]. Here, we provide a further explanation of the basic principle of MADLI-MSI and elucidate the workflow for MALDI.

The basic working principle of MALDI is that target analytes on the surface of tissue are crystallized with matrix (e.g., α-cyano-4-hydroxycinnamic acid and 2,5-dihydroxyacetophenone) to form a complex. When the complex is exposed by infrared laser at 2.94 or 10.6 μm and/or ultraviolet laser at 337, 355, or 266 nm, it absorbs the laser energy and converts these analytes into a phase of gas, which causes molecule ionization. The ionized molecules automatically enter a mass spectrometer where the molecules are detected and mapped [19].

A typical experimental workflow for MALDI is as follows (**Figure 1**):

a. Insect tissues are flash-frozen (with or without fixation) in an embedding media with gelatin, carboxymethylcellulose, or water;

b. Each sample is cryo-sectioned at 10–20 μm thickness and mounted onto glass slides coated with indium tin oxide, which is then lyophilized for tissue imaging;

c. The lyophilized slide is subject to three irregular fiducial markings on the surface of each sample for localization;

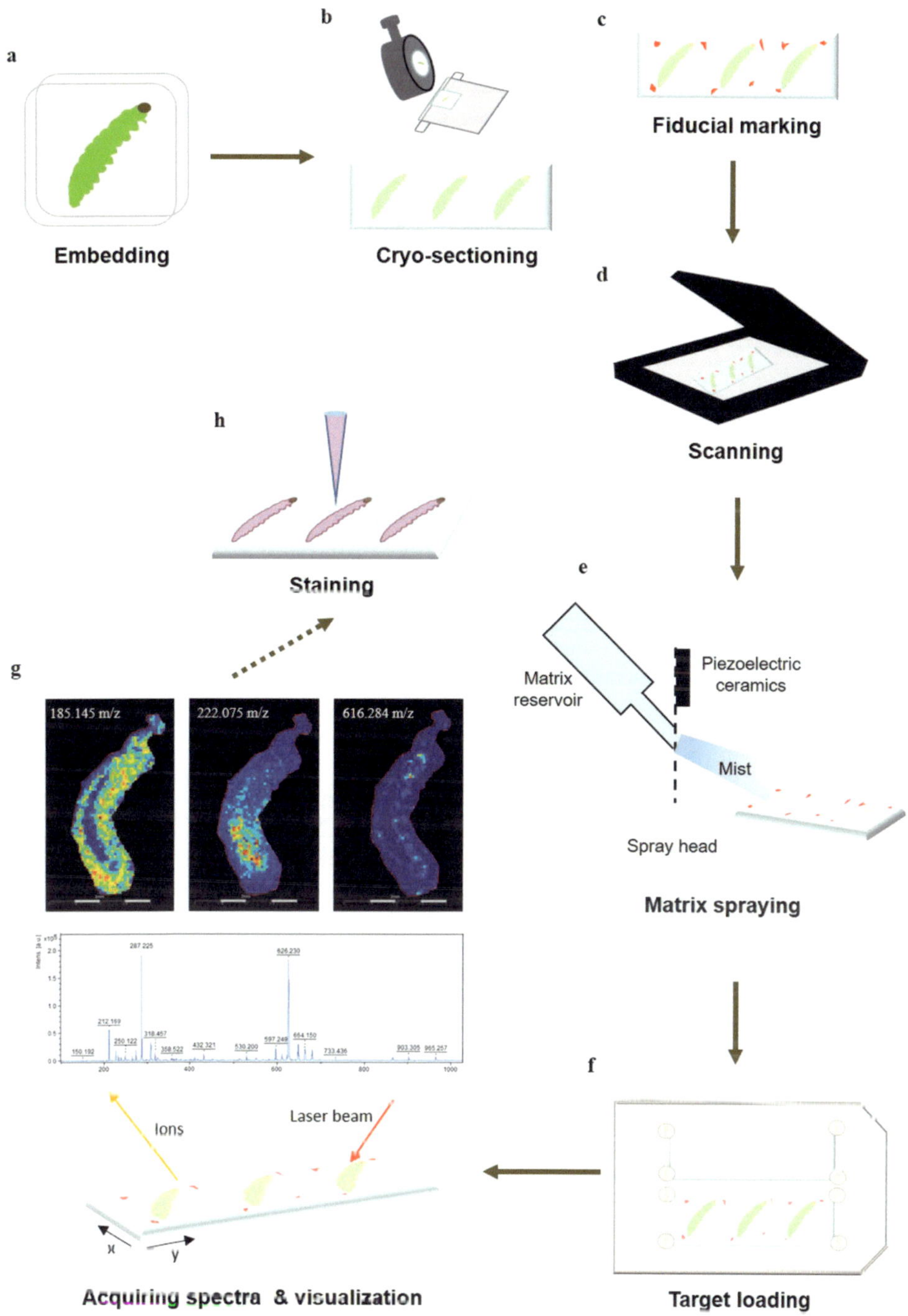

Figure 1.
MALDI-MSI imaging workflow.

d. A digital image of the sample with fiducials is acquired;

e. A chemical matrix is applied to promote desorption and ionization. Matrix is coated by a sprayer/nebulizer or by solvent-free sublimation to acquire homogeneous matrix coverage over the entire tissue surface;

f. After matrix deposition, the target is inserted into the instrument, for which experimental parameters (e.g., laser energy, step size of plate movement, and a selected region of the tissue) are optimized to scan the image;

g. A laser beam is emitted for desorption to acquire mass spectra at every x and y grid points within the scanning area, so to visualize target ions and convert the ion's intensity into a color scheme;

h. Hematoxylin-eosin staining is optional for displaying tissue localization.

3. Application of mass spectrometry imaging in entomological researches

MSI can visualize the spatial and temporal distributions of molecules. Endogenous metabolites, exogenous metabolites, and insect-plant interactions are three main aspects of MSI application to insect tissue section for *in situ* characterization. Endogenous metabolites refer to lipids, neuropeptides, proteins, and defense compounds [32–35]; exogenous metabolites are drugs and insecticides [36, 37]; insect-plant interactions are associated with the fate of plant secondary defense compounds in insects [38]. We summarize the major applications of MSI for a better understanding of insect physiology and metabolism (**Table 1**).

3.1 Insect endogenous metabolites

3.1.1 Lipids

Lipids are basic cell components and play important roles in insect development and reproduction, such as maintenance of cell membrane structure and intra or extracellular signaling [39–41]. For example, glycerophospholipids, phosphatidylcholines, and phosphatidylethanolamines are basic components of cell and lysophospholipids have an important function in inflammation, abiotic stress, and biotic stress signal transmit [42]. MSI has been widely applied in many aspects in model insect *Drosophila melanogaster*, such as the neutral lipids three-dimensional spatial distribution on the surface adults [43, 44], body lipid distribution [45], brain lipid structure [46, 47], wing lipids [34, 44], Malpighian tubule phospholipid distribution [48], and phospholipids in the brain treated with cocaine [49]. Moreover, MSI detected and localized the composition and distribution of triacylglyceride in *Aedes aegypti*, phospholipid and phosphatidylcholine in *Anopheles stephensi* [42], and phospholipids in *Schistocerca gregaria* [37].

3.1.2 Neuropeptides

Neuropeptides, a kind of structurally diverse signaling molecules, can control and regulate fundamental physiological functions such as growth, reproduction, and environmental stress tolerance in animals [50]. MSI detected and localized the distribution of 14 neuropeptides in coronal brain sections in all development stages

Species	Tissue	Major analyte	Method	Embedding medium	Thickness (μm)	Matrix	Ref.
Endogenous metabolites							
Anopheles stephensi	Whole-body	Lipids	AP-SMALDI	5% CMC	20	DHB	[42]
Aedes aegypti	Ovarian follicles	Lipids	3D-SIMS	/	100	/	[41]
Apis mellifera	Brain	Neuropeptides	MALDI	/	14	CHCA	[51, 52]
	Brain	Protein	MALDI	/	12	CA	[32]
	Brain	L-arginine	MALDI	/	12	DHB	[53]
	Venom	Venom toxins	MALDI	/	10	CHCA	[54]
Drosophila melanogaster	Body	Peptide	MALDI	Agarose	10	CHCA	[55]
	Brain	Lipids	SIMS	10% Gelatin	15	/	[47]
	Brain	Phospholipid	SIMS	10% Gelatin	12	/	[49]
	Collar	Lipids	SIMS	10% Gelatin	10	/	[46]
	Brain & head	GABA	MALDI	4% CMC	15	CHCA	[56]
	Malpighian tubule	Lipids	MALDI	5% CMC	12	DHB、DAN	[48]
	Surface	Lipids	MALDI	/	/	LiDHB	[43]
	Whole-body	Neuropeptides	AP-SMALDI	5% CMC	20	DHB	[29]
	Whole-body	Lipids	MALDI	10% Gelatin	20	DHB	[45]
	Wing	Lipids	SIMS	PBS	/	DHB	[34, 44]
Graphosoma lineatum	Head to abdomen	Non-polar compounds	DAPPI	/	/	/	[57]
Paederus riparius	Whole-body	Defensive compounds	AP-SMALDI	10% Tragacanth gum	16	DHB	[35]

Species	Tissue	Major analyte	Method	Embedding medium	Thickness (μm)	Matrix	Ref.
Periplaneta americana	Brain	Neuropeptides	MALDI	Gelatin	14	CHCA	[33]
	Neuro-endocrine tissues	Neuropeptides	MALDI	Paraffin	20	DHB	[58]
Prorhinotermes simplex	Head to abdomen	Non-polar compounds	DAPPI	/	/	/	[57]
Solenopsis invicta	Venom	Venom proteins	MALDI	Gelatin	14	DHB	[59]
Exogenous metabolites							
Drosophila melanogaster	Whole-body	Insecticide	MALDI	10% Gelatin	15	DHB	[36]
Schistocerca gregaria	Whole-body	Drugs	DESI	5% CMC	50	/	[37, 60]
Helicoverpa armigera	Whole-body	Biopesticide	MALDI	/	16	DHB	[61]
Insect-plant interaction system							
Aphis glycines	Feeding leaf	Metabolites	MALDI	/	/	DHB、DAN	[62, 63]
Athalia rosae	Whole-body	Glucosinolates	MALDI	Water	15	CHCA	[64]
Chorthippus dorsatus	Gut, feces	Metabolites	LDI	1% PBS	12	DHB、DAN、CHCA	[38]
Others							
Acromyrmex echinatior	Nest	Microbial	MALDI	/	/	DHB、CHCA	[65]
Ants	Propleural plate	Fungus	MALDI	/	/	DHB	[66]
Bombus terrestris	Whole-body	/	MALDI	pHPMA	12	DHB、SA	[67]

Table 1.
Overview of the application of MSI in insect sciences.

of *D. melanogaster* [29]. These neuropeptides play important roles in physiological processes (e.g., allatostatins and tachykinin-like peptides participate in odor perception and locomotor activity). Neuropeptides can act as transmitters or neuromodulators in the central nervous system [33]. Neuropeptides in the brain of *Apis mellifera* are related to the functional division of the population and their activities. Worker bees' neuropeptide levels at the age of 0–15 d increased with the in-hive activities but decreased with out-hive activities (guarding and foraging) at 15–25 d [51]. Further study proved that allatostatin and tachykinin-related neuropeptides in the brain of worker bees were related to aggressiveness behaviors [52]. Neuropeptides distribution in the retrocerebral complex of *Periplaneta Americana* revealed the differentiation of prohormone processing and the distinctness of neuropeptides-based compartmentalization [33]. These studies proved that MSI has the advantages of sensitivity, which can facilitate to detect peptides in low abundance.

3.1.3 Proteins

As a kind of macromolecules, proteins are fundamental compounds of organisms and take part in important cellular processes, such as DNA replication and metabolisms. MSI can simultaneously and specifically detect the spatial distribution of massive proteins and overcome antibody cross-contamination. MSI system has been used to evaluate the negative impacts in the brain of *A. mellifera* exposing to a sublethal concentration of imidacloprid. The system has successfully visualized the distribution of 24 proteins (e.g., cytochrome P450s, glutathione S-transferases, and heat shock protein 70s). Besides, 8-day exposure to imidacloprid triggered biochemical changes in *A. mellifera* brain (e.g., up-regulated acetylcholinesterase and amyloid precursor-like protein and down-regulated cytochrome P450 and disulfide-isomerase protein). This could influence the well-being of *A. mellifera* (e.g., learning and memory acquisition, maintaining neuronal integrity, detoxification, and apoptosis) [32].

3.1.4 Others

In addition to lipids, neuropeptides, and proteins, MSI can also be used to visualize the distributions of defensive compounds, special proteins (e.g., venom allergens and toxins) and other small molecules (e.g., betaine and amino acids). Defensive compounds (e.g., pederin, pseudopederin, and pederon) were detected and localized in the organs of *Paederus riparius* [35]. Three venom allergens and two venom toxins were mapped in the honeybee [54]. Poison sac was the lactation of main venom proteins in *Solenopsis invicta* [59]. Nonpolar compounds (e.g., (*E*)-1-nitropentadec-1-ene and (*E*)-hex-2-enal) can be detected from the head to the abdomen in two model insects, *Prorhinotermes simplex* and *Graphosoma lineatum*. Gland openings and gland reservoirs were the most active areas in *P. simplex* and *G. lineatum* [57]. Other small molecules (e.g., betaine and amino acids) were detected in *Schistocerca gregaria* [37]. Semiochemicals were mapped on the surface of the adults of *D. melanogaster* [43]. Two male-specific sex pheromones were localized in the ejaculatory bulb of *D. melanogaster* [45]. MSI can also be used as a novel *in situ* metabonomic tool to study the metabolism of L-arginine of the honeybee brain in response to proboscis extension [53].

3.2 Insect exogenous metabolites

3.2.1 Insecticides

MSI can be applied to visualize the distribution of insecticides in insects and their negative influence on the target insects. Imidacloprid was used to study its distribution and accumulation in *D. melanogaster*. Based on laser irradiation, imidacloprid was found to be converted to guanidine-imidacloprid. It eventually accumulated and spread in the abdominal region [36]. *Azadirachta indica* is an economical tree that can be used to distract a biopesticide component, azadirachtin-A. It was only presented in the midgut of *Helicoverpa armigera* after application [61].

3.2.2 Drug/pharmacological test

Pharmacology model animals are crucial for scientists or pharmacologists to test the side effects of newly developed drugs before clinical trials on human beings. Common pharmacology model animal species include mice, rabbits, dogs, and monkeys. Insects, compared with the above-mentioned animals, have pros such as low costs, high fertility, and moral constraints. Locust has become a new model species for pharmacology test because of its high similarity with mammals. Antihistamine drug terfenadine was tested in locust to study the distribution of secondary metabolites. Terfenadine was gradually degraded from hemolymph to stomach and intestinal wall. However, terfenadine-related chemical compounds such as terfenadine acid, terfenadine glucoside, and terfenadine phosphate were distributed in the unexcreted feces in the intestine, which revealed a rapid discharge of metabolites through defecation [37]. Besides, the spatial and temporal distribution of midazolam was tested in locust. The results showed that midazolam was abundant in 30-min but only found in the feces after a 2-hour application. Midazolam glucoside was found in gut, gastric caeca, and feces after a 2-hour application, indicating that glucose conjugates are a major detoxification pathway to neutralize the effects caused by midazolam in locusts [60].

In addition, *D. melanogaster* was used to test how cocaine, drug removal, and methylphenidate influence the brain lipids. The results showed that cocaine increased the level of phosphatidylcholines and decreased the levels of phosphatidylethanolamines and phosphatidylinositols. Methylphenidate-treated flies failed to rescue the levels of phosphatidylethanolamines and phosphatidylinositols, but enhanced the reversal of phosphatidylcholine levels [49].

3.3 Insect-plant interaction

Plants and herbivorous insects are co-evolved in nature. Plants activate defense reaction by releasing secondary metabolites when they are under attack by herbivorous insects, while herbivores trigger anti-defense systems for adapting and overcoming the side effects of secondary metabolites produced by plants [68]. Illuminating the changes of secondary metabolites during the interactions between insects and plants can contribute to a better understanding of plant resistance and insect adaptability.

Chemical interaction between soybean (*Glycine max*) and aphid (*Aphis glycines*) was studied. The metabolite changes (e.g., phosphorylcholine and amino acid) were detected in the aphid-infested soybean leaves. The results suggested that secondary metabolites were produced by dead cells after aphid infestation. Moreover, other compounds such as pipecolic acid, salicylic acid, formononetin, and dihydroxyflavone were consistently detected in the plant regions infested by aphids [62]. It was also found that isoflavones can be accumulated in mesophyll cells or epidermis but were not present in the vasculature. The results indicated that isoflavones take part in non-phloem defense response [63].

In addition, MSI can be used for physiological studies such as annihilation of the plant secondary metabolites by herbivores. Glucosinolate gradually changes in the distribution and metabolic sequestration were detected in the body of *Athalia rosae* that fed on host plants after different periods. The glucosinolate sinalbin was accumulated in the hemolymph and eventually circulated the Malpighian tubules. The results indicated that the insect gut plays a crucial role as a regulatory functional organ [64].

Moreover, MSI can be applied to the entire metabolic process of secondary metabolites in the plant-insect-soil system. The fate of the secondary metabolites produced by *Dactylis glomerata* was tracked in the different organs of herbivore *Chorthippus dorsatus*, and finally in the soil solution. After infestation by herbivores, levels of quinic acid, apigenin, and luteolin decreased, while those of flavonoids and rosmarinic acid increased in the leaf wounds of plants in 1 d. Quinic acid can be detected during the digestion process in the grasshoppers' gut and unexcreted feces [38]. Overall, MSI is a useful tool to visualize plant defense and insect defense processes from the responses of plants infested by herbivores to insect defense systems responding to plant-derived metabolites.

4. Conclusion

MSI has been proved to be an effective and powerful tool to visualize molecules' spatial distribution and temporal changes. In this chapter, we introduce the major types of MSI methodologies and describe the typical experimental workflow for MALDI-MSI. We also retrospect three major applications of MSI in insect physiology, for example, endogenous metabolites, exogenous metabolites, and insect-plant interaction. However, MSI still has some technical cons with limited application range that need to be optimized. In addition, MSI can cooperate with other genetic tools (e.g., proteomics, metabolomics, or lipidomics) for a better understanding of sophisticated insect biology.

Acknowledgements

The work was supported by National Key R&D Program of China (2017YFD0200400), Special Key Project of Fujian Province (2018NZ01010013), Natural Science Foundation of Fujian Province (2019J01369) in China, and Innovation Fund of Fujian Agriculture and Forestry University (CXZX2018092, CXZX2016128, CXZX2017321 and 324-1122yb059).

Conflict of interest

The authors declare no conflict of interest.

Author details

Fei-Ying Yang[1,2,3], Wei-Yi He[1,2,3]* and Min-Sheng You[1,2,3]

1 State Key Laboratory for Ecological Pest Control of Fujian and Taiwan Crops, College of Plant Protection, Fujian Agriculture and Forestry University, Fuzhou, China

2 International Joint Research Laboratory of Ecological Pest Control, Ministry of Education, Fujian Agriculture and Forestry University, Fuzhou, China

3 Institute of Applied Ecology, Fujian Agriculture and Forestry University, Fuzhou, China

*Address all correspondence to: wy.he@fafu.edu.cn

References

[1] Chandrasekar R, Brintha PG, Park EY, Pelsoi P, Liu F, Goldsmith M, et al. Introduction to insect molecular biology. In: Chandrasekar R, Tyagi BK, Gui ZZ, Reeck GR editors. Short Views on Insect Biochemistry and Molecular Biology. 1st ed. Manhattan, Academic Publisher; 2014:3-56

[2] Li Y, Wang X, Chen Q, Hou Y, Xia Q, Zhao P. Metabolomics analysis of the larval head of the silkworm, *Bombyx mori*. International Journal of Molecular Sciences. 2016;**17**(9):1460. DOI: 10.3390/ijms17091460

[3] Chintapalli VR, Al Bratty M, Korzekwa D, Watson DG, Dow JAT. Mapping an atlas of tissue-specific *Drosophila melanogaster* metabolomes by high resolution mass spectrometry. PLoS One. 2013;**8**(10):1-13. DOI: 10.1371/journal.pone.0078066

[4] Shi T, Burton S, Wang Y, Xu S, Zhang W, Yu L. Metabolomic analysis of honey bee, *Apis mellifera* L. response to thiacloprid. Pesticide Biochemistry and Physiology. 2018;**152**:17-23. DOI: 10.1016/j.pestbp.2018.08.003

[5] Koinuma H, Maejima K, Tokuda R, Kitazawa Y, Nijo T, Wei W, et al. Spatiotemporal dynamics and quantitative analysis of phytoplasmas in insect vectors. Scientific Reports. 2020;**10**:4291. DOI: 10.1038/s41598-020-61042-x

[6] Ban FX, Yin TY, Guo Q, Pan LL, Liu YQ, Wang XW. Localization and quantification of begomoviruses in whitefly tissues by immunofluorescence and quantitative PCR. Journal of Visualized Experiments. 2020;**156**:1-8. DOI: 10.3791/60731

[7] Jia HR, Sun YF, Luo SP, Wu KM. Characterization of antennal chemosensilla and associated odorant binding as well as chemosensory proteins in the *Eupeodes corollae* (Diptera: Syrphidae). Journal of Insect Physiology. 2019;**113**:49-58. DOI: 10.1038/s41598-018-25996-3

[8] Ho YN, Shu LJ, Yang YL. Imaging mass spectrometry for metabolites: Technical progress, multimodal imaging, and biological interactions. Wiley Interdisciplinary Reviews. Systems Biology and Medicine. 2017;**9**(5):e1387. DOI: 10.1002/wsbm.1387

[9] Longuespée R, Casadonte R, Kriegsmann M, Pottier C, Picard de Muller G, Delvenne P, et al. MALDI mass spectrometry imaging: A cutting-edge tool for fundamental and clinical histopathology. Proteomics. Clinical Applications. 2016;**10**(7):701-719. DOI: 10.1002/prca.201500140

[10] Yalcin EB, de la Monte SM. Review of matrix-assisted laser desorption ionization-imaging mass spectrometry for lipid biochemical histopathology. The Journal of Histochemistry and Cytochemistry. 2015;**63**(10):762-771. DOI: 10.1369/0022155415596202

[11] Schubert KO, Weiland F, Baune BT, Hoffmann P. The use of MALDI-MSI in the investigation of psychiatric and neurodegenerative disorders: A review. Proteomics. 2016;**16**(11-12):1747-1758. DOI: 10.1002/pmic.201500460

[12] Bowrey HE, Anderson DM, Pallitto P, Gutierrez DB, Fan J, Crouch RK, et al. Imaging mass spectrometry of the visual system: Advancing the molecular understanding of retina degenerations. Proteomics - Clinical Applications. 2016;**10**(4):391-402. DOI: 10.1002/prca.201500103

[13] Liu X, Hummon AB. Mass spectrometry imaging of therapeutics from animal models to three-dimensional

cell cultures. Analytical Chemistry. 2015;**87**(19):9508-9519. DOI: 10.1021/acs.analchem.5b00419

[14] Kwon HJ, Kim Y, Sugihara Y, Baldetorp B, Welinder C, Watanabe KI, et al. Drug compound characterization by mass spectrometry imaging in cancer tissue. Archives of Pharmacal Research. 2015;**38**(9):1718-1727. DOI: 10.1007/s12272-015-0627-2

[15] Yoshimura Y, Goto-Inoue N, Moriyama T, Zaima N. Significant advancement of mass spectrometry imaging for food chemistry. Food Chemistry. 2016;**210**:200-211. DOI: 10.1016/j.foodchem.2016.04.096

[16] Dong Y, Li B, Aharoni A. More than pictures: When MS imaging meets histology. Trends in Plant Science. 2016;**21**(8):686-698. DOI: 10.1016/j.tplants.2016.04.007

[17] Sturtevant D, Lee YJ, Chapman KD. Matrix assisted laser desorption/ionization-mass spectrometry imaging (MALDI-MSI) for direct visualization of plant metabolites *in situ*. Current Opinion in Biotechnology. 2016;**37**:53-60. DOI: 10.1016/j.copbio.2015.10.004

[18] Boughton BA, Thinagaran D, Sarabia D, Bacic A, Roessner U. Mass spectrometry imaging for plant biology: A review. Phytochemistry Reviews. 2016;**15**(3):445-488. DOI: 10.1007/s11101-015-9440-2

[19] Qin L, Zhang Y, Liu Y, He H, Li Y, Zeng M, et al. Recent advances in matrix-assisted laser desorption/ionisation mass spectrometry imaging (MALDI-MSI) for *in situ* analysis of endogenous molecules in plants. Phytochemical Analysis. 2018;**29**(4):351-364. DOI: 10.1002/pca.2759

[20] Shih CJ, Chen PY, Liaw CC, Lai YM, Yang YL. Bringing microbial interactions to light using imaging mass spectrometry. Natural Product Reports. 2014;**31**(6):739-755. DOI: 10.1039/c3np70091g

[21] Yang JY, Phelan VV, Simkovsky R, Watrous JD, Trial RM, Fleming TC, et al. Primer on agar-based microbial imaging mass spectrometry. Journal of Bacteriology. 2012;**194**(22):6023-6028. DOI: 10.1128/JB.00823-12

[22] Wang J, Qiu S, Chen S, Xiong C, Liu H, Wang J, et al. MALDI-TOF MS imaging of metabolites with a N-(1-Naphthyl) ethylenediamine dihydrochloride matrix and its application to colorectal cancer liver metastasis. Analytical Chemistry. 2014;**87**(1):422-430. DOI: 10.1021/ac504294s

[23] Hoshi T, Kudo M. High resolution static SIMS imaging by time of flight SIMS. Applied Surface Science. 2003;**203**:818-824. DOI: 10.1016/S0169-4332(02)00834-6

[24] Phan NTN, Mohammadi AS, Dowlatshahi Pour M, Ewing AG. Laser desorption ionization mass spectrometry imaging of drosophila brain using matrix sublimation versus modification with nanoparticles. Analytical Chemistry. 2016;**88**(3):1734-1741. DOI: 10.1007/s00216-017-0336-4

[25] Le Pogam P, Legouin B, Geairon A, Rogniaux H, Lohézic-Le Dévéhat F, Obermayer W, et al. Spatial mapping of lichen specialized metabolites using LDI-MSI: Chemical ecology issues for Ophioparma ventosa. Scientific Reports. 2016;**6**:37807. DOI: 10.1038/srep37807

[26] Ifa DR, Wiseman JM, Song Q, Cooks RG. Development of capabilities for imaging mass spectrometry under ambient conditions with desorption electrospray ionization (DESI). International Journal of Mass Spectrometry. 2007;**259**(1-3):8-15. DOI: 10.1016/j.ijms.2006.08.003

[27] Nemes P, Vertes A. Laser ablation electrospray ionization for atmospheric pressure, *in vivo*, and imaging mass spectrometry. Analytical Chemistry. 2007;**79**(21):8098-8106. DOI: 10.1021/ac071181r

[28] Coello Y, Jones AD, Gunaratne TC, Dantus M. Atmospheric pressure femtosecond laser imaging mass spectrometry. Analytical Chemistry. 2010;**82**(7):2753-2758. DOI: 10.1021/ac9026466

[29] Khalil SM, Pretzel J, Becker K, Spengler B. High-resolution AP-SMALDI mass spectrometry imaging of *Drosophila melanogaster*. International Journal of Mass Spectrometry. 2017;**416**:1-19. DOI: 10.1016/j.ijms.2017.04.001

[30] Tang F, Bi Y, He J, Li T, Abliz Z, Wang X. Application of super-resolution reconstruction of sparse representation in mass spectrometry imaging. Rapid Communications in Mass Spectrometry. 2015;**29**(12):1178-1184. DOI: 10.1002/rcm.7205

[31] Zaima N, Hayasaka T, Goto-inoue N, Setou M. Matrix-assisted laser desorption/ionization imaging mass spectrometry. International Journal of Molecular Sciences. 2010;**11**(12):5040-5055. DOI: 10.3390/ijms11125040

[32] Catae AF, da Silva Menegasso AR, Pratavieira M, Palma MS, Malaspina O, Roat TC. MALDI-imaging analyses of honeybee brains exposed to a neonicotinoid insecticide. Pest Management Science. 2019;**75**(3):607-615. DOI: 10.1002/ps.5226

[33] Ly A, Ragionieri L, Liessem S, Becker M, Deininger SO, Neupert S, et al. Enhanced coverage of insect neuropeptides in tissue sections by an optimized mass-spectrometry-imaging protocol. Analytical Chemistry. 2019;**91**(3):1980-1988. DOI: 10.1021/acs.analchem.8b04304

[34] Marty F, Rago G, Smith DF, Gao X, Eijkel GB, MacAleese L, et al. Combining time-of-flight secondary ion mass spectrometry imaging mass spectrometry and CARS microspectroscopy reveals lipid patterns reminiscent of gene expression patterns in the wing imaginal disc of *Drosophila melanogaster*. Analytical Chemistry. 2017;**89**(18):9664-9670. DOI: 10.1021/acs.analchem.7b00125

[35] Bhandari DR, Schott M, Römpp A, Vilcinskas A, Spengler B. Metabolite localization by atmospheric pressure high-resolution scanning microprobe matrix-assisted laser desorption/ionization mass spectrometry imaging in whole-body sections and individual organs of the rove beetle *Paederus riparius*. Analytical and Bioanalytical Chemistry. 2015;**407**(8):2189-2201. DOI: 10.1007/s00216-014-8327-1

[36] Ohtsu S, Yamaguchi M, Nishiwaki H, Fukusaki E, Shimma S. Development of a visualization method for imidacloprid in *Drosophila melanogaster* via imaging mass spectrometry. Analytical Sciences. 2018;**34**(9):991-996. DOI: 10.2116/analsci.18scp04

[37] Olsen LR, Hansen SH, Janfelt C. Distribution of terfenadine and its metabolites in locusts studied by desorption electrospray ionization mass spectrometry imaging. Analytical and Bioanalytical Chemistry. 2015;**407**(8):2149-2158. DOI: 10.1007/s00216-014-8292-8

[38] Crecelius AC, Michalzik B, Potthast K, Meyer S, Schubert US. Tracing the fate and transport of secondary plant metabolites in a laboratory mesocosm experiment by employing mass spectrometric imaging. Analytical and Bioanalytical Chemistry. 2017;**409**(15):3807-3820. DOI: 10.1007/s00216-017-0325-7

[39] Beenakkers AT, van der Horst DJ, Van Marrewijk WJA. Role of lipids in

energy metabolism. In: Downer RGH, editor. Energy Metabolism in Insects. 1st ed. New York: Plenum; 1981. p. 53-100

[40] Gilbert LI, Chino H. Transport of lipids in insects. Journal of Lipid Research. 1974;**15**(5):439-456

[41] Castellanos A, Ramirez CE, Michalkova V, Nouzova M, Noriega FG, Fernández-Lima F. Three dimensional secondary ion mass spectrometry imaging (3D-SIMS) of: *Aedes aegypti* ovarian follicles. Journal of Analytical Atomic Spectrometry. 2019;**34**(5):874-883. DOI: 10.1039/c8ja00425k

[42] Khalil SM, Römpp A, Pretzel J, Becker K, Spengler B. Phospholipid topography of whole-body sections of the *Anopheles stephensi* mosquito, characterized by high-resolution atmospheric-pressure scanning microprobe matrix-assisted laser desorption/ionization mass spectrometry imaging. Analytical Chemistry. 2015;**87**(22):11309-11316. DOI: 10.1021/acs.analchem.5b02781

[43] Kaftan F, Vrkoslav V, Kynast P, Kulkarni P, Böcker S, Cvačka J, et al. Mass spectrometry imaging of surface lipids on intact Drosophila melanogaster flies. Journal of Mass Spectrometry. 2014;**49**(3):223-232. DOI: 10.1002/jms.3331

[44] Vrkoslav V, Muck A, Cvačka J, Svatoš A. MALDI imaging of neutral cuticular lipids in insects and plants. Journal of the American Society for Mass Spectrometry. 2010;**21**(2):220-231. DOI: 10.1016/j.jasms.2009.10.003

[45] Niehoff AC, Kettling H, Pirkl A, Chiang YN, Dreisewerd K, Yew JY. Analysis of drosophila lipids by matrix-assisted laser desorption/ionization mass spectrometric imaging. Analytical Chemistry. 2014;**86**(22):11086-11092. DOI: 10.1021/ac503171f

[46] Le MUT, Son JG, Shon HK, Park JH, Lee SB, Lee TG. Comparison between thaw-mounting and use of conductive tape for sample preparation in ToF-SIMS imaging of lipids in *Drosophila* microRNA-14 model. Biointerphases. 2018;**13**(3):03B414. DOI: 10.1116/1.5019597

[47] Phan NTN, Munem M, Ewing AG, Fletcher JS. MS/MS analysis and imaging of lipids across drosophila brain using secondary ion mass spectrometry. Analytical and Bioanalytical Chemistry. 2017;**409**(16):3923-3932. DOI: 10.1007/s00216-017-0336-4

[48] Yang E, Gamberi C, Chaurand P. Mapping the fly Malpighian tubule lipidome by imaging mass spectrometry. Journal of Mass Spectrometry. 2019;**54**(6):557-566. DOI: 10.1002/jms.4366

[49] Philipsen MH, Phan NTN, Fletcher JS, Ewing AG. Interplay between cocaine, drug removal, and methylphenidate reversal on phospholipid alterations in *Drosophila* brain determined by imaging mass spectrometry. ACS Chemical Neuroscience. 2020;**11**(5):806-813. DOI: 10.1021/acschemneuro.0c00014

[50] Schoofs L, de Loof A, van Hiel MB. Neuropeptides as regulators of behavior in insects. Annual Review of Entomology. 2017;**62**(1):35-52. DOI: 10.1146/annurev-ento-031616-035500

[51] Pratavieira M, Ribeiro A, Maria A, Garcia C, Simo D, Gomes PC, et al. MALDI imaging analysis of neuropeptides in the Africanized honeybee (*Apis mellifera*) brain: Effect of ontogeny. Journal of Proteome Research. 2014;**13**(6):3054-3064. DOI: 10.1021/pr500224b

[52] Pratavieira M, Menegasso ARDS, Esteves FG, Sato KU, Malaspina O, Palma MS. MALDI imaging analysis of neuropeptides in Africanized

honeybee (*Apis mellifera*) brain: Effect of aggressiveness. Journal of Proteome Research. 2018;**17**(7):2358-2369. DOI: 10.1021/acs.jproteome.8b00098

[53] Pratavieira M, da Silva Menegasso AR, Roat T, Malaspina O, Palma MS. *In situ* metabolomics of the honeybee brain: The metabolism of l -arginine through the polyamine pathway in the proboscis extension response (PER). Journal of Proteome Research. 2020;**19**(2):832-844. DOI: 10.1021/acs.jproteome.9b00653

[54] Francese S, Lambardi D, Mastrobuoni G, la Marca G, Moneti G, Turillazzi S. Detection of honeybee venom in envenomed tissues by direct MALDI MSI. Journal of the American Society for Mass Spectrometry. 2009;**20**(1):112-123. DOI: 10.1016/j.jasms.2008.09.006

[55] Zhong Y, Shobo A, Hancock MA, Multhaup G. Label-free distribution of anti-amyloid D-AIP in *Drosophila melanogaster*: Prevention of Aβ42-induced toxicity without side effects in transgenic flies. Journal of Neurochemistry. 2019;**150**(1):74-87. DOI: 10.1111/jnc.14720

[56] Enomoto Y, Phan NTAN, Yamaguchi M, Fukusaki E, Shimma S. Mass spectrometric imaging of GABA in the *Drosophila melanogaster* adult head. Analytical Sciences. 2018;**34**(9):1055-1059. DOI: 10.2116/analsci.18SCN01

[57] Rejšek J, Vrkoslav V, Hanus R, Vaikkinen A, Haapala M, Kauppila TJ, et al. The detection and mapping of the spatial distribution of insect defense compounds by desorption atmospheric pressure photoionization Orbitrap mass spectrometry. Analytica Chimica Acta. 2015;**886**:91-97. DOI: 10.1016/j.aca.2015.06.007

[58] Paine MRL, Ellis SR, Maloney D, Heeren RMA, Verhaert PDEM. Digestion-free analysis of peptides from 30-year-old formalin-fixed, paraffin-embedded tissue by mass spectrometry imaging. Analytical Chemistry. 2018;**90**(15):9272-9280. DOI: 10.1021/acs.analchem.8b01838

[59] Das T, Alabi I, Colley M, Yan F, Griffith W, Bach S, et al. Major venom proteins of the fire ant *Solenopsis invicta*: Insights into possible pheromone-binding function from mass spectrometric analysis. Insect Molecular Biology. 2019;**27**(4):505-511. DOI: 10.1111/imb.12388

[60] Olsen LR, Gabel-Jensen C, Wubshet SG, Kongstad KT, Janfelt C, Badolo L, et al. Characterization of midazolam metabolism in locusts: The role of a CYP3A4-like enzyme in the formation of 1′-OH and 4-OH midazolam. Xenobiotica. 2016;**46**(2):99-107. DOI: 10.3109/00498254.2015.1051604

[61] Dawkar VV, Barage SH, Barbole RS, Fatangare A, Grimalt S, Haldar S, et al. Azadirachtin-a from azadirachta indica impacts multiple biological targets in cotton bollworm *Helicoverpa armigera*. ACS Omega. 2019;**4**(5):9531-9541. DOI: 10.1021/acsomega.8b03479

[62] Klein AT, Yagnik GB, Hohenstein JD, Ji Z, Zi J, Reichert MD, et al. Investigation of the chemical interface in the soybean-aphid and rice-bacteria interactions using maldi-mass spectrometry imaging. Analytical Chemistry. 2015;**87**(10):5294-5301. DOI: 10.1021/acs.analchem.5b00459

[63] Hohenstein JD, Studham ME, Klein A, Kovinich N, Barry K, Lee YJ, et al. Transcriptional and chemical changes in soybean leaves in response to long-term aphid colonization. Frontiers in Plant Science. 2019;**10**:310. DOI: 10.3389/fpls.2019.00310

[64] Abdalsamee MK, Giampà M, Niehaus K, Müller C. Rapid incorporation of glucosinolates as a

strategy used by a herbivore to prevent activation by myrosinases. Insect Biochemistry and Molecular Biology. 2014;**52**(1):115-123. DOI: 10.1016/j.ibmb.2014.07.002

[65] Cristopher ABP, Christian MH, Fernandez-Marin H, Gutierrez M. Fungus-growing ant's microbial interaction of *streptomyces sp.* and *escovopsis sp.* through molecular networking and MALDI imaging. Natural Product Communications. 2019;**14**(1):63-66. DOI: 10.1177/1934578X1901400117

[66] Gemperline E, Horn HA, Delaney K, Currie CR, Li L. Imaging with mass spectrometry of bacteria on the exoskeleton of fungus-growing ants. ACS Chemical Biology. 2019;**12**(8):1980-1985. DOI: 10.1021/acschembio.7b00038

[67] Strohalm M, Strohalm J, Kaftan F, Krásný L, Volný M, Novák P, et al. Poly[N -(2-hydroxypropyl) methacrylamide]-based tissue-embedding medium compatible with MALDI mass spectrometry imaging experiments. Analytical Chemistry. 2011;**83**(13):5458-5462. DOI: 10.1021/ac2011679

[68] Voelckel C, Jander G, editors. Insect-plant interactions. Annual Plant Reviews. 1st ed. Vol. 47. West Sussex: Wiley-Blackwell; 2014

Chapter 6

The Role of Irrigation and Nitrogen Fertilization on the Feeding Behavior of European Corn Borer

Ankica Sarajlić, Emilija Raspudić, Zdenko Lončarić, Marko Josipović and Ivana Majić

Abstract

European corn borer (ECB) creates tunnels inside the plant stalks, causing damage, which could significantly decrease yield loss. This study aimed to determine the relationship between damage caused by ECB larval feeding and different irrigation and nitrogen fertilization rates on different maize genotypes. We conducted a field experiment in Croatia from 2012 to 2014. Increased plant nitrogen adsorption was observed under irrigation only in drought years, and it was decreased in optimal or extremely wet years. We found a weak or a moderate relationship between ECB damage and nitrogen concentration, but the greatest ECB damage was in all years recorded in treatments with the highest fertilization rates. However, the highest plant nitrogen concentration was observed in the hybrid with the lowest damage from ECB larvae. The maize damage caused by ECB larval feeding was negatively affected by high plant nitrogen concentrations only when plants were under drought stress. Nitrogen uptake was increased in irrigated plots. We did not find a strong relationship between the C/N ratio or irrigation and intensity of ECB damage. In 2012, when the narrowest C/N ratio was calculated, the greatest damage by ECB was measured. Further studies are needed since we detected the significant impact of drought on intensive ECB larval feeding.

Keywords: *Ostrinia nubilalis*, nitrogen, irrigation, C/N ratio, fertilization

1. Introduction

European corn borer (ECB) is a polyphagous insect able to attack more than 200 different plant species. This pest mostly damages maize plants; however, a yield of peppers can be decreased up to 60% due to the ECB feeding activity [1]. The intensity of the ECB attack in Croatia often reaches 100% of maize plants in the last 20 years, with few exceptions, which could be related to recent global changes in climate [2]. The cultural and biological controls are mostly used for the management of ECB in maize, while chemicals are mostly applied in sweet maize and production of maize for seeds. Tolerant maize hybrids are a very important tool in integrated pest management programs against ECB. Plant breeders successfully developed hybrids with stronger stalk, a larger root system, and such plants are more tolerant to this pest. Tolerant hybrids are also characterized by the low potential of plant lodging and yield losses in conditions of the highest ECB infestation and stalk damage [3, 4]. One ECB larvae per plant on commercial hybrids can reduce yield up to 6.05% [5]. The amount of plant-available water has been studied as a potentially important factor for ECB infestation and damage. Irrigated plants have a higher yield, and grain quality is better [6]. If the water supply is satisfactory, plants will adsorb more micro and macro elements. Nitrogen is the most important essential element, and plants adsorb it in a mineral form. Agricultural soils usually contain 0.1–0.3% of nitrogen, and from this, only 1–3% is available for plants. Maize plants under nitrogen fertilization treatment are attractive for ECB females for laying eggs. Such plants contain more quality ingredients for larval nutrition. The positive correlation between nitrogen fertilization and ECB oviposition and larval feeding was found by many researchers [7–9]. C/N ratio is important for immobility and mineralization processes and varies with levels of nitrogen fertilization. If the C/N ratio is wider than 1:25, nitrogen would become immobile. However, if the ratio is narrower, the mineralization is possible. Plants alone are complex systems in which a lot of factors interact; therefore, we should elucidate damage potential of ECB in changing environments and techniques that will enable better and sustainable protection of plants against this pest.

2. European corn borer

European corn borer (*Ostrinia nubilalis* Hübner) (Crambidae, Lepidoptera) (**Figure 1**) is a destructive pest on maize, but it also occurs on many other plant species such as peppers and tomatoes. In eastern Croatia, ECBs have been studied for many years [10, 11], population dynamics and intensity of attack of ECB have been monitored since 1965 (Ivezić, personal commun.). In Croatia, ECB usually develops two generations per year; however, the third generation may occur in years with favorable agro-climatic conditions [12]. Larvae feed on aboveground parts of a plant and pass through five larval stages before attaining the pupal stage. The first larval stage tends to move toward higher moisture conditions in

Figure 1.
Female and male of European corn borer (Photo: Sarajlić, 2010).

plants, and for this reason, they feed within the whorls of leaves. When all leaves are developed, typical symptoms of larval feeding can be observed as small holes on leaves, lined in one row (**Figure 2**). Third stage larvae bore into the plant stalk or fruits and disrupt physiological processes inside the plant. Larvae pupate into the stalk holes (**Figure 3**). Larvae are the most abundant in the tasseling and silking stage of maize. During these stages, maize plants are the most attractive for insects to feed on due to the low content of flavonoids [13].

2.1 Influence of agroclimatic conditions on European corn borer

High-temperature stress has a negative impact on plant growth, productivity, and metabolism, and plants are the most vulnerable in the reproductive phase [14]. Climatic stress has a significant impact on ECB oviposition and larvae mortality. The first and second larval stages are the most sensitive to stress, and mortality rates can reach up to 62% [15, 16]. Quantity of available water can be regulated by the irrigation system, but other climatic factors, such as air temperatures, cannot be manipulated in field crops. Excessive rainfalls and lower temperatures can delay ECB appearance from 10 and more days, and the intensity of attack is lower; consequently, the damage is also lowered [2]. We are already facing often deviations from average multiyear precipitation and air temperature; therefore, models to predict the occurrence of a new generation of ECB have been proposed [17].

2.2 Potential of ECB natural enemies

During the vegetation, ECB has been exposed to different species of natural enemies, such as wasps and flies that feed on eggs or larval stages of this pest. The most investigated natural enemies of ECB are *Trichogramma* spp. Contradicting results on their effectiveness have been reported [18]. *Trichogramma* wasps are insect egg parasitoids. We have detected the natural parasitism of ECB eggs by these wasps in Croatia on the second generation of ECB. We reported low populations of natural enemies in the field crop, insufficient to significantly reduce the population of ECB and keep it under the economic threshold. Only artificial release of

Figure 2.
Leaf damage from ECB larvae (Photo: Sarajlić, 2010).

Figure 3.
ECB pupae inside maize stalk (Photo: Sarajlić, 2010).

these wasps may be considered as a reliable suppressant of ECB [19]. *Trichogramma* spp. has a great potential in biological control of sweet maize, where ECB has been controlled by chemicals that pose an environmental and toxicological threat [20].

3. The role of plant defense mechanisms against herbivory

Plants have developed direct and indirect defense mechanisms against herbivores. Direct plant defense is reflected through primary and secondary metabolites (silica, trichomes, proteinase inhibitors, polyphenol oxidases, toxic compounds, and other). The indirect defense is plants' ability to attract insect predators by emitting different volatile organic compounds [21]. Blend structure of volatiles is specific and very complex and depends on plant and herbivore species, plant developmental stage, and environmental conditions [22]. Many plants induce volatile emissions only during the photo phase, while many lepidopteran larvae are nocturnal insects and consume plant material during the night. Damaged plants may emit volatiles for the attraction of natural enemies. Adult insects may avoid these plants for egg deposition; however, lepidopteran larvae could be attracted to induced volatiles [23]. To reveal and better understand signaling pathways between plants and insects, plant defense mechanisms post herbivory attack have been in the focus of many researchers, so far. In a study of the Asian Corn Borer (*Ostrinia furnacalis*), it is found that females laid fewer eggs on damaged plants by insect larvae than mechanically damaged or healthy plants [24]. The blends of plant volatiles are tested to attract gravid females and manipulate with a population of pests as well [25]. The ECB is considered a model insect for studying the sex pheromone communication system [26, 27]. Several strains of ECB (E, Z, and E/Z) are recognized in the world [28], while in eastern Croatia Z-strain is present only [29].

4. The role of nitrogen and C/N ratio on the intensity of the ECB attack

A high amount of nitrogen in the soil system is unavailable for plants. Nitrogen fertilization is applied according to the plans of plant nutrition and expected yields. Nitrogen also rinses in groundwater, so it is necessary to ensure plants with a sufficient amount of nitrogen fertilizer. Plant dry matter contains

2–5% of nitrogen. In soils with higher pH, plants prefer ammonium form of nitrogen, and in soils with lower pH, nitrate form.

The quality of host plants influences the feeding behavior of insects. Plant nitrogen and carbon concentrations as well as other metabolites directly affect insect fertility (oviposition, size, and quality of egg masses). The quality of the host plant is being changed due to damage caused by insect feeding [30]. Plants in treatments with nitrogen fertilization have been damaged more intensively from lepidopteran larvae than plants without fertilization. The plants with the highest growth potential attract lepidopteran pests. This attractive behavior of pests is mostly affected by the plants' nitrogen concentration [31]. High nitrogen supply usually increases protein production and decreases carbohydrates, so plants have softer tissue, and they are more susceptible to the ECB attack [32]. Contradictory results are reported on the effect of sulfur and calcium concentration in maize leaves concerning plant defense mechanisms. The consensus has been made, and their interaction is very important in plant protection against herbivores [33]. Mixtures of lignin, proteins, minerals, and carbohydrates are organic compounds that contain a different ratio of carbon and nitrogen, which is usually abbreviated to the C/N ratio. Preferably, ECB attacks plants that contain higher nitrogen concentration and a narrower C/N ratio. The susceptible plants, those with the highest damages, are characterized by high lignin content and decreased C/N ratio [34]. We aimed to evaluate the role of irrigation and nitrogen fertilization on the ECB attack on maize in Croatia.

5. Experimental design

The open field experiment was set up at the Agricultural Institute Osijek, Croatia (45° 33′27.11N, 18°40′46.52E) during three vegetation seasons: 2012, 2013, and 2014. We tested the efficacy of three treatments of irrigation and nitrogen fertilization on ECB attack on four different maize genotypes. Soybean was used in crop rotation. A factorial experimental design, the split-split plot, was used with three replications. Each independent variable was one factor in experimental design. Three levels of irrigation and nitrogen fertilization are applied for each tested maize genotype 3 × 3 × 4 (**Table 1**).

Factor	**A**	**B**	**C**
	Irrigation	**Nitrogen fertilization**	**Hybrid**
Treatment	A1—natural rainfall (control)	B1—soil nitrogen (control)	C1—OSSK 596 C2—OSSK 613
	A2—60–100% FWC	B2—100 kg N ha^{-1}	C3—OSSK 602
	A3—80–100% FWC	B3—200 kg N ha^{-1}	C3—OSSK 552

FWC, field water capacity.

Table 1.
Factors and treatments in the experiment.

The area of the experimental field was 0.5 ha. The basic plot consisted of two maize rows 10 m long. Row spacing was 70 cm. The first three rows at the edges of fertilization plots were omitted in evaluations. Maize was sown with hand planters with two seeds per together and thinned in the phase of 4–6 leaves plants.

1. Plot size was: factor (a), irrigation, 4 hybrids 56 m^2 × 3 level of nitrogen fertilization = 168 m^2;

2. Subfactor (B), nitrogen fertilization, 4 hybrids × 14 m^2 = 56 m^2; and

3. Subsubfactor (C), maize hybrid, 14 m^2 (10 m × 2 rows × 0.7 m = 14 m^2).

5.1 Irrigation

Maize was irrigated by Typhon system (sprinkling irrigation). The range of the system was 20–25 m. Intensity and amount of water were regulated by the selection of nozzles and the speed system movement. Watermark 200SS device was used to determine the start of irrigation and to monitor the soil moisture condition. The water quality analysis was satisfactory [35].

5.2 Nitrogen fertilization

Nitrogen fertilization was applied four times per vegetation. One-third of urea (46% N) was applied in the autumn with the basic soil cultivation, and the two-thirds were added pre-sowing. Top dressing was done twice with KAN (calcium-ammonium nitrate; 27% N). The first top dressing, one sixth of N was done in phase 6–8 leaves, and the last application, one sixth of N was done in phase 8–10 leaves.

5.3 The maize genotypes

Maize genotypes used in this experiment are developed at the Agricultural Institute Osijek and presented in **Table 2**.

Hybrid	Factor	FAO group	Insect tolerance
OSSK 596	C1	590	+
OSSK 617	C2	610	+
OSSK 602	C3	620	++
OSSK 552	C4	580	—

+—Increased tolerance to diseases and insects. ++—Increased tolerance to ECB. – Not specified (Source: http://www.poljinos.hr) [36].

Table 2.
Maize genotypes.

5.4 The evaluation of ECB larval damage

Stalks were dissected at the end of the vegetation period [2], at the beginning of September in 2012 and 2013, and October in 2014 due to the unfavorable environmental conditions. In total, 1080 maize plants were evaluated for tunnel length (cm) [TL].

5.5 Chemical analysis of the maize leaves

For plant analysis, we took 10 randomly chosen leaves below the ear in the silking stage, from each plot from 1080 maize plants. Leaves were cleaned from dust and other debris, stored in paper bags, and dried at 70°C to decrease in elasticity and then further dried at 40°C for 24 hours. Dried leaves were crushed on Retsch Gmbh Germany, SM 100 mill, and from 10 crushed leaves, subsample was taken. Organic carbon was determined by oxidation of dry samples by a wet process at 135°C [37] and nitrogen content by the Kjeldahl method. The C/N ratio was calculated.

5.6 Statistics

The data were evaluated by analysis of variance after the data were subjected to a normality test using the SAS software [38]. Log transformation (log [n + 1]) was used to normalize the data. Least square means with the Tukey adjustment for multiple comparisons were calculated and reported for significance at the 95% confidence level. Back transformation was done for original values. Data in figures are presented with standard error (SE) bars. Pearson correlation coefficient is used to test relationships between variables.

5.7 Influence of irrigation on nitrogen concentration and C/N ratio

In the formation of maize reproductive tissues, heat stress occurs at 32.5°C air temperatures, and as a consequence, pollen viability is decreased and pollen tube elongated [14]. We have observed the highest air temperatures (on average 19.95°C) in vegetation season in 2012 with peaks above 33°C in July and August [2] and low natural rainfall (**Figure 4**). In 2012, plants undergone drought stress since 22% less precipitation was recorded compared to the multi-year average (62.41 mm; https://meteo.hr). In subsequent years, we noticed over multi-year average precipitations (**Figure 4**).

The data on the damage of maize stalks caused by ECB larval feeding were previously analyzed and reported [2]. The greatest damage of maize stalks was observed in 2012, ranging from 49.13 cm (A3) to 79.22 cm (A1) tunnel length per stalk (**Figure 4**). The lowest damage was recorded at the highest level of irrigation (A3). Drought in 2012 probably affected ECB survival and consequently larval damage. Excessive rainfall in 2013 and 2014 could have caused eggs to rinse from maize leaves, thus preventing the penetration of larvae inside the stalk and less plant damage.

Available nitrogen and water supply are the most important factors for plant growth and quality. Plant nitrogen concentration in maize leaves significantly differed between the irrigation plots and the control plots in 2012. Treatment with the highest nitrogen concentration was also with the highest irrigation level A3 (2.93%); compared to the control plot, it was higher for 8.87%. In the subsequent years, plant nitrogen concentrations were not affected by irrigation treatments (**Figure 5**). Soil moisture level and texture are the major factors influencing the root uptake of nitrogen [2]. Our results revealed that nitrogen uptake was not only increased under irrigation in drought year, which was characterized by high temperature and low rainfall, but also decreased in optimal or extremely wet years with a large amount of rainfall. Some authors reported that nitrogen concentration was increased in the aboveground part of irrigated plants and roots nitrogen concentration decreased in irrigation plots [39].

In 2012, the C/N ratio was the widest in control plots of irrigation (A1) (15:1), and it was significantly higher for 1.17 than in the plots with the highest irrigation level (A3). However, the C/N ratio in 2012 was narrowest comparing to the other 2 years. In subsequent years, there were no statistical differences between irrigation treatments concerning the C/N ratio (**Figure 6**). Due to the decreased nitrogen concentration during drought, the values of the C/N ratio were wider in a drought year (2012) and narrower in the other 2 years. In our study, the greatest damage by

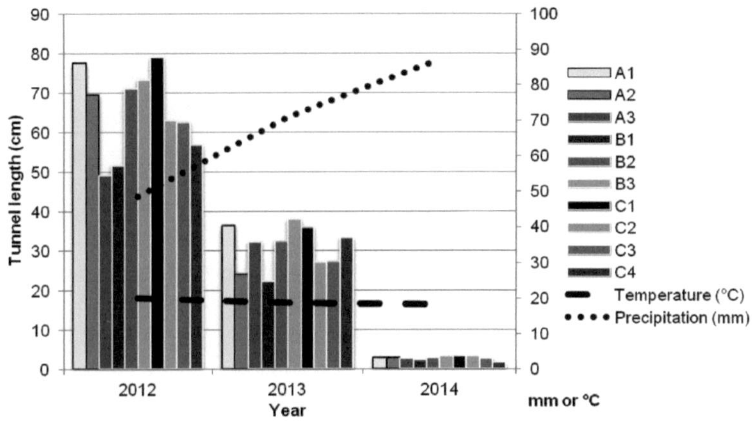

Figure 4.
Maize damage caused by ECB larval feeding and agroclimatic conditions 2012–2014.

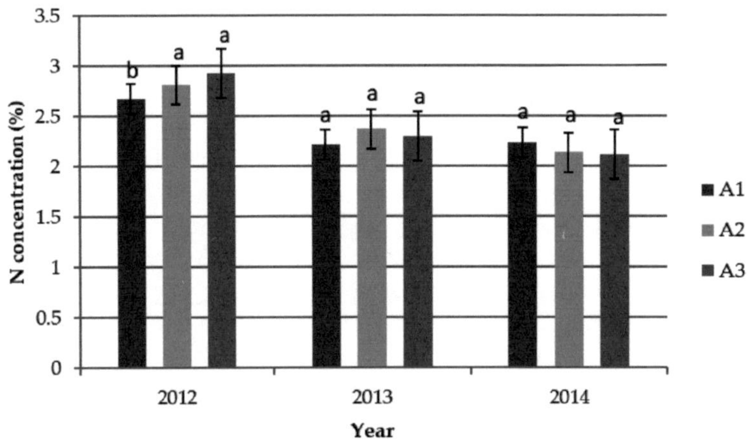

Figure 5.
Plant nitrogen concentrations in irrigation treatments presented by years of investigation.

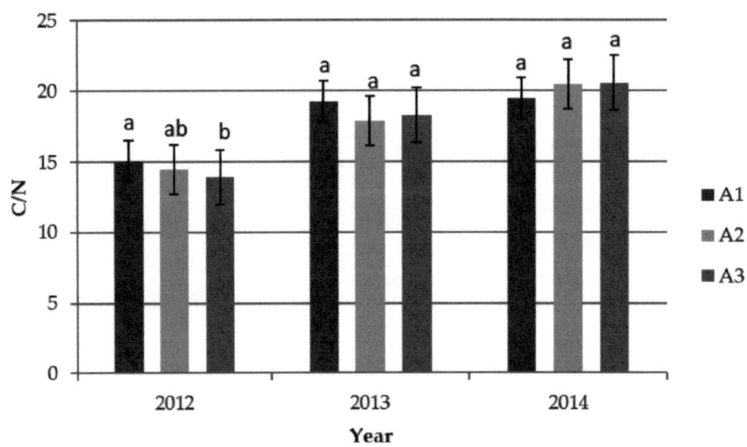

Figure 6.
The C/N ratio in irrigation treatments presented by years of investigation.

ECB was found in 2012 when the narrowest C/N ratio was calculated compared to the other 2 years.

Plant nitrogen concentration increased with an increase in fertilization rates, while the C/N ratio was narrowed. This problem has been studied by many researchers, and they also obtained similar results [40–43]. Upon herbivores attack, maize plants differently react, and it can be observed as translocation of sugars in stalk and root, increase in nutrition and photosynthesis, and other processes. All these changes can affect the C/N ratio in plant tissue [44]. The significantly highest concentration of nitrogen occurred, as it was expected in fertilization treatments with the highest rates (B3) in all years of research, compared to the control (B1) (**Figure 7**).

The B2 and B3 treatments did not differ statistically in plant nitrogen concentration. The lowest nitrogen concentrations were detected in 2012 in control plots (B1), and compared to the plots with the highest level of fertilization (B3), it was lower for 2.87%, in 2013 for 31.41%, and in 2014 for 22.26% (**Figure 7**). On average, 20.7% higher nitrogen concentration was found in plants in the B3 treatment compared to the control. Nitrogen fertilization increases ear weight and yields [45]. In all years, the greatest damage, tunnel length created by ECB larvae, was in the B3 treatment and the lowest in the control. Our data are similar to previously reported studies [8, 9]. By increasing the level of nitrogen fertilization, the C/N ratio was significantly reduced. Significant differences in C/N ratio were found in all years between the control (B1) and the other two treatments of nitrogen fertilization (B2 and B3), with the exception in 2012, when significantly differed only B1 and B3. The rate of applied nitrogen was not a significant factor for the C/N ratio since the B2 and B3 treatments did not differ statistically. In 2014, the C/N ratio was the widest (15:1) on the control treatment (B1), and it was wider for 5.47 than the B3 treatment. Similar results are obtained in 2013 (23:1) when the C/N ratio was wider for 8.38 and in 2012 (15:1) for 1.27 (**Figure 8**). On average, the widest C/N ratio (21:1) was recorded at the treatment B1, and compared to the B3 treatment, it was significantly wider for 5.04.

The highest nitrogen concentration in this research and the lowest damage from ECB larvae were observed in maize hybrid C4 (**Figure 4**) [2]. These results are contrary to the studies who reported a positive relationship between plant nitrogen and ECB damage. The insects' interactions are complex, and other compounds

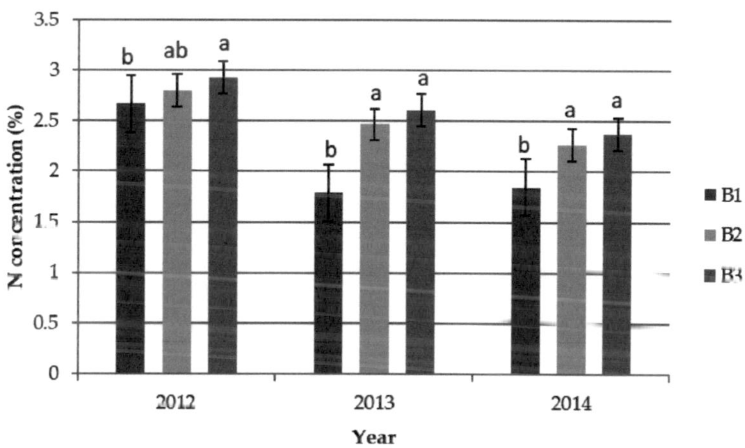

Figure 7.
Nitrogen concentrations in fertilization treatments presented by years of investigation.

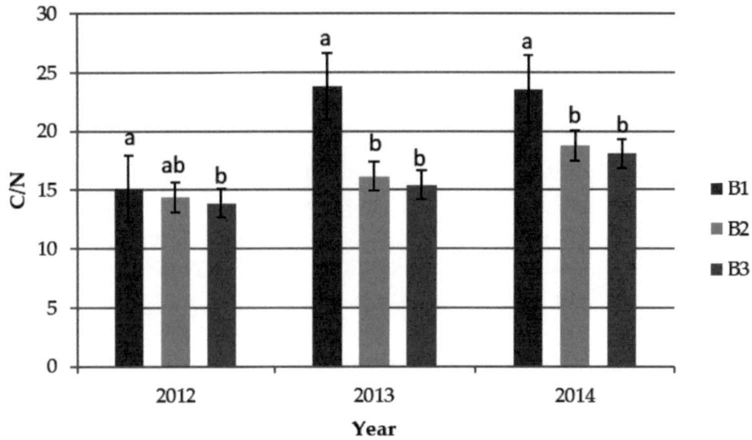

Figure 8.
The C/N ratio in nitrogen fertilization treatments presented by years of investigation.

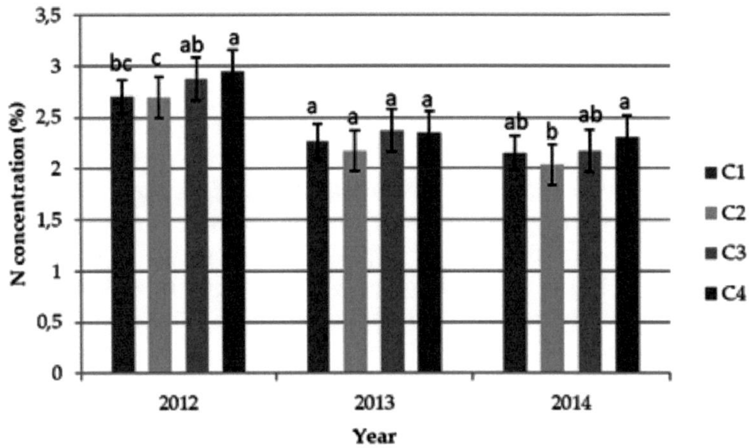

Figure 9.
Nitrogen concentrations in different maize hybrids presented by years of investigation.

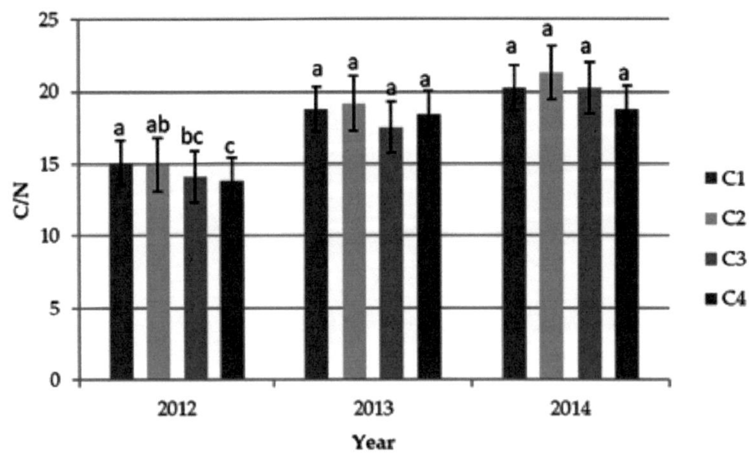

Figure 10.
The C/N ratio in different maize hybrids presented by years of investigation.

	Irrigation			Nitrogen fertilization			Maize hybrids			
	A1	A2	A3	B1	B2	B3	C1	C2	C3	C4
2012										
$R_{tl;N}$	−0.04	0.16	0.07	−0.43**	0.09	−0.26	−0.16	0.09	−0.21	0.07
$R_{tl;C/N}$	0.01	−0.15	−0.01	0.36*	−0.18	0.30	0.19	−0.08	0.24	−0.11
$R_{N;C/N}$	−0.89**	−0.91**	−0.93**	−0.85**	−0.86**	−0.95**	−0.91**	−0.91**	−0.91**	−0.89**
2013										
$R_{tl;N}$	0.09	0.27	0.09	0.07	−0.32	0.11	0.23	−0.17	0.47*	−0.14
$R_{tl;C/N}$	−0.11	−0.26	−0.09	−0.05	0.28	−0.11	−0.24	0.10	−0.49**	0.13
$R_{N;C/N}$	−0.98**	−0.99**	−0.98**	−0.99**	−0.95**	−0.99**	−0.98**	−0.99**	−0.99**	−0.99**
2014										
$R_{tl;N}$	0.01	−0.01	0.03	0.23	0.02	0.14	−0.26	0.22	0.19	−0.05
$R_{tl;C/N}$	0.01	−0.03	−0.05	−0.28	−0.05	−0.07	0.32	−0.26	−0.15*	0.02
$R_{N;C/N}$	−0.96**	−0.98**	−0.97**	−0.97**	−0.93**	−0.96**	−0.96**	−0.98**	−0.98**	−0.97**

A1—control, A2—from 60 to 100% WFC, A3—from 80 to 100% WFC; B1—control, B2—100 kg N ha^{-1}, B3—200 kg N ha^{-1}; C1—OSSK 596, C2—OSSK 617, C3—OSSK 602, C4—OSSK 552; tl, tunnel length; N, nitrogen; C/N ratio.
$*P < 0.05$.
$**P < 0.01$.

Table 3.
Correlation coefficients between ECB feeding and nitrogen concentrations and C/N ratio among tested years and treatments.

may have an impact on larval feeding such as phenols, carbohydrates, and other components [46]. The highest nitrogen concentrations on average were determined in 2012, while the lowest was found in 2014 (**Figure 9**). In both years, a statistically significant difference occurred between C2 and C4 hybrid. On average, hybrid C4 had the highest nitrogen concentration, and it was higher from 2.75 to 9.45% than observed in other hybrids.

The C/N ratio was found to be significantly different among several hybrids only in 2012. The hybrid C1 had significantly wider value than the hybrid C4 (**Figure 10**). On average, no significant difference occurred between the hybrids; however, the widest C/N ratio was measured for hybrid C2 (18:1).

The relationship between nitrogen concentration and C/N ratio was strong negative in all years and all treatments. The relationship between tunnel length in stalks caused by the ECB larvae and nitrogen concentrations was weak or moderately strong but inconsistent over the years of investigation. Plants require carbon and nutrients for growth. If nutrients are limited, plants tend to accumulate more carbohydrates that can be immediately used. When the ratio of carbon is increased to nutrients, some carbohydrates can be incorporated into secondary metabolism of plant. Secondary metabolites have a defensive role in plants [47]. Carbon, water, and mineral nutrient allocation in a plant depend on genotype and plant environment [48]. The concentration of secondary metabolites increases with drought stress [49]. Nitrogen fertilization leads to a high concentration of nitrogen in plant tissue and a lower concentration of secondary metabolites, but drought stress limits nitrogen adsorption, and such plants are not attractive to herbivores. Our investigation did not give strong evidence that nitrogen concentration and the C/N ratio impact the feeding behavior of ECB larvae in maize stalks (**Table 3**).

6. Conclusion

Nitrogen uptake was increased in irrigation treatments in drought year characterized by high temperatures and a small amount of rainfall. Decreased plant nitrogen concentrations were observed in optimal or extremely wet years with a large amount of rainfall. By increasing the level of nitrogen fertilization, the C/N ratio was significantly reduced. The highest nitrogen concentration in this research and the lowest damage from ECB larvae were observed in maize hybrid C4. The relationship between nitrogen concentration and C/N ratio was strongly negative. We found a weak or moderately strong relationship between damage caused by the ECB larva and nitrogen concentration. Our results indicate that maize damage caused by ECB is negatively affected by plant nitrogen concentrations only when plants are under drought stress. However, the relationship between ECB larval damage and plant nitrogen concentration depends on the nitrogen fertilization rates. We did not find strong evidence for this hypothesis and did not prove that plant nitrogen concentration or more quality plants would be more damaged by European corn borer. Further studies, in controlled environments, are needed since our results were inconsistent over the years and indicate the great impact of agroclimatic conditions (drought) on the potential of ECB to create damage.

Acknowledgements

The research was financed by the Ministry of Science, Education, and Sports of the Republic of Croatia (Project: 079-0790570-2208).

Conflict of interest

All authors saw and approved this book chapter. We warrant that the chapter is the authors' original work and is not under consideration for publication elsewhere. All coauthors agree that the corresponding author will be responsible for the submission. We warrant that all authors have contributed significantly to the work.

Author details

Ankica Sarajlić[1*], Emilija Raspudić[1], Zdenko Lončarić[1], Marko Josipović[2] and Ivana Majić[1]

1 Faculty of Agrobiotechnical Sciences Osijek, Department of Phytomedicine Section of Entomology and Nematology, Osijek, Croatia

2 Agricultural Institute Osijek, Osijek, Croatia

*Address all correspondence to: ankica.sarajlic@fazos.hr

References

[1] Welty C. Monitoring and control of European corn borer, *Ostrinia nubilalis* (Lepidoptera: Pyralidae), on bell peppers in Ohio. Journal of Agriculture Entomology. 1995;**12**(2-3):145-161

[2] Sarajlić A, Raspudić E, Lončarić Z, Josipović M, Brmež M, Ravlić M, et al. Significance of irrigation treatments and weather conditions on European corn borer appearance. Maydica. 2017;**62**(2):1-8

[3] Melchinger AE, Kreps R, Späth R, Klein D, Schulz B. Evaluation of early maturing European maize inbreds for resistance to the European corn borer. Euphytica. 1998;**99**:115-125. DOI: 10.1023/A:1018305316853

[4] Raspudić E, Ivezić M, Brmež M, Majić I. Susceptibility of Croatian maize hybrids to European corn borer. Cereal Research Communications. 2009;**37**:177-180. DOI: 10.1556/CRC.37.2009.Suppl.2

[5] Bohn M, Kreps RC, Klein D, Melchinger AE. Damage and grain yield losses caused by European corn borer (Lepidoptera: Pyralidae) in early maturing European maize hybrids. Journal of Economic Entomology. 1999;**92**(3):723-731. DOI: 10.1093/jee/92.3.723

[6] Kresovic B, Gajić B, Tapanarova A, Dugalić G. How irrigation water affects the yield and nutritional quality of maize (*Zea mays* L.) in a temperate climate. Polish Journal of Environmental Studies. 2018;**27**(3):1123-1131. DOI: 10.15244/pjoes/76674

[7] Haq M, Alvi SM. Efect of nitrogenous and phosphatic fertilizers on the susceptibility of maize crop to different insect pests and on the grain yield. Pakistan Entomologists. 1982;**4**:1-2

[8] Martin RC, Arnason TJ, Lambert JDH, Isabelle P, Voldeng HD, Smith DL. Reduction of European corn borer (Lepidoptera: Pyralidae) damage by intercropping corn with soybean. Journal of Economic Entomology. 1998;**82**:1455-1459. DOI: 10.1093/jee/82.5.1455

[9] Altieri MA, Nicholls CI. Soil fertility management and insect pests: Harmonizing soil and plant health in agroecosystems. Soil and Tillage Research. 2003;**72**:203-211. DOI: 10.1016/S0167-1987(03)00089-8

[10] Ivezić M, Raspudić E. Intensity of attack of the corn borer (*Ostrinia nubilalis* Hübner) on the territory of Baranja in the period 1971-1990. Natura Croatica. 1997;**6**(1):137-142

[11] Ivezić M, Raspudić E. The European corn borer (*Ostrinia nubilalis* Hübner) – review of results from Croatia. Agriculture. 2001;**7**(1):15-17

[12] Raspudić E, Sarajlić A, Ivezić M, Majić I, Brmež M, Gumze A. Efficiency of the chemical treatment against the European corn borer in seed maize production. Agriculture. 2013;**19**(1):11-15

[13] Fadamiro HY, Baker TC. Reproductive performance and longevity of female European corn borer, *Ostrinia nubilalis*: Effects of multiple mating, delay in mating, and adult feeding. Journal of Insect Physiology. 1999;**45**:385-392. DOI: S0022-1910(98)00137-1

[14] Kaushal N, Bhandari K, Siddique KHM, Nayyar H. Food crops face rising temperatures: An overview of responses, adaptive mechanisms, and approaches to improve heat tolerance. Cogent Food and Agriculture. 2016;**2**:1-42. DOI: 10.1080/23311932.2015.1134380

[15] Lee DA. Factors affecting mortality of the european corn borer, *Ostrinia nubilalis* (Hübner), in

Alberta. The Canadian Entomologist. 1998;**120**(10):841-853. DOI: 10.4039/Ent120841-10

[16] Ross SE, Ostlie KR. Dispersal and survival of early instars of European corn borer (Lepidoptera: Pyralidae) in field corn. Journal of Economic Entomology. 1990;**83**(3):831-836. DOI: 10.1093/jee/83.3.831

[17] Trnka M, Muška M, Semerádová D, Dubrovský M, Kocmánková E, Žalud Z. European corn borer life stage model: Regional estimates of pest development and spatial distribution under present and future climate. Ecological Modelling. 2007;**207**(2-4):61-84. DOI: 10.1016/j.ecolmodel.2007.04.014

[18] Meissle M, Mouron P, Musa T, Bigler F, Pons X, Vasileiadis VP, et al. Pests, pesticide use and alternative options in European maize production: Current status and future prospects. Journal of Applied Entomology. 2010;**134**(5):357-375. DOI: 10.1111/j.1439-0418.2009.01491.x

[19] Sarajlić A, Raspudić E, Majić I, Ivezić M, Brmež M, Josipović M. Efficasy of natural population of *Trichogramma* wasps against european corn borer in field maize. Agriculture. 2014;**20**(2):18-22

[20] Gagnon A-E, Audette C, Duval B, Boisclair J. Can the use of *Trichogramma ostriniae* (Hymenoptera: Trichogrammatidae) to control *Ostrinia nubilalis* (Lepidoptera: Crambidae) be economically sustainable for processing sweet corn? Journal of Economic Entomology. 2017;**110**(1):59-66. DOI: 10.1093/jee/tow293

[21] Kessler A, Baldwin IT. Plant responses to insect herbivory: The emerging molecular analysis. Annual Review of Plant Biology. 2002;**53**(1):299-328. DOI: 10.1146/annurev.arplant.53.100301.135207

[22] Takabayashi J, Takahashi S, Dicke M, Posthumus MA. Developmental stage of herbivore *Pseudaletia* separate affects production of herbivore-induced synomone by corn plants. Journal of Chemical Ecology. 1995;**21**:273-287. DOI: 10.1007/BF02036717

[23] Von Mérey GE, Veyrat N, D'Alessandro M, Turlings TCJ. Herbivore-induced maize leaf volatiles affect attraction and feeding behavior of *Spodoptera littoralis* caterpillars. Frontiers in Plant Science. 2013;**4**:209. DOI: 10.3389/fpls.2013.00209

[24] Huang C-H, Yan F-M, Byers JA, Wang R-J, Xu C-R. Volatiles induced by the larvae of the Asian corn borer (*Ostrinia furnacalis*) in maize plants affect behavior of conspecific larvae and female adults. Insect Sci. 2009;**16**(4):311-320. DOI: 10.1111/j.1744-7917.2009.01257.x

[25] Molnár BP, Tóth Z, Fejes-Tóth A, Dekker T, Kárpáti Z. Electrophysiologically-active maize volatiles attract gravid female European corn borer, *Ostrinia nubilalis*. Journal of Chemical Ecology. 2015;**41**(11):997-1005. DOI: 10.1007/s10886-015-0640-4

[26] Dopman EB, Robbins PS, Seaman A. Components of reproductive isolation between north American pheromone strains of the European corn borer. Evolution. 2010;**64**:881-902. DOI: 10.1111/j.1558-5646.2009.00883.x

[27] Lassance JM. Journey in the Ostrinia world: From pest to model in chemical ecology. Journal of Chemical Ecology. 2010;**36**:1155-1169. DOI: 10.1007/s10886-010-9856-5

[28] Kochansky J, Cardé RT, Liebherr J, Roelofs WL. Sex pheromone of the european corn borer,*Ostrinia nubilalis* (Lepidoptera: Pyralidae), in New York. Journal of Chemical Ecology. 1975;**1**(2):225-231. DOI: 10.1007/BF00987871

[29] Sarajlić A. The influence of abiotic factors on the presence of European corn borer (*Ostrinia nubilalis* Hübner) [thesis]. University of Josip Juraj Strossmayer; 2015

[30] Awmack CS, Leather SR. Host plant quality and fecundity in herbivorous insects. Annual Review of Entomology. 2002;**47**:817-844. DOI: 10.1146/annurev.ento.47.091201.145300

[31] Berry SD, Leslie GW, Spaull VW, Cadet P. Within-field damage and distribution patterns of the stalk borer, *Eldana saccharina* (Lepidoptera: Pyralidae), in sugarcane and a comparison with nematode damage. Bulletin of Entomological Research. 2010;**100**:373-385. DOI: 10.1017/S0007485309990319

[32] Sétamou M, Schulthess F, Bosque-Pérez NA, Thomas-Odjo A. Effect of plant nitrogen and silica on the bionomics of *Sesamia calamistis* (Lepidoptera: Noctuidae). Bulletin of Entomological Research. 1993;**83**(3):405-411. DOI: 10.1017/S000748530002931X

[33] Bastos CS, Caldoso Galvao JC, Picaneo MC, Gomes Pereira PR, Cecon PR. Nutrient content affecting *Spodoptera frugiperda* and *Dalbulus maidis* occurrence in corn. Insect Sci. 2007;**14**:117-123. DOI: 10.1111/j.1744-7917.2007.00133.x

[34] Yanni SF, Whalen JK, Ma B, Gelinas Y. European corn borer injury effects on lignin, carbon and nitrogen in corn tissues. Plant and Soil. 2011;**341**:165-177. DOI: 10.1007/s11104-010-0632-7

[35] Rhoades JD, Kandiah A, Mashali AM. The use of saline waters for crop production. Rome, Italy: FAO Irrigation and Drainage; 1992

[36] Zdunić Z, Brkić I, Andrić L, Plavšić H, Šimić B, Šimenić J, et al. New catalogue of maize in 2018. Available from: http://www.poljinos.hr [Accessed: 15 May 2020]

[37] ISO. Soil quality, determination of organic carbon by sulfochromic oxidation. International Standard. ISO 14235:1998. Geneva, Switzerland: International Organization for Standardization; 1998

[38] SAS Institute Inc. SAS/STAT® 9.2 User's Guide. 2nd ed. Cary, NC: SAS Institute Inc.; 2009

[39] Hammad HM, Farhad W, Abbas F, Fahad S, Saeed S, Nasim W, et al. Maize plant nitrogen uptake dynamics at limited irrigation water and nitrogen. Environmental Science and Pollution Research. 2017;**24**(3):2549-2557. DOI: 10.1007/s11356-016-8031-0

[40] Melillo JM, Aber JD, Muratore JF. Nitrogen and lignin control of hardwood leaf litter decomposition dynamics. Ecology. 1982;**63**:621-626. DOI: 10.2307/1936780

[41] Cadisch G, Giller K. Driven by Nature: Plant Litter Quality and Decomposition. UK: CAB International; 1997

[42] Vanlauwe B, Diels J, Sanginga N, Merckx R. Residue quality and decomposition: An unsteady relationship? In: Cadisch G, Giller K, editors. Driven by Nature: Plant Litter Quality and Decomposition. UK: CAB International; 1997. pp. 157-166

[43] Johnson ML, Zalucki MP. Foraging behaviour of *Helicoverpa armigera* first instar larvae on crop plants of different developmental stages. Journal of Applied Entomology. 2005;**129**:239-245. DOI: 10.1111/j.1439-0418.2005.00958.x

[44] Howe G, Schaller A. Direct defenses in plants and their induction by wounding and insect herbivores.

In: Schaller A, editor. Induced Plant Resistance to Herbivory. The Netherlands: Springer; 2008. pp. 7-29

[45] El Hallof N, Sarvary M. Relationship between yield quality and quantity of maize hybrids and fertilizer. Cereal Research Communications. 2007;**35**(2):369-372

[46] Gotoa M, Fujiia M, Suzukib K, Sakaib M. Factors affecting carbohydrate and free amino acid content in overwintering larvae of *Enosima leucotaeniella*. Journal of Insect Physiology. 1997;**44**(1):87-94. DOI: 10.1016/S0022-1910(97)00098-X

[47] Ayres MP. Global change, plant defense and herbivory. In: Kareiva PM, Kingsolver JG, Huey RB, editors. Biotic Interactions and Global Change. Sunderland, MA: Sinauer Associates; 1993. pp. 79-94

[48] Bloom AJ, Chapin FS, Mooney HA. Resource limitation in plants—An economic analogy. Annual Review of Ecology and Systematics. 1985;**16**:363-392. DOI: 10.1146/annurev.es.16.110185.002051

[49] Mattson WJ, Haack RA. The role of drought in outbreaks of plant-eating insects. Bioscience. 1987;**37**:110-118

Section 2

Herbicides: Scientific Background and Food Safety

Chapter 7

A Review of the Analytical Methods Based on Chromatography for Analyzing Glyphosate in Foods

Pasquale Avino, Ivan Notardonato and Mario Vincenzo Russo

Abstract

Glyphosate is a pesticide widely used in agriculture, horticulture, and silviculture as well as around homes and gardens. It was introduced by Monsanto in the early 1970s, and it is a broad spectrum, nonselective, post-emergence herbicide that inhibits plants' shikimic acid pathway. Glyphosate is considered as "difficult herbicide" in terms of trace analysis. It has low molecular weight, low volatility, thermal lability, and good water solubility. These properties cause problems in its extraction, purification, and detection. The determination often requires additional processes that may allow quantification by chromatographic methods. Several analytical procedures have been developed based on solid-phase extraction, ion-exchange chromatography, or matrix solid phase dispersion. Most published methods involve liquid extraction followed by clean-up. This review would like to revise the literature on this issue discussing the relevant chromatographic methods reported in the literature in terms of analytical parameters for analyzing such compound in food chain.

Keywords: glyphosate, pesticide, herbicide, chromatography, GC, LC, MS, LOD/LOQ, food, recovery, human health

1. Introduction

Glyphosate (GLYP) (or, less commonly, but still used, glyphosphate), a broad-spectrum herbicide, is one of the most used pesticides in the world [1], nearly $5 billion in sales and an annual global production about 825,800,000 kg [2]. Glyphosate is a nonselective herbicide; therefore, it is a molecule that eliminates all weeds without distinction.

Glyphosate [IUPAC N-(phosphonomethyl)glycine; CAS registry number 1071-83-6] is an aminophosphoric analogue of glycine and an important amino acid. It was discovered in the early 1950s by Henri Martin and was patented by Monsanto and sold as a Roundup® product for about 20 years; after 2001 (patent expiration date), free production of glyphosate was legally permitted [3, 4]. As of 2010, more than 750 glyphosate products have been on the market [5, 6]. The first important worldwide warning about the GLYP occurred in 2017: the Canadian Food Inspection Agency (CFIA) confirmed that 36.6% of the Canadian wheat samples had a high presence of GLYP (3.9% above the legal limits, which in Canada is 5 ppm) [7]. In Canada, GLYP-based products are widely used for improving the wheat ripening and drying. Such occurrence has created a big supply problem in Europe where this practice is prohibited: for instance, Italy imported large amounts of wheat to make flour for pasta from Canada (and from the United States as well).

GLYP inhibits the 5-enolpyruvylshikimate-3-phosphate (EPSP) enzyme produced by plants, which is involved in the synthesis of three essential amino acids such as tyrosine, tryptophan, and phenylalanine. The mechanism of action is absorption through the foliage, and to a small extent through the roots, and transport to growth points. Since this enzyme is present only in the plant kingdom, glyphosate acts only on plant organisms.

GLYP is a leaf herbicide (it is absorbed by the leaves of the plant), systemic (once absorbed, it passes toward the growth points, causing the death of the plant), nonselective (in fact, it is active on all plants, if not genetically modified). Glyphosate-based products are activated by the addition of a surfactant, polyoxyethylene amine (POEA), which promotes penetration through the leaf surface of plants; other additives used are sulfuric acid and phosphoric acid. Its main metabolite is aminomethylphosphonic acid (AMPA). It should be noted that a fraction of AMPA could be due to degradation processes of the detergents/surfactants rather than from glyphosate. GLYP does not penetrate deeply into the soil (maximum 20 cm) and is easily degraded by bacteria. This means that the probability that it reaches the aquifers is very low and that its presence is certainly lower than that of other dangerous pollutants.

The half-life of GLYP in the soil is between 2 and 197 days, a typical half-life of 47 days has been suggested. The soil and climate conditions on the persistence of glyphosate in the soil are very important. The average half-life of GLYP in water varies from few to 91 days. The AMPA metabolite of glyphosate has been found in Swedish forest soils for up to 2 years after a glyphosate application. In this case, the persistence of AMPA has been attributed to frozen soil for most of the year. The adsorption of glyphosate into the soil, and then its release from the soil, varies according to the type of soil. GLYP is generally less persistent in water than in land, with 12–60 days persistence observed in Canadian ponds, although persistence of more than a year has been recorded in American lake sediments.

GLYP (**Figure 1**) is a weak acid commonly used in the form of salt, distributed as a powder or as a water-soluble concentrate. At room temperature, it appears as a colorless crystalline solid, is completely soluble in water, and is highly insoluble in common organic solvents such as benzene and dichloromethane. GLYP is a nonvolatile and photo-resistant molecule, and its dissolution in water generates four chemical equilibria represented by the respective acid dissociation constants (K_a). In logarithmic form, pKa acquires the following values: 2.0, 2.6, 5.6, and 10.6. This aspect makes the molecule highly polar and amphoteric [8].

During the reactions involving the enzymes glyphosate oxidase and glyphosate N-acetyl transferase, glyphosate can form different metabolites: the main is

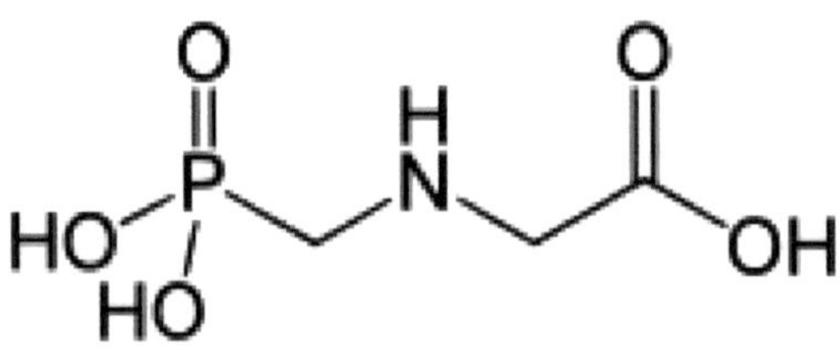

Figure 1.
Glyphosate [N-(phosphonomethyl)glycine; CAS number 1071-83-6].

considered the amino-methylphosphonic acid (AMPA), whereas the others are glyoxylate, N-acetyl glyphosate, N-acetyl-AMPA, methylphosphonic acid, sarcosine, N-methyl-aminomethylphosphonic acid (MAMPA), hydroxymethylphosphonic acid, and phosphonoformic acid [9]. This behavior is important: these compounds should be considered when toxicity and environmental studies are performed for the risk assessment. Similarly, compounds used as adjuvants in commercial glyphosate formulations should be considered: for instance, polyoxyethylene amine (POEA), used as a surfactant in Roundup [10] or isopropylamine, ammonium and trimesium salts, or formulation impurities such as N-(phosphonomethyl)iminodiacetic acid and bis(phosphonomethyl)amine. This occurrence is really important because the adjuvants can modify the toxicity of pesticides based on glyphosate as active ingredient; so, the result is the need of a novel toxicological evaluation [11].

All these considerations play an important role in the GLYP toxicity. The toxicity of a substance is assessed according to its median lethal dose (lethal dose, 50% – LD_{50}), that is, the dose that causes the death of 50% of the individuals taking the test substance: Class 1, high acute toxicity, LD_{50} less than 50 mg per kg of live weight; Class 2, moderate toxicity, LD_{50} between 50 and 500; Class 3, mild toxicity, LD_{50} between 500 and 5000; and Class 4, harmless, LD_{50} of over 5000 mg. The GLYP is in Class 3, while in Class 2, we find, for example, caffeine, aspirin, and boiling chloride, and in Class 1, the vitamin D_3. In **Table 1**, acute toxicity assessment is reported.

Also, important is the concept of daily limit dose (expressed in milligrams per kilogram of body weight considered) definable as the maximum amount of herbicide that can be consumed daily without causing damage. Based on this concept, the glyphosate content of a food or a drink should be correctly evaluated using the milligrams of glyphosate per kilogram of body weight that can be taken per day as a unit of measurement. In this way, the European Food Safety Authority (EFSA) has set a daily limit dose of 0.5 mg kg^{-1} of weight per day [12].

A tumor associated with glyphosate would be the non-Hodgkin lymphoma (NHL). In 2013, the German Federal Institute for Risk Assessment (BfR) found that "the available data are contradictory and far from convincing" in terms of the relationship between exposure to glyphosate formulations and the risk of various cancers, including the NHL [13–18]. A meta-analysis published in 2014 identified an increased risk of NHL in workers exposed to glyphosate formulations [19, 20].

	High toxicity	Moderate toxicity	Low toxicity	Very low toxicity
Acute oral[a]	≤50 mg kg^{-1}	>50–500 mg kg^{-1}	>500–5000 mg kg^{-1}	>5000 mg kg^{-1}
Inhalation[b]	≤0.05 mg L^{-1}	>0.05–0.5 mg L^{-1}	>0.5–2.0 mg L^{-1}	>2.0 mg L^{-1}
Dermal[a]	≤200 mg kg^{-1}	>200–2000 mg kg^{-1}	>2000–5000 mg kg^{-1}	>5000 mg kg^{-1}
Primary eye irritation	Corrosive or corneal involvement	Corneal involvement (8–21 days)	Corneal involvement (7 days)	Minimal effects clearing in 24 hours
Primary skin irritation	Corrosive	Severe irritation at 72 hours	Moderate irritation at 72 hours	Mild or slight irritation at 72 hours

[a]*As LD_{50}.*
[b]*As lethal concentration, 50% (LC_{50}).*

Table 1.
Relationship between GLYP levels and toxicity.

In March 2015, the International Agency for Research on Cancer (IARC) classified glyphosate "probably carcinogenic to humans" (Group 2a) based on epidemiological studies, animal studies, and in vitro studies: in particular, GLYP has been defined genotoxic through at least two mechanisms known to be associated with human carcinogens [21–23]. In contrast, EFSA concluded in November 2015 that "the substance is unlikely to be genotoxic (i.e., harmful to DNA), or pose a threat to humans." Subsequently, EFSA itself states that while there may be formulations containing glyphosate that are carcinogenic, studies relating only to glyphosate as an active ingredient do not show this effect [24, 25]. The European Chemicals Agency (ECHA), on the basis of "the scientific evidence available at the moment," classified GLYP, according to the CLP Regulation, as a chemical causing eye damage (H318) and being toxic to aquatic life with long-lasting effects (H411), but "the available scientific evidence did not meet the criteria in the CLP Regulation to classify glyphosate for specific target organ toxicity, or as a carcinogen, as a mutagen or for reproductive toxicity" [26]. The United States Environmental Protection Agency (US EPA) has classified glyphosate as a Group E chemical, meaning the agency has determined that there is "evidence of noncarcinogenicity to humans" [27, 28]. In any case, US EPA has established tolerances for GLYP residues in different commodities [29]. The difference of point of views depends on the fact that IARC and US EPA have analyzed different studies and applied different statistics. Further, EFSA analyses concern only the glyphosate molecule, whereas the studies considered by IARC also concern glyphosate-based products placed on the market [30].

This brief analysis shows that, in any case, international pesticide regulatory agencies and scientific organizations agree that there is no evidence that GLYP as an active substance is carcinogenic to humans, only IARC has classified glyphosate as "probably carcinogenic."

Finally, it should be considered an interesting hypothesis by Samsel and Seneff [31]: they propose a relationship between celiac disease and imbalances in gut bacteria generated by the known GLYP effects on them.

The EFSA has renewed the authorization for GLYP, establishing the acute reference dose (ARfD) at 0.5 mg kg^{-1} of body weight, while the acceptable operator exposure level (AOEL) was set at 0.1 mg kg^{-1} body weight per day and the acceptable daily intakes (ADIs) for consumers are in line with the ARfD threshold, 0.5 mg kg^{-1} body weight per day.

There are several exposure sources of humans to GLYP in the environment, for example, air, water, application to crops and target weeds, and food [32–34]. Solomon deeply reviewed the exposure data from the literature (PubMed and Google Scholar) and unpublished reports in different papers [35, 36]: in both papers, he reaches a similar conclusion: "In all cases, measured and estimated systemic exposures to glyphosate in humans and animals were less than the ADIs and the RfD. Based on this large dataset, these exposures represent a *de minimis* risk." The conclusion reached by Gillezeau et al. [33] is instead intermediate by reviewing the same literature (PubMed and Google Scholar): they state that "additional studies are urgently needed to evaluate levels of glyphosate and related metabolites in the general population and in workers." Further, they observe the great differences in the analyzed papers: they detected some bias such as the few studies on potential occupational GLYP exposure, or no study designed to address the hypothesis of seasonality in exposure, or the use of a few populations of farmers and relative collection of one-time spot urine. They rise serious doubts about the data generalizability, which they consider rather limited.

This paper would like to critically revise the literature on chromatographic methods developed for analyzing GLYP and AMPA in food matrices, specifically grains (e.g., rice, wheat, soybean, and maize), honey, olive and oil, vegetables, fruit, beverages (e.g., drinking water, milk, tea, and coffee), cheese, and meat/fish products. In literature (source: Scopus database), there are 2666 papers using keywords "glyphosate" and "analysis" by the end of April 2020 and 361 using "chromatography" as third keyword.

2. Glyphosate determination in different food matrices

Starting from the Canadian study performed in 2017, the scientific attention on GLYP has become stronger, and several papers are annually published dealing the determination of such compound, along with its main metabolite AMPA, on different agricultural and food matrices. For avoiding dispersive information due to the big amount of studies aimed to this determination, the authors have focused their attention on the main innovative analytical methods based on chromatographic methods for determining both compounds in such matrices. It is also necessary to advise the reader that different matrices could be determined with same analytical protocols, at least showing different analytical parameters (multiresidue analyses), as well as in literature are present papers dealing important toxicological studies with no analytical information.

2.1 Approaching the determination

Before approaching the discussion on the different analytical methodologies developed for analyzing GLYP and AMPA in agricultural and food matrices, it should be necessary to resume some toxicological information on it along with some chemical characteristics to be taken into account for evaluating the analytical process.

First, a maximum residue level (MRL) is defined as the highest level of a pesticide residue legally tolerated in or on food or feed when pesticides are applied correctly [37]. For each product, an MRL of GLYP has been determined [38]. An example of this database is reported in **Table 2**.

A preliminary important information comes from the EU Reference Laboratories for Residues of Pesticides (EURL-SRM): for all the analytical steps, it is highly recommended the use of plastic vials because there is an interaction between the pesticide and the glass surface, especially when aprotic solvents are used. These interactions greatly affect the precision and accuracy, especially at low GLYP concentration. This statement is important because it influences its stability and degradation as well. Among the different solvents, water with 10% acetonitrile is considered a good storage solvent, whereas the compound is not stable in water and methanol. At room temperature, the degradation is very low within 14 days, whereas if extract is stored in the refrigerator, it is stable over 7 months [39].

Finally, the authors would like to remember some definitions regarding the parameter of an analytical method. Recovery is the term used in analytical and preparative chemistry to denote the fraction of the total quantity of a substance recoverable following a chemical procedure [40]. Accuracy is the difference between the mean of some measurements and the value considered as the true or correct value for the quantity measured, whereas precision is the measurement reproducibility, that is, the dispersion around a central value. In regard to the chromatographic separation, a signal-to-noise (S/N) ratio of 3 is acceptable for determining the limit of detection (LOD), that is, the lowest amount of analyte in a sample, which can be detected, whereas a ratio of 10 for the limit of quantification (LOQ), that is, the lowest amount of analyte in a sample, which can be quantitatively determined with precision and accuracy [41–44]. The S/N definition for chromatography is the ratio

Product	MRL
Tangerines, clementines, oranges, and grapes	0.5
Lemons, grapefruits, cedars, kumquats, apples, pears, peaches, apricots, cherries, plums, almonds, hazelnuts, strawberries, and table olives	0.1
Oil olives	1
Potatoes	0.5
Wild mushrooms	50
Other vegetables	0.1
Baked beans	2
Grain peas, lupines, and lentils	10
Other leguminous vegetables	0.1
Flax seeds, rapeseed, mustard, and cotton	10
Sunflower and soybeans	20
Other oil seeds	0.1
Wheat and rye	10
Barley, oats, and sorghum	20
Corn	1
Other cereals	0.1
Sugar beets (roots)	15
Forage from meadows and pastures, and alfalfa	0.1

Table 2.
Maximum residue levels for some food products.

of the peak height relative to the middle of the noise range (S) to the difference between the maximum and minimum baseline signal values for the noise (N) [45].

2.2 Cereal grains

Grain is the largest and well-studied matrix in this field. Many papers deal the glyphosate determination in cereals and legumes due to the worldwide use of such herbicide in the relative cereal crops. We must remember that, as said at the beginning, the first warning came precisely by analyzing several Canadian wheat samples and finding almost 37% of them with high presence of the pesticide. So, after this occurrence, scientific and health attention has been very high and focused on cereals in general, for example, maize corn, millet, barley, oats, rice, wheat wild rice, amaranth, and quinoa.

The literature analysis for the GLYP determination in such matrix is very large; for this reason, the authors focused their attention on the main publications starting from the last deep review, that is, by Tadeo et al. [46]. The same method will be applied to the revision of the analytical methods for GLYP determination in vegetables and fruit matrices.

A routine control method based on extraction with water by ultrasonication was developed by Granby et al. [47] for analyzing several Danish mill products. It was one of the first studies based on green chemistry, that is, the authors used no organic solvents or chemicals except diluted solutions of $NaHCO_3$ (as eluent) and, in some cases, H_2SO_4. The samples (rye or wheat in grain and flour) were subjected to online clean-up and separation by in-series system of ion chromatography

(IC) and high performance liquid chromatography (HPLC) with detection by electrospray ionization mass spectrometry in the negative-ion mode. The method, investigated in the range of 0.03–0.33 mg kg^{-1}, shows a GLYP recovery of 85%, a repeatability between 1 and 14%, a reproducibility from 4 to 16%, and a LOD of 0.02 mg kg^{-1} (LOQ was not reported).

A very interesting paper was published by Tseng et al.: they used the gas chromatography coupled with a pulsed flame photometric detector (PFPD) for a simultaneous determination of GLYP and glufosinate (DL-homoalanin-4-yl-(methyl)phosphinic acid, GLUF) along with their main metabolites including AMPA [48] after a single-step derivatization with trimethyl orthoacetate (TMOA). In particular, the authors studied the influence of the heating temperature (70–90°C) and time (90–120 min) on the AMPA and 3-(methylphosphinico) propionic acid (3-MPPA, a GLUF metabolite) derivatization. They optimized the method on soybean sprouts and rice samples and determined the different analytical parameters (recoveries 72–81, 71–86, 101–119, and 83–90%; LOD of 0.02, 0.03, 0.02, and 0.01 μg g^{-1}; and LOQ of 0.06, 0.10, 0.06, and 0.04 μg g^{-1} for glyphosate, AMPA, GLUF, and 3-MPPA, respectively; RSD < 10%). On the other hand, Li et al. used fluorenylmethyl chloroformate (FMOC-Cl) as derivatizating agent followed by HPLC-MS/MS for analyzing GLYP and AMPA residues in different matrices such as rice, wheat, vegetables, fruits and tea, pig and chicken muscles, aquatic products, chestnut, and honey [49]. Further, they also used an isotope-labeled 1,2-$1^{3}C^{15}N$ GLYP for increasing the accuracy and the precision of the measurements at low GLYP concentration. In this way, they obtained recoveries between 80.0 and 104% and RSDs from 6.7 and 18.2% with a LOQ of 0.05 mg kg^{-1} for both compounds and a correlation of 0.998 in the linear range of 0.20–10 μg L^{-1}.

In 2007, Granby's group published a paper on the (six) laboratory intercomparison for determining GLYP, chlormequat, and mepiquat (these two are plant growth regulators, also used for the growth reduction of the lowest straw part) residues in cereals [50]. GLYP was analyzed by treating the samples twice with MilliQ water by ultrasonication followed by centrifugation, filtration, clean-up on polystyrene-based reverse phase column, and separation by IC-HPLC-MS/MS, whereas the other two compounds were extracted by Ultra-Turrax and cleaned-up by SPE-C_{18}. The results showed very different LOQs and recoveries reached by the six laboratories (0.01 and 0.3 mg kg^{-1} and 29 and 109% for GLY) with a good within-laboratory precision and a poor between-laboratory precision [51]. For glyphosate, the authors stated the presence of a systematic component between laboratories to be the reason of such large data variability.

Simple sample preparation and fast chromatographic analysis are the main features of the paper by Martins-Júnior et al. [52]. They analyzed GLYP and AMPA in soybean samples by means of liquid-liquid partition with dichloromethane and protein precipitation followed by HPLC-MS/MS determination (in positive and negative electrospray ionization, ESI, mode). This paper highlights the choice of the liquid-liquid partition and protein precipitation. Particularly, the paper evidences the importance of the second step, that is, the protein precipitation for eliminating the matrix interference: different solvents, that is, acetone, acetonitrile, and methanol, were tested, and methanol was found the best for reducing it (but it does not eliminate it). The authors took advantage of the great performance of the tandem mass spectrometry (MS/MS) and reached LODs of 0.09 and 0.1 mg kg^{-1} and LOQs of 0.30 and 0.31 mg kg^{-1} for GLYP and AMPA, respectively, with recoveries between 79.6 and 109.1% and RSD below 12.2%. Further, the authors suggested to apply this analytical protocol to other crop matrices, where GLYP is largely used, for instance, corn and cotton.

As just noted above and especially using the LC-MS/MS as GLYP detection, the matrix effect is not negligible. In literature, different possibilities have been studied for reducing this artifact: for instance, sample dilution [53], injection of smaller volumes [54], the optimization of sample preparation and/or chromatographic parameters [55], or the use of expensive internal standard (IS). Ding et al. [56] developed a combination of C18 and SAX cartridge for reducing the matrix effect. After to have optimized the analytical conditions, the authors used hydrophilic interaction chromatography (HILIC)/WAX mixed-mode stationary phases for glyphosate retention and LC-MS/MS in negative ion mode for the detection. They used this methodology for analyzing soybean, corn, spicy cabbage, apple, and carrot samples. GLYP is investigated in a linear range between 0.02 and 10 mg kg^{-1}, with a $R^2 > 0.999$, and the intra- and inter-day errors are 2.7 and 1.8%, respectively, whereas the precision as RSD is below 7%. Using the developed analytical procedure, the authors reached good LOD and LOQ, 0.02 and 0.005 mg kg^{-1}, respectively, and recoveries ranging between 83.1 and 100.8% according to the different matrices analyzed, specifically 89–96% for soybean, 84–101% for corn, 86–94% for carrot, 85–93% for spicy cabbage, and 83–100% for apple. Quite interesting in this paper are both the possibilities to quantify such herbicides in different plant-derived or processed foods (this is not so common in the literature) and to use solution calibration curves instead of matrix-match calibration curve for the analysis.

Botero-Coy et al. explored for first the possibility to analyze GLYP in rice, maize, and soybean without derivatization step but just direct LC–MS/MS with a triple quadrupole instrument after water extraction and SPE using Oasis HLB cartridge [57]. The method has allowed to reach high correlation coefficients (<0.99) in the range of 1–250 μg L^{-1}, recoveries between 77 and 100% with RSDs below 17%, and good LODs and LOQs (0.007–0.12 mg kg^{-1} and 0.1 and 2 m kg^{-1}, respectively) for all matrices.

A Chinese-French scientific paper in 2018 dealt the determination of GLYP and GLUF in 136 food samples, of which 34% of samples with high (banana, apple, orange, potato, carrot, and juice) and low (biscuits or bread) water contents and 66% of animal origin samples (milk-based foods included, e.g., milk, cheese, and butter) [58]. After a solvent extraction (acidified water, methanol, and dichloromethane), the authors performed a derivatization by means of FMOC and a solid phase extraction (SPE) C_{18} for purifying and concentrating the extract and a HPLC-MS/MS analysis for determining the two compounds. Using these conditions, recoveries between 82 and 112%, LODs and LOQs of 1.7 and 5 μg kg^{-1}, respectively, and RSDs below 20% for both compounds were achieved.

An ion chromatography-tandem mass spectrometry-based method was developed by Adams et al. for analyzing 14 polar pesticides including GLYP in cereal and grape samples [59]. The extraction is based on quick polar pesticide (QuPPe). Although the method is interesting, not all the analytical parameters are reported except the recoveries for cereals (specifically, oat flour) ranging between 85 and 104%.

A simple method based on acidified methanol solution extraction followed by centrifugation and filtration and LC-MS/MS analysis was developed by Santilio et al. for analyzing GLYP in rice and maize [60]. The authors highlighted the importance of using GLYP isotope labeled in the matrix effect reduction. LODs of 2 μg kg^{-1} for rice and 4 μg kg^{-1} for maize and a LOQ of 10 μg kg^{-1} for both matrices were reached in a linearity range of 0.01–1.5 mg kg^{-1} (R^2 0.9982) with recoveries ranging between 74 and 98% and RSD < 20%. Finally, it should be reported that the authors' principal aim was to develop a method to be routinely used for analyzing rice and maize, taking into account the relative MRLs established, and to extend it to other matrices.

Finally, Herrera López et al. set up a multiresidue analysis for determining 14 highly polar pesticides (parents and metabolites) in 352 samples including oat and soya beans, lettuce, grapes, and oranges [61]. After a solvent extraction step, a LC system coupled with a hybrid quadrupole/linear ion trap mass spectrometer system (with ESI source) (LC-ESI-QTRAP-MS) was used for reaching high analytical performances: linearity range between 0.01 and 10 mg kg^{-1} with $r^2 > 0.99$, recoveries between 70 and 120% with RSD < 22% (specifically, GLYP between 83 and 118% with RSD < 22%), LOQs between 20 and 500 μg kg^{-1} (particularly, for GLYP 500 and 20 μg kg^{-1}) for all the investigated matrices. The clean-up procedure was not involved because no appropriate sorbent was found to increase the protocol, and the derivatization step was not necessary, whereas these authors also stated that the use of an isotopically labeled internal standard helps in the matrix effect correction.

2.3 Meat, fish, and cheese

The scientific attention on GLYP contamination in this food class is on the rise recently. Only few papers are available on such matrices. In fact, if the GLYP behavior in the aquatic environment is studied since many years [62–64], poor information is presented on its presence in foods.

Starting from the paper by Botero-Coy et al. [57], Chiesa et al. developed a method based on IC coupled to high resolution mass spectrometry (IC-HRMS) for determining GLYP, GLUF, and AMPA in different foods of animal origin without a derivatization step [65]. The authors focused their attention on the matrix, particularly on the lipid composition, which is the major interfering group because co-extracted with the analytes. The main contribution of this study was to identify the best extraction solvent: among different assays, the best solution is 30% of methanol and 70% of acidified water (1% formic acid). Thirty samples among fish (bass), bovine muscle, and organic honey were analyzed. The detector, an orbitrap quipped with heated electrospray ionization (HESI) source, allowed to reach very low LOQs (4.26–5.38 ng g^{-1}, 6.25–6.47 ng g^{-1}, and 4.30–9.26 ng g^{-1} for fish, bovine, and honey, respectively), good recoveries (96.9, 76.1, and 97.0%, respectively), RSDs <13.1%, and good correlation coefficients ($R^2 > 0.992$).

Actually, in literature, there are other few papers showing the determination of GLYP and AMPA in muscle meat (bovine, cow, pig, and chicken), but the LODs are higher (50 ng g^{-1}) [49, 66, 67], whereas the only paper on fish does not report any information on LOQ [68].

A communication dealing with the determination of GLYP and GLUF in animal feeds shows linearity more than 0.999, instrumental detection limits (IDLs) of 8.3 μg kg^{-1} and 1.1 μg kg^{-1}, respectively, accuracy between 102 and 112%, and precision below 6% in both matrices [69].

Finally, about the GLYP determination in cheese or, basically, in milk-based foods, the authors just discussed above the only paper present in the literature [58]. Please note that the milk as beverage will be discussed in other section.

2.4 Vegetables

Some papers dealing with the GLYP determination in such food matrices are just discussed previously [46, 49, 56, 58, 61]: here the attention is focused on papers showing novelty or improvements in the analytical methodology or large studies on the herbicide content. The first interesting paper is dated in 1992: Tanaka and coauthors developed a very easy method employing routinely available instrumentation, that is, HPLC with a fluorescence detection [70]. The analytical parameters are quite weak (recoveries >68% and >88% for GLYP and AMPA, respectively, and LOD 0.05 ppm for both), but it is to be appreciated the use of common equipment. In 1996, two papers investigated the GLYP presence in green lentils, fresh beans [71], and carrot [72]. The first paper introduced a post-column reaction, a denitrozation, for obtaining a N-nitroso-GP (NGP) derivate to be analyzed by HPLC coupled with thermal energy analyzer (TEA), that is, a chemiluminescence detector. Over vegetables, the authors also analyzed beverages (water and beer) and cereals (rice flour, corn, barley, and rye). They obtained recoveries between 83 and 97% for vegetables, 70–100% for beverages, and 67–100% for cereals with LODs ranging between 0.005 and 1 μg g^{-1}. On the other hand, the second paper presents a GC analysis coupled with flame photometric detection (FPD) for analyzing GLY, AMPA, and GLU. The use of instrumentation commonly present in each laboratory is to be appreciated also in this case. The three compounds were derivatized with N-isopropoxycarbonyl (isoPOC) for obtaining the relative isoPOC methyl ester derivatives: 0.5–1 μL of this solution wax injected in the GC-FPD. The authors determined the LODs (12, 8, and 20 pg injected for GLY, AMPA, and GLU, respectively), the recoveries (91–104, 94–104, and 91–100%, respectively), and the correlation coefficients (R^2 0.9992, 0.9982, and 0.9992, respectively) in a linearity range of 5–200 ng.

Hooijschuur and coauthors explored the possibility to use the microcolumn liquid chromatography with FPD detection (μLC-FPD) and compared these results with those obtained by capillary electrophoresis (CE) with FPD (CE-FPD) [73]. They used a silica column (25 cm × 320 μm ID, 450 μm OD) with 5 μm LiChrosorb RP-1 bonded silica. Although CE-FPD was faster than μLC-FPD, this is more sensitive for the GLYP and AMPA analysis: LODs are 15 and 7.5 ng mL^{-1}, respectively, versus LOD of 1.0 μg mL^{-1} for both compounds by CE-FPD. Grey et al. applied the LC-ESI/MS analysis after the derivatization with FMOC-Cl of GLYP and AMPA [74]. They evaluated the use of isotope-labeled compounds: their conclusions were positive in the GLYP determination (LODs 0.11 μg g^{-1} and 0.06 μg L^{-1} for lettuce and water samples, respectively), whereas they did not find any contribution for the accurate AMPA analysis (LODs 0.53 μg g^{-1} and 0.3 μg L^{-1}, respectively). Finally, the recoveries increased from 23.2 to 98.4% for GLYP and from 33.8 to 99.4% using the isotope dilution mass spectrometry (IDMS)-based glyphosate analytical method. Finally, Takahashi et al. determined GLYP and GLUF in cabbage, Chinese cabbage, carrot, onion, strawberry, lemon, kiwi fruit, over soybean, corn, and brown rice after derivation with FMOC-Cl and analysis by HPLC with fluorescence detection [75]. Another interesting paper came from Japan in 2004: Watanabe set up a rapid method for determining GLYP, GLUF, and 3-MPPA in vegetables (cucumber and spinach) and fruits (apple, mandarin, and orange) using an anion exchange resin and elution with acetic acid, followed by derivatization with trimethyl orthoacetate and clean-up on SPE Florisil cartridge and GC-FPD analysis [76]. The method allows to reach LODs of 0.01, 0.01, and 0.005 μg g^{-1} and recoveries of 83.5–89.8, 77.9–92.2, and 75.0–87.2% for GLYP, GLUF, and 3-MPPA, respectively.

A Chinese group proposed an original method for determining GLYP in apple samples [77]: after clean-up with SPE-C18, a derivatization step was performed using 4-chloro-3,5-dinitrobenzotrifluoride (CNBF). The quantification occurred by reverse ion-pair liquid chromatography using cetyltrimethylammonium bromide (CTAB) as ion-pair reagent. The strengths of the method are the formation of a stable derivative (5% degradation after 7 storage days at room temperature) and the easy pretreatment procedure. LOD of 0.01 μg g^{-1}, recoveries from 86.0 to 99.5%, and RSDs from 1.43 to 6.32 were achieved applying this method to apple samples.

Rembisz and coauthors started from a different idea: GLYP (as well GLU) is an aminophosphonic acid, analogous of the amino acid. So, they proposed a derivatization with phenyl isothiocyanate (PITC) for obtaining phenylthiocarbamyl derivatives (PTC derivatives): a thin-layer chromatography (TLC) with iodine-azide detection allowed to detect such compounds in parsley and lettuce samples [78]. The method was sensitive, accurate, and inexpensive showing recoveries between 95 and 104%, LODs 0.99–4 μg per spot, LOQs 1.78–8.45 μg per spot, and RSDs <7.7 for both compounds.

A fast routine analysis was developed by Boušová et al. for routinely determining the polar pesticides, including GLYP, AMPA, GLUF, and 3-MPPA, in lettuce, orange, and flour samples [79]. The coupling of ion chromatography to a triple quadrupole mass spectrometer allowed the authors to reach very good LODs and LOQs (1–10 μg kg^{-1} and 10–20 μg kg^{-1}), recoveries ranging between 71 and 116% according to the matrix, and RSD < 18%. Rajski et al. implemented this procedure using an orbitrap detector and validating the method for aubergine, zucchini, cabbage, orange, and watermelon samples [80], achieving good recoveries (70–120%) and LOQ (0.01 mg kg^{-1}) for GLY, AMPA, and GLU. Melton et al. still used the ion chromatography but coupled with the tandem mass spectrometry (IC-MS/MS) for determining highly polar pesticides (including GLY, AMPA, and GLU) in 288 samples of melon, peas without pods, and pineapple [81]. Finally, a paper by Savini et al. worth to be mentioned: the authors used the UHPLC coupled with a orbitrap detector for analyzing GLPY, AMPA, GLUF, and other polar pesticides in 98 samples (83 processed fruits and vegetables and 15 infant foods) [82]. Using the developed method, the authors obtained LOQ of 0.003 mg kg^{-1} for all three compounds, recoveries 75–113% in all matrices, RSDs below 18.5%, and a R^2 between 0.9954 and 0.9998 in the linear range of 0.001–0.1 mg L^{-1}. Another important advantage of this method is the simultaneous determination of six polar pesticides (i.e., AMPA, glyphosate, phosphonic acid, chlorate, fosetyl-Al, and perchlorate) in 25 min.

2.5 Olives and olive oil

Two papers dealt with the determination of GLYP in olives and olive oil [83, 84]. Both papers deal the difficulty of analyzing such matrices, and there is strong matrix effect. In the first paper, two different methods were developed, that is, UHPLC-TOFMS and UHPLC-MS/MS using HILIC separation: in this way, the authors reached LOQ of 0.3 μg kg^{-1} and 0.1 μg kg^{-1}, respectively, and recoveries between 57.2 and 117.6% with a linearity >0.99 and an RSD < 3.9%. The two different LOQs were calculated using time-of-flight mass spectrometry (TOFMS) and triple quadrupole instruments: as expected, the MS/MS shows lower quantifiable levels. The second paper presented a green fast-analytical method based on vortexing (1 min with acidified water) and centrifugation (10 min at 3700 rpm) and extract injection in UHPLC–MS/MS for determining GLYP, AMPA, and GLUF in different olive oils, that is, extra virgin olive oil, virgin olive oil, olive pomace oil, and soy oil. Particularly, the paper reported the determination with no internal standards nor matrix-matched calibration. The authors tested the linearity in the concentration range of 5–250 μg L^{-1}; they fixed LOQs at 5 μg kg^{-1} for AMPA and at 10 μg kg^{-1} for GLYP and GLUF and determined recoveries between 81.4 and 119.4% with intra and inter-day precision lower than 19%.

2.6 Honey

During the past few years, the important question has emerged about GLYP contamination in natural honey samples. Different papers have been published dealing this issue. Some of them have already been discussed previously [49, 65].

A first interesting paper dealing with such of matrix was this of Karise and coauthors [85]. They set up a multiresidue method for analyzing GLYP along with other

47 pesticides in 33 honey samples collected from beehives of Estonia. The paper was focused on the detection of the pesticide concentration and the relative maximum residue levels and the possible impact of the agriculture on the product. In any case, the authors largely used the analytical methodology based on using QuEChERS (acronym of Quick, Easy, Cheap, Effective, Rugged and Safe) extraction methodology followed by detection using GC-MS and ultra-high-performance liquid chromatography-MS/MS (UHPLC-MS/MS): the method shows recovery between 78 and 115%, repeatability from 3.0 to 16%, LOQ for GLYP of 0.050 mg kg^{-1} (and 0.010 mg kg^{-1} for the other pesticides), and correlation coefficients >0.990 for all compounds.

In 2018, Zoller et al. found GLYP at very low levels in 15 of 16 honey samples analyzed; in addition, they also analyzed pulses (tofu and soy sauce), breakfast cereals (corn flakes and pops), durum wheat, pastry and snacks (crisps, etc.), bread, flour and baking mixtures, and beverages (beer, wine, milk, fruit juices, and mineral water) for a total of 243 samples [86]. The authors applied a well-tested analytical method based on solvent extraction with methanol and LC-MS/MS analysis for determining GLYP and AMPA (LODs 0.2–0.4 and 0.5–1 μg kg^{-1}, respectively; LOQs 0.5–1 and 1–2.5 μg kg^{-1}; recoveries 92–103 and 92–115%; RSDs <9.5 and <13.9%). Further, in this paper, the authors assessed a dietary risk of each food for a child of 15 kg body weight and for an adult of 60 kg body weight. The first findings of this work were that the GLYP maximum residue levels did not exceed more than the legally tolerated ones (0.1 mg kg^{-1} for plant products and 0.05 mg kg^{-1} for animal products). So, the scores reported by authors for the risk assessment highlighted a low exposure only for the pulses (5% of the acceptable daily intake, ADI, and acute reference dose, ARfD), whereas in all the other cases, honey samples included, the exposure to GLYP is less than 1% of the ADI/ARfD, meaning there is no any human health issue in all samples. Further, the authors, simulating a daily ingestion of the different investigated foods, estimated the probable GLYP content in urine. They found levels in agreement with those found by other authors in German [3, 17] and Swiss [87] populations, whereas some differences could be expected in AMPA concentration comparison [17].

A pilot study for monitoring GLYP and AMPA in 32 honey samples was set up by Pareja et al. based on IC coupled to a Q-Orbitrap accurate high-resolution mass spectrometry [88]. It is still confirmed that the use of IC simplifies the polar pesticide determination, whereas the use of an orbitrap detector allows to reach a GLYP LOQ of 5 μg kg^{-1} (20 μg kg^{-1} for AMPA), less than the allowed EU MRL (50 μg kg^{-1}) and recoveries ranging between 80 and 110% with RSDs <20% in the linearity range of 5–500 μg kg^{-1}.

Still in 2019, a Canadian group developed an easy method for analyzing GLYP, AMPA, and GLUF at low μg kg^{-1} levels based on both the derivatization with FMOC-Cl in acetonitrile solution and online SPE(C18)-LC-MS/MS analysis [89] and the use of isotopically labeled internal standards (as just evidenced previously). In particular, for all the investigated compounds, the authors obtained accuracies ranging between 95.2 and 105.3% (intraday precision 1.6–7.2%) and LOQ 1 μg kg^{-1}. By this method, 200 honey samples were analyzed: GLYP was found in 196 samples at maximum level of 49.8 μg kg^{-1} with a 95th percentile of 14.2 μg kg^{-1}, evidencing no risks for the consumers. Further, the authors performed a survey between their data with others from worldwide studies (the United States, Estonia, Switzerland, some just cited in this review) [85, 86, 90–92].

A 2020 paper evaluated the exposure risk of bees and humans to GLYP and AMPA residues in three different bee matrices, that is, beebread, wax, and paired samples of wax/honey collected from 379 Belgian apiaries using an analytical method based on clean-up on SPE-C18 followed by derivatization step with FMOC-Cl and

HPLC-ESI-MS/MS analysis [93]. LOD and LOQ of 1 ng g^{-1} and 10 ng g^{-1}, respectively, were achieved for both compounds in all matrices with recoveries ranging between 72.2 and 112.9% and RSDs from 0.1 to 4.5%. The authors stated that the GLYP levels were below the EU regulation in all samples. In any case, they suggest particular attention because recent studies deal the effects of GLYP [94] and AMPA [95] below the allowed concentrations.

2.7 Beverages

This last matrix is really important considering the large use of beverages in the daily dietary intake. Beverages such as water, beer, milk, and fruit juices are under strict attention by the different national authorities. For instance, in 2019, a study by Cook of the CalPIRG Education Fund (available at https://uspirg.org/sites/pirg/files/reports/WEB_CAP_Glyphosate-pesticide-beer-and-wine_REPORT_022619.pdf?_ga=2.33097086.1581849178.1551185850-857148262.1551185850) reported that 19 of wine (5) and beer (14) brands contained GLYP at levels ranging between 4.8 and 51.4 ppb. Several papers have been published in recent years, some of which have already been mentioned in this review [49, 58, 71, 73, 86].

The first interesting paper by Hao et al. describes a method for analyzing GLYP, AMPA, and GLUF in drinking water, surface water, and groundwater samples [96], that is, a LC-MS/MS method with reversed-phase and weak anion-exchange mixed-mode Acclaim® WAX-1 column. Good analytical parameters were obtained: LODs of 1.5, 3.9, and 1.7 μg L^{-1} for GLYP, AMPA, and GLUF, respectively; LOQs of 4.5, 11.6, and 5.3 μg L^{-1}; and recoveries between 62 and 102%. The main aspect is the analysis by direct injection of aqueous samples without derivatization or clean-up procedures with the risk of artifacts.

In 2015, a Chinese group developed a procedure for analyzing GLYP and GLUF in tea samples by means of FMOC-Cl derivatization and UPLC–MS/MS analysis [97]. The method shows good linearity (r > 0.990) in the range of 0.003–0.1 mg L^{-1}, LODs of 0.03 mg kg^{-1} for both compounds, and recoveries between 81.4 and 99.1% with RSDs <2.3%.

Two papers published in 2015 reported the GLYP, AMPA, and GLUF determination in milk and milk-based products. Ehling and Reddy carried out a derivatization with FMOC-Cl followed by means of LC-MS/MS in different nutritional milk matrices such as cow's milk, human breast milk, soy milk, and whole milk powder [98]. This study is important because the reported analytical method does not require any analytical treatment such as clean-up, evaporation, or concentration; so, the possible artifact formation is drastically reduced. Further, the importance of the use of a triple-quadrupole mass spectrometry is still confirmed in terms of selectivity and fragment analysis. This occurrence gives good analytical parameters: R^2 > 0.99 in the entire investigated linearity range (5–500 ng mL^{-1}); recoveries between 91.1 and 115.2%; LODs of 0.012 and 0.01 μg g^{-1} for GLY and AMPA, respectively; LOQ of 0.05 μg g^{-1} for both; high intra-day (<4.0 and <7.7% for GLYP and GLUF, respectively) and inter-day (<8.4 and <3.0, respectively) precision. The second paper investigates the direct injection of milk extract after deproteination and SPE on Oasis cartridge [99]: the LC–MS/MS analysis under the negative ion-spray ionization mode allowed to reach low method detection limits (MDLs), that is, 0.3, 1.4, and 0.4 ng mL^{-1} for GLYP, AMPA, and GLUF, respectively, and low method quantification limits (MQLs), 1, 4, and 1 ng mL^{-1}, respectively, with recoveries ranging between 81 and 107% and RSDs 2.04–8.36%. A LC-MS/MS method (6 min chromatographic run) was successfully applied to a sample of fortified milk with a very low herbicides concentration (0.025 μg mL^{-1}). Further, the use of negative mode ion spray offers high sensitivity and selectivity. According to the study's authors (and these authors agree), this methodology could be competitive with the enzyme-linked immunosorbent assay (ELISA) method.

Steinborn et al. reported of a survey on the GLYP content in 114 breast milk samples collected in Bavaria and Lower Saxony, Germany, by comparing the data obtained by LC-MS/MS and GC-MS/MS analyses [100]. The two analyses required (a) an ultrafiltration and chromatography on an anion exchange column for LC–MS/MS and (b) a clean-up step on a cation exchange column and derivatization with trifluoroacetic acid anhydride (TFAA) and heptafluorobutanol (HFB) for the GC–MS/MS. The authors deeply investigated the difference between the chromatograms obtained with the two methods, especially for evaluating parameters such as precision, accuracy, LOD, and LOQ. Basically, GC–MS/MS allowed to reach instrumental detection limit (IDL) lower than that found in LC–MS/MS (0.02 vs. 0.5 ng mL^{-1}), but they detected an interference on a GLYP peak, which they did not manage to identify (all reagents, ultrapure water, all components were tested). Therefore, they fixed the LOQ at 1 ng mL^{-1}, the same concentration determined by LC–MS/MS (whose LOD is 0.5 ng mL^{-1}). The recoveries ranged between 83 and 128% with RSD < 17% for LC–MS/MS and between 71 and 102% with RSD < 13% for GC–MS/MS. Resuming, the GC–MS/MS is powerful at lower concentrations, but it simultaneously gives more bias than LC–MS/MS; both methods manage to investigate concentration above 1 ng mL^{-1} with high precision and accuracy.

Two papers investigated the GLYP and AMPA content in human milk and urine samples. In the first, a high-throughput LC–MS/MS method using stable isotope labeled internal standard and clean-up with methylene chloride allowed to reach very low LODs (0.92 and 1.2 for GLYP and AMPA in human milk samples and 0.023 and 0.033 μg mL^{-1} in human urine samples) and LOQs (10 μg mL^{-1} for both in breast human milk samples and 0.1 μg mL^{-1} in human urine samples), high recoveries (GLYP ranging between 92 and 107% in both matrices, AMPA between 89 and 107%) with low RSDs (<7.4 and <11.6% in human milk and urine samples, respectively) [101]. The authors also studied the matrix stability over a storage in 5°C (refrigerator) and at –20°C (freezer): in the first case, the recoveries were acceptable also after 24 hours, whereas in the second case, they were good also after 3 months. On the other hand, the second paper investigated the presence of GLYP and AMPA in milk (41 samples) and urine (40 samples) from healthy lactating women from Russia and the United States [102]. The authors used the same analytical procedure as reported above (i.e., LC-MS/MS, the use of stable isotope labeled internal standard and two fragments, such as precursor and product ion transitions, for the quantification) for the analysis, that is, the same analytical parameters. The results showed GLYP and AMPA in milk samples at levels below the LODs, whereas at low concentrations (<LOD and 1.93 μg mL^{-1} and <LOD and 1.33 μg mL^{-1}, respectively, in urine samples). The authors extrapolated the maximum intake of milk containing 1 μg mL^{-1} of GLYP for a 5-kg infant: their conclusions were that the expected levels should be 12,000 times lower than the health concern.

The presence of MRLs for GLYP in barley, wheat, rye, and hops is regulated by EU Regulation (EC) No. 396/2005 (i.e., 20, 10, and 0.1 mg kg^{-1}) [37, 38]. These are the raw agricultural commodities for beer beverage. Jansons et al. (2018) analyzed 100 beer samples from 24 different producers and distributors in Latvia with LC–MS/MS method (R^2 > 0.999 in the range of 0.2–25 μg kg^{-1}; LOD 0.2 μg kg^{-1}; LOQ 0.5 μg kg^{-1}; RSD < 4.1%) [103]. Among the numerous samples analyzed, 8 samples showed levels below the LOD and 9 samples below the LOQ, whereas 80 samples reported a GLYP concentration below 15 μg kg^{-1} and 1 sample reached a GLYP content of 150 μg kg^{-1}. The authors pointed out the attention on beer brands of "undisclosed" origin, that is, no country production reported on the labeling (it sounds strange to the authors of this review considering the restrictions on food labeling in the EU, but we reported the authors' considerations), which could have higher GLYP content than the locally produced beer. Further, they also compared beers by malt type (barley or combined/other), color (light or dark), packaging (canned or bottled), the presence of precipitate (precipitate or no precipitate), filtration (filtered or not filtered), and pasteurization (pasteurized or not pasteurized), finding no significant differences in these cases.

Over these papers, it should be underlined two other paper dealing the GLYP determination in river water and soil samples. This particular occurrence regards the analytical protocol used by authors. In the first paper, Kudzin et al. developed a procedure based on derivatization with TEA-trifluoroacetic anhydride (TFAA)-trimethyl orthoacetate reagent and analyses by GC-CI(or EI)/MS (LOD 2.5–5.0 pmol) and GC-flame ionization detection (GC-FID; LOD 30–80 pmol, recovery 97%) [104]. In the second paper, Hu et al. investigated the performance of a method based on GC with nitrogen-phosphorus detector (GC-NPD): they estimated a LOD of 9×10^{-12} g and a LOQ of 0.01 mg kg^{-1} in samples, recoveries between 84.4 and 94.0%, and RSDs between 8.1 and 13.7% [105]. These two papers deserve to be mentioned for having introduced the possibility to analyze GLYP by two very easy, cheap, and worldwide available detectors such as FID and NPD.

3. Conclusion

This long *excursus* wanted to cover the novel or advanced methodologies based on chromatographic analysis reported in the literature. The GLYP determination in foods is a really important issue, even if the different international agencies still do not totally agree on the human health concern. The importance of a continuous monitoring of such compound (and its main metabolite, AMPA), and GLUF as well, is well known by scientists and politics worldwide due to its large use in agriculture. The suggestion is to continuously develop new methods, more accurate and sensitive, based on GC-MS/MS or LC-MS/MS analysis but also routine method based on inexpensive or use-friendly detectors (FID, FPD, or NPD).

In any case, the fear for the future is that the refinement of analytical methods increasingly leads to alarmist attitudes based on the discovery of very low quantities of GLYP, which is possible for a very wide range of products, even extremely toxic, without forgetting that in nature the zero residue does not exist.

Acknowledgements

The authors would like to thank Giuseppe Ianiri for his help in the database revision and data analysis.

Conflict of interest

The authors declare no conflict of interest.

Author details

Pasquale Avino*, Ivan Notardonato and Mario Vincenzo Russo
Department of Agricultural, Environmental and Food Sciences, University of Campobasso, Campobasso, Italy

*Address all correspondence to: avino@unimol.it

References

[1] Duke SO. Glyphosate: The world's most successful herbicide under intense scientific scrutiny. Pest Management Science. 2018;**74**:1025-1026

[2] Benbrook CM. Trends in glyphosate herbicide use in the United States and globally. Environmental Sciences Europe. 2016;**28**:1-15

[3] Conrad A, Schroter-Kermani C, Hoppe HW, Ruther M, Pieper S, Kolossa-Gehrig M. Glyphosate in German adults—Time trend (2001 to 2015) of human exposure to a widely use herbicide. International Journal of Hygiene and Environmental Health. 2017;**200**:8-16

[4] Bai SH, Ogbourne SM. Glyphosate: Environmental contamination, toxicity and potential risks to human health via food contamination. Environmental Science and Pollution Research. 2017;**23**:18988-19001

[5] Duke SO, Powles SB. Glyphosate: A once-in-a-century herbicide. Pest Management Science. 2008;**64**:319-325

[6] Duke SO. The history and current status of glyphosate. Pest Management Science. 2018;**74**:1027-1034

[7] Canadian Food Inspection Agency (CFIA). Safeguarding with science: glyphosate testing in 2015-2016. Science Branch Survey Report; Ottawa, Canada, 2017. Available from: https://www.inspection.gc.ca/food-safety-for-industry/chemical-residues-microbiology/food-safety-testing-bulletins/2017-04-13/executive-summary/glyphosate-testing/eng/1491846907641/1491846907985 [Accessed: 19 April 2020]

[8] Stalikas CD, Konidari CN. Analytical methods to determine phosphonic and amino acid group-containing pesticides. Journal of Chromatography A. 2001;**907**:1-19

[9] Van Bruggen AHC, He MM, Shin K, Mai V, Jeong KC, Finckh MR, et al. Environmental and health effects of the herbicide glyphosate. The Science of Total Environment. 2018;**616**:255-268

[10] Cuhra M, Bohn T, Petr C. Too much of a good thing? Frontiers in Environmental Science. 2016;**4**:28

[11] Mesnage R, Bernay B, Séralini GE. Ethoxylated adjuvants of glyphosate-based herbicides are active principles of human cell toxicity. Toxicology. 2013;**313**:122-128

[12] EFSA (European Food Safety Authority). Conclusion on the peer review of the pesticide risk assessment of the active substance glyphosate. EFSA Journal. 2015:13, 4302, 107 pp. DOI: 10.2903/j.efsa.2015.4302

[13] European Commission. Review report for the active substance glyphosate. Health & Consumer Protection Directorate-General, 6511/VI/99-final; 2002

[14] Acquavella JF, Alexander BH, Mandel JS, Gustin C, Baker B, Chapman P, et al. Glyphosate biomonitoring for farmers and their families: Results from the farm family exposure study. Environmental Health Perspective. 2004;**112**:321-326

[15] BfR. The BfR has finalised its draft report for the re-evaluation of glyphosate. In: Press Release. Berlin: Bundesinstitut fu€r Risikobewertung (BfR); 2014. Available from: http://www.bfr.bund.de/en/the_bfr_has finalised_its_draft_report_for_the_re_evaluation_of_glyphosate-188632.html [Accessed: 08 April 2020]

[16] BfR. Does glyphosate cause cancer? In: BfR Communication No. 007/2015, 23 March 2015. Berlin: Bundesinstitut fu€r Risikobewertung (BfR); 2015.

Available from: http://www.bfr.bund.de/cm/349/does-glyphosate-cause-cancer.pdf [Accessed: 08 April 2020]

[17] Niemann L, Sieke C, Pfeil R, Solecki R. A critical review of glyphosate findings in human urine samples and comparison with the exposure of operators and consumers. Journal of Consumer Protection and Food Safety. 2015;**10**:3-12

[18] Glyphosate [Internet]. 2016. Available from: https://www.bfr.bund.de/en/a-z_index/glyphosate-193962.html [Accessed: 21 April 2020]

[19] Schinasi L, Leon ME. Non-Hodgkin lymphoma and occupational exposure to agricultural pesticide chemical groups and active ingredients: A systematic review and meta-analysis. International Journal of Environmental Research and Public Health. 2014;**11**:4449-4527

[20] Alavanja MC, Hofmann JN, Lynch CF, Hines CJ, Barry KH, Barker J, et al. Non-Hodgkin lymphoma risk and insecticide, fungicide and fumigant use in the Agricultural Health Study. PLoS One. 2014;**9**:e109332

[21] IARC. Glyphosate. Leon, France: International Agency for Research on Cancer Report; 2015. pp. 1-92

[22] Williams GM, Aardema M, Acquavella J, Berry SC, Brusick D, Burns MM, et al. A review of the carcinogenic potential of glyphosate by four independent expert panels and comparison to the IARC assessment. Critical Reviews in Toxicology. 2016;**46**:3-20

[23] Davoren MJ, Schiestl RH. Glyphosate-based herbicides and cancer risk: A post-IARC decision review of potential mechanisms, policy and avenues of research. Carcinogenesis. 2018;**39**:1207-1215

[24] Tarazona JV, Court-Marques D, Tiramani M, Reich H, Pfeil R, Istace F, et al. Glyphosate toxicity and carcinogenicity: A review of the scientific basis of the European Union assessment and its differences with IARC. Archives of Toxicology. 2017;**91**:2723-2743

[25] European Food Safety Authority (EFSA). Review of the existing maximum residue levels for glyphosate according to Article 12 of Regulation (EC) No, 396/2005-revised version to take into account omitted data. EFSA Journal. 2019;**17**:e05862

[26] ECHA. Committee for Risk Assessment. 2016. Available from: https://echa.europa.eu/it/-/glyphosate-not-classified-as-a-carcinogen-by-echa [Accessed: 02 April 2020]

[27] Greim H, Saltmiras D, Mostert V, Strupp C. Evaluation of carcinogenic potential of the herbicide glyphosate, drawing on tumor incidence data from fourteen chronic/carcinogenicity rodent studies. Critical Reviews in Toxicology. 2015;**45**:185-208

[28] US EPA. Revised glyphosate issue paper: Evaluation of carcinogenic potential. 2017. Available from: https://www.regulations.gov/document?D=EPA-HQ-OPP-2009-0361-0073 [Accessed: 07 April 2020]

[29] Environmental Protection Agency. Tolerances and exemptions for pesticide chemical residues in food. §180.364 Glyphosate; tolerances for residues. 45 FR 64911. 1980. Available from: https://www.ecfr.gov/cgi-bin/text-idx?SID=2c85909360c7c5aff63ddd1447545d6a&mc=true&node=se40.24.180_1364&rgn=div8 [Accessed: 27 March 2020]

[30] Benbrook CM. How did the US EPA and IARC reach diametrically opposed conclusions on the genotoxicity of glyphosate-based herbicides? Environmental Sciences Europe. 2019;**31**:2

[31] Samsel A, Seneff S. Glyphosate, pathways to modern diseases. II: Celiac sprue and gluten intolerance. Interdisciplinary Toxicology. 2013;**6**:159-184

[32] Li Z, Jennings A. Worldwide regulations of standard values of pesticides for human health risk control: A review. International Journal of Environmental Research and Public Health. 2017;**14**:826

[33] Gillezeau C, Van Gerwen M, Shaffer RM, Rana I, Zhang L, Sheppard L, et al. The evidence of human exposure to glyphosate: A review. Environmental Health: A Global Access Science Source. 2019;**18**:2

[34] Ledoux ML, Hettiarachchy N, Yu X, Howard L, Sun-Ok L. Penetration of glyphosate into the food supply and the incidental impact on T the honey supply and bees. Food Control. 2020;**109**:106859

[35] Solomon KR. Glyphosate in the general population and in applicators: A critical review of studies on exposures. Critical Reviews in Toxicology. 2016;**46**:21-27

[36] Solomon KR. Estimated exposure to glyphosate in humans via environmental, occupational, and dietary pathways: An updated review of the scientific literature. Pest Management Science. 2019. DOI: 10.1002/ps.5717

[37] Regulation (EC) NO 396/2005 of the European Parliament and of the Council of 23 February 2005. Official Journal of the European Union. 2005;**70**:1-16

[38] EU Database on current MRL values. Part A of Annex I to Reg. 396/2005, 2005. Available from: https://ec.europa.eu/food/plant/pesticides/eu-pesticides-database/public/?event=pesticide.residue.CurrentMRL&language=EN [Accessed: 07 April 2020]

[39] Anastassiades M, Kolberg DI, Eichhorn E, Benkenstein A, Wachtler A-K, Zechmann S, et al. Quick method for the analysis of numerous highly polar pesticides in foods of plant origin via LC-MS/MS involving simultaneous extraction with methanol (QuPPe-Method). I. Food of Plant Origin. Version 10. 2019. Available from: https://www.eurl-pesticides.eu/userfiles/file/EurlSRM/meth_QuPPe-PO_EurlSRM.pdf [Accessed: 28 March 2020]

[40] Duffus JH. Glossary for chemists of terms used in toxicology. Pure & Applied Chemistry. 1993;**65**:2003-2122

[41] Knoll JK. Estimation of the limit of detection in chromatography. Journal of Chromatographic Science. 1985;**23**:422-425

[42] Armbruster DA, Tillman MD, Hubbs LM. Limit of detection (LQD)/limit of quantitation (LOQ): Comparison of the empirical and the statistical methods exemplified with GC-MS assays of abused drugs. Clinical Chemistry. 1994;**40**:1233-1238

[43] Armbruster DA, Pry T. Limit of blank, limit of detection and limit of quantitation. The Clinical Biochemist Reviews. 2008;**29**:S49-S52

[44] Boqué R, Heyden YV. The limit of detection. LC-GC Europe. 2009;**22**:82-85

[45] Dolan JW. Method reproducibility—A case study. LC-GC North America. 2006;**24**:1274-1278

[46] Tadeo JL, Sánchez-Brunete C, Pérez RA, Fernández MD. Analysis of herbicide residues in cereals, fruits and vegetables. Journal of Chromatography A. 2000;**882**:175-191

[47] Granby K, Johannesen S, Vahl M. Analysis of glyphosate residues in

cereals using liquid chromatography-mass spectrometry (LC-MS/MS). Food Additives and Contaminants. 2003;**20**:692-698

[48] Tseng SH, Lo YW, Chang PC, Chou SS, Chang HM. Simultaneous quantification of glyphosate, glufosinate, and their major metabolites in rice and soybean sprouts by gas chromatography with pulsed flame photometric detector. Journal of Agricultural and Food Chemistry. 2004;**52**:4057-4063

[49] Li B, Deng X, Guo D, Jin S. Determination of glyphosate and aminomethylphosphonic acid residues in foods using high performance liquid chromatography-mass spectrometry/mass spectrometry. Chinese Journal of Chromatography. 2007;**25**:486-490

[50] Andersen JH, Bille RL, Granby K. An intercomparison study of the determination of glyphosate, chlormequat and mepiquat residues in wheat. Food Additives and Contaminants. 2007;**24**:140-148

[51] Horwitz W. Evaluation of analytical methods used for regulation of foods and drugs. Analytical Chemistry. 1982;**54**:67A-76A

[52] Martins-Júnior HA, Lebre DT, Wang AY, Pires MAF, Bustillos OV. An alternative and fast method for determination of glyphosate and aminomethylphosphonic acid (AMPA) residues in soybean using liquid chromatography coupled with tandem mass spectrometry. Rapid Communications in Mass Spectrometry. 2009;**23**:1029-1034

[53] Van Eeckhaut A, Lanckmans K, Sarre S, Smolders I, Michotte Y. Validation of bioanalytical LC-MS/MS assays: Evaluation of matrix effects. Journal of Chromatography B. 2009;**877**:2198-2207

[54] Heller DN. Ruggedness testing of quantitative atmospheric pressure ionization mass spectrometry methods: The effect of coinjected matrix on matrix effects. Rapid Communications in Mass Spectrometry. 2007;**21**:644-652

[55] Hernandez F, Sancho JV, Pozo OJ. Critical review of the application of liquid chromatography/mass spectrometry to the determination of pesticide residues in biological samples. Analytical Bioanalytical Chemistry. 2005;**382**:934-946

[56] Ding J, Jin G, Jin G, Shen A, Guo Z, Yu B, et al. Determination of underivatized glyphosate residues in plant-derived food with low matrix effect by solid phase extraction-liquid chromatography-tandem mass spectrometry. Food Analytical Methods. 2016;**9**:2856-2863

[57] Botero-Coy AM, Ibáñez M, Sancho JV, Hernández F. Direct liquid chromatography–tandem mass spectrometry determination of underivatized glyphosate in rice, maize and soybean. Journal of Chromatography A. 2013;**1313**:157-165

[58] Liao Y, Berthion J-M, Colet I, Merlo M, Nougadère A, Hu R. Validation and application of analytical method for glyphosate and glufosinate in foods by liquid chromatography-tandem mass spectrometry. Journal of Chromatography A. 2018;**1549**:31-38

[59] Adams S, Guest J, Dickinson M, Fussell RF, Beck J, Schoutsen F. Development and validation of ion chromatography–tandem mass spectrometry-based method for the multiresidue determination of polar ionic pesticides in food. Journal of Agricultural and Food Chemistry. 2017;**65**:7294-7304

[60] Santilio A, Pompili C, Giambenedetti A. Determination of

glyphosate residue in maize and rice using a fast and easy method involving liquid chromatography–mass spectrometry (LC/MS/MS). Journal of Environmental Science and Health Part B. 2019;**54**:1-6

[61] Herrera López S, Scholten J, Kiedrowska B, de Kok A. Method validation and application of a selective multiresidue analysis of highly polar pesticides in food matrices using hydrophilic interaction liquid chromatography and mass spectrometry. Journal of Chromatography A. 2019;**1594**:93-104

[62] Grunewald K, Schmidt W, Unger C, Hanschmann G. Behavior of glyphosate and aminomethylphosphonic acid (AMPA) in soils and water of reservoir Radeburg II catchment (Saxony/Germany). Journal of Plant Nutrition and Soil Science. 2001;**164**:65-70

[63] Laitinen P, Rämö S, Siimes K. Glyphosate translocation from plants to soil—Does this constitute a significant proportion of residues in soil? Plant and Soil. 2007;**300**:51-60

[64] Mercurio P, Flores F, Mueller JF, Carter S, Negri AP. Glyphosate persistence in seawater. Marine Pollution Bulletin. 2014;**85**:385-390

[65] Chiesa LM, Nobile M, Panseri S, Arioli F. Detection of glyphosate and its metabolites in food of animal origin based on ion-chromatography-high resolution mass spectrometry (IC-HRMS). Food Additives & Contaminants: Part A. 2019;**36**:592-600

[66] Alferness PL, Iwata Y Determination of glyphosate and (aminomethyl) phosphonic acid in soil, plant and animal matrixes, and water by capillary gas chromatography with mass-selective detection. Journal of Agricultural Food Chemistry. 1994;**42**:2751-2759

[67] Picó Y, Fernández M, Ruiz MJ, Font G. Current trends in solid-phase-based extraction techniques for the determination of pesticides in food and environment. Journal of Biochemical and Biophysical Methods. 2007;**70**:117-131

[68] Wang YS, Jaw CG, Chen YL. Accumulation of 2,4-D and glyphosate in fish and water hyacinth. Water Air Soil Pollution. 1994;**74**:397-403

[69] Lee J, Kim W, Lee J, Park S, Yang H, Kho Y, et al. Development of analytical method for glyphosate and glufosinate in animal feeds using LC-MS/MS. Environmental Epidemiology. 2019;**3**:225

[70] Tanaka T, Imou M, Tamase K, Umoto F, Okayama A, Oobayashi H, et al. Simple determination of glyphosate and its metabolite (aminomethylphosphonic acid) in fruits and vegetables by high performance liquid chromatography with fluorescence detection. Shokuhin Eiseigaku Zasshi. 1992;**33**:125-132 (in Japanese)

[71] Sen NP, Baddoo PA. Determination of glyphosate as N-nitroso derivative by high performance liquid chromatography with chemiluminescence detection. International Journal of Environmental Analytical Chemistry. 1996;**63**:107-117

[72] Kataoka H, Ryu S, Sakiyama N, Makita M. Simple and rapid determination of the herbicides glyphosate and glufosinate in river water, soil and carrot samples by gas chromatography with flame photometric detection. Journal of Chromatography A. 1996;**726**:253-258

[73] Hooijschuur EWJ, Kientz CE, Dijksman J, Brinkman UAT. Potential of microcolumn liquid chromatography and capillary electrophoresis with flame photometric detection for determination

of polar phosphorus-containing pesticides. Chromatographia. 2001;**54**:295-301

[74] Grey L, Nguyen B, Yang P. Liquid chromatography/electrospray ionization/isotopic dilution mass spectrometry analysis of n-(phosphonomethyl) glycine and mass spectrometry analysis of aminomethyl phosphonic acid in environmental water and vegetation matrixes. Journal of AOAC International. 2001;**84**:1770-1780

[75] Takahashi K, Horie M, Aoba N. Analysis of glyphosate and its metabolite, aminomethylphosphonic acid, in agricultural products by HPLC. Shokuhin Eiseigaku Zasshi. 2001;**42**:304-308 (in Japanese)

[76] Watanabe S. Simultaneous analysis of glyphosate and glufosinate in vegetables and fruits by GC-FPD. Shokuhin Eiseigaku Zasshi. 2004;**45**:38-43 (in Japanese)

[77] Qian K, Tang T, Shi T, Li P, Li J, Cao Y. Solid-phase extraction and residue determination of glyphosate in apple by ion-pairing reverse-phase liquid chromatography with pre-column derivatization. Journal of Separation Science. 2009;**32**:2394-2400

[78] Rembisz Z, Zakrzewski R, Skowron M, Ciesielski W. Image analysis of phenylisothiocyanate derivatised and charge-couple device-detected glyphosate and glufosinate in food samples separated by thin-layer chromatography. International Journal of Environmental Analytical Chemistry. 2016;**96**:320-331

[79] Boušová K, Bruggink C, Godula M. Fast routine analysis of polar pesticides in foods by suppressed ion chromatography and mass spectrometry. ThermoScientific Application Note. 2016;**661**

[80] Rajski L, Díaz Galiano FJ, Cutillas V, Fernádez-Alba AR. Coupling ion chromatography to Q-orbitrap for the fast and robust analysis of anionic pesticides in fruits and vegetables. Journal of AOAC International. 2018;**101**:352-359

[81] Melton LM, Taylor MJ, Flynn EE. The utilisation of ion chromatography and tandem mass spectrometry (IC-MS/MS) for the multi-residue simultaneous determination of highly polar anionic pesticides in fruit and vegetables. Food Chemistry. 2019;**298**:125028

[82] Savini S, Bandini M, Sannino A. An improved, rapid, and sensitive ultra-high-performance liquid chromatography-high-resolution orbitrap mass spectrometry analysis for the determination of highly polar pesticides and contaminants in processed fruits and vegetables. Journal of Agricultural and Food Chemistry. 2019;**67**:2716-2722

[83] Nortes-Méndez R, Robles-Molina J, López-Blanco R, Vass A, Molina-Díaz A, Garcia-Reyes JF. Determination of polar pesticides in olive oil and olives by hydrophilic interaction liquid chromatography coupled to tandem mass spectrometry and high resolution mass spectrometry. Talanta. 2016;**158**:222-228

[84] Chiarello M, Jiménez-Medina ML, Marín Saéz J, Moura S, Garrido Frenich A, Romero-González R. Fast analysis of glufosinate, glyphosate and its main metabolite, aminomethylphosphonic acid, in edible oils, by liquid chromatographycoupled with electrospray tandem mass spectrometry. Food Additives & Contaminants: Part A. 2019;**36**:1376-1384

[85] Karise R, Raimets R, Bartkevics V, Pugajeva I, Pihlik P, Keres I, et al. Are pesticide residues in honey related to oilseed rape treatments? Chemosphere. 2017;**188**:389-396

[86] Zoller O, Rhyn P, Rupp H, Zarn JA, Geiser C. Glyphosate residues in Swiss market foods: Monitoring and risk evaluation. Food Additives & Contaminants: Part B. 2018;**11**:83-91

[87] RTS. Du glyphosate, herbicide contesté, découvert dans l'urine des Romands. 2015. Available from: https://www.rts.ch/info/suisse/7125072-du-glyphosate-herbicide-conteste-decouvert-dans-l-urine-des-romands.html [Accessed: 12 April 2020]

[88] Pareja L, Jesús F, Heinzen H, Hernando MD, Rajski L, Fernández-Alba AR. Evaluation of glyphosate and AMPA in honey by water extraction followed by ion chromatography mass spectrometry. A pilot monitoring study. Analytical Methods. 2019;**11**:2123-2128

[89] Thompson TS, van den Heever JP, Limanowka RE. Determination of glyphosate, AMPA, and glufosinate in honey by online solid-phase extraction-liquid chromatography-tandem mass spectrometry. Food Additives & Contaminants: Part A. 2019;**36**:434-446

[90] Rubio F, Guo E, Kamp L. Survey of glyphosate residues in honey, corn and soy products. Journal of Environmental & Analytical Toxicology. 2014;**4**:249

[91] Chamkasem N, Vargo JD. Development and independent laboratory validation of an analytical method for the direct determination of glyphosate, glufosinate, and aminomethylphosphonic acid in honey by liquid chromatography/tandem mass spectrometry. Journal of Regulatory Science. 2017;**5**:1-9

[92] Berg CJ, King HP, Delenstarr G, Kumar R, Rubio F, Glaze T. Glyphosate residue concentrations in honey attributed through geospatial analysis to proximity of large-scale agriculture and transfer off-site by bees. PLoS One. 2018;**13**:e0198876

[93] El Agrebi N, Tosi S, Wilmart O, Scippo M-L, de Graaf DC, Saegerman C. Honeybee and consumer's exposure and risk characterisation to glyphosate-based herbicide (GBH) and its degradation product (AMPA): Residues in beebread, wax, and honey. The Science of the Total Environment. 2020;**704**:135312

[94] Mesnage R, Defarge N, Spiroux de Vendômois J, Séralini GE. Potential toxic effects of glyphosate and its commercial formulations below regulatory limits. Food and Chemical Toxicology. 2015;**84**:133-153

[95] Mañas F, Peralta L, Raviolo J, García Ovando H, Weyers A, Ugnia L, et al. Genotoxicity of AMPA, the environmental metabolite of glyphosate, assessed by the Comet assay and cytogenetic tests. Ecotoxicology and Environmental Safety. 2009;**72**:834-837

[96] Hao C, Morse D, Morra F, Zhao X, Yang P, Nunn B. Direct aqueous determination of glyphosate and related compounds by liquid chromatography/tandem mass spectrometry using reversed-phase and weak anion-exchange mixed-mode column. Journal of Chromatography A. 2011;**1218**:5638-5643

[97] Wu X, Chen X, Xiao H, Liu B. Simultaneous determination of glyphosate and glufosinate-ammonium residues in tea by ultra performance liquid chromatography-tandem mass spectrometry coupled with pre-column derivatization. Se Pu. 2015;**33**:1090-1096 (in Chinese)

[98] Ehling S, Reddy TM. Analysis of glyphosate and aminomethylphosphonic acid in nutritional ingredients and milk by derivatization with fluorenylmethyloxycarbonyl chloride and liquid chromatography-mass spectrometry. Journal of Agricultural and Food Chemistry. 2015;**63**:10562-10568

[99] Chamkasema N, Morrisa C, Harmon T. Direct determination of glyphosate, glufosinate, and AMPA in milk by liquid chromatography/tandem mass spectrometry. Journal of Regulatory Science. 2015;**2**:20-26

[100] Steinborn A, Alder L, Michalski B, Zomer P, Bendig P, Martinez SA, et al. Determination of glyphosate levels in breast milk samples from Germany by LC-MS/MS and GC-MS/MS. Journal of Agricultural and Food Chemistry. 2016;**64**:1414-1421

[101] Jensen PK, Wujcik CE, McGuire MK, McGuire MA. Validation of reliable and selective methods for direct determination of glyphosate and aminomethylphosphonic acid in milk and urine using LC-MS/MS. Journal of Environmental Science and Health, Part B. 2016;**51**:254-259

[102] McGuire MK, McGuire MA, Price WJ, Shafii B, Carrothers JM, Lackey KA, et al. Glyphosate and aminomethylphosphonic acid are not detectable in human milk. The American Journal of Clinical Nutrition. 2016;**103**:1285-1290

[103] Jansons M, Pugajeva I, Bartkevičs V. Occurrence of glyphosate in beer from the Latvian market. Food Additives & Contaminants: Part A. 2018;**35**:1767-1775

[104] Kudzin ZH, Gralak DK, Drabowicz J, Luczak J. Novel approach for the simultaneous analysis of glyphosate and its metabolites. Journal of Chromatography A. 2002;**947**:129-141

[105] Hu JY, Chen C-L, Li J-Z. A simple method for the determination of glyphosate residues in soil by capillary gas chromatography with nitrogen phosphorus. Journal of Analytical Chemistry. 2008;**63**:371-375

Chapter 8

Glyphosate Residues in Soil and Air: An Integrated Review

Evagelia Tzanetou and Helen Karasali

Abstract

Glyphosate [N-(phosphonomethyl) glycine] (GPS) is currently the most commonly applied herbicide worldwide. Given the widespread use of glyphosate, the investigation of the relationship between glyphosate and soil ecosystem is critical and has great significance for its valid application and environmental safety evaluation. However, although the occurrence of glyphosate residues in surface and groundwater is rather well documented, only few information are available for soils and even fewer for air. Due to this, the importance of developing methods that are effective and fast to determine and quantify glyphosate and its major degradation product, aminomethylphosphonic acid (AMPA), is emphasized. Based on its structure, the determination of this pesticide using a simple analytical method remains a challenge, a fact known as the "glyphosate paradox." In this chapter a critical review of the existing literature and data comparison studies regarding the occurrence and the development of analytical methods for the determination of pesticide glyphosate in soil and air is performed.

Keywords: glyphosate, AMPA, soil, air, extraction, analytical methods, quantification

1. Introduction

After World War II, the world was in the need to overcome food scarcity. Therefore, several pest and weed management techniques were adopted by farmers all over the world using various synthetic herbicides. The invention of glyphosate (GLY; N-(phosphonomethyl) glycine) was a big breakthrough in that era. GLY with CAS No. 1071-83-6 is a broad-spectrum, postemergent, nonselective, and synthetic universal herbicide, whose commercial formulations are referred to as glyphosate-based herbicides (GBHs) [1, 2]. Glyphosate was first synthesized in 1950 by Swiss chemist Henry Martin, who worked for the Swiss company Cilag. The work was never published. Its herbicidal activity was not discovered until GBHs were resynthesized and tested in 1970, being used for this purpose since 1974. It was the Monsanto Corporation in 1974 that introduced and made commercially available the herbicidal formulation Roundup containing GLY as active substance. Farmers quickly adopted glyphosate for agricultural weed control, gaining the potential to kill weeds without killing their crops. Indeed, glyphosate proved able to kill weeds without killing their crops, especially annual broadleaf weeds and grasses known to compete with commercial

IntechOpen

crops grown around the globe by interfering with the synthesis of the aromatic amino acids phenylalanine, tyrosine, and tryptophan [3].

Since then, its use in agricultural and nonagricultural settings has steadily increased from a total of 0.6 Mg applied in 1974 to a total of 125.5 Mg applied in 2014, and it is currently the most widely used herbicide in the United States and throughout the world [4, 5]. Monsanto's last commercially relevant US patent expired in 2000. Nowadays, GLY formulations that are used as a broad-spectrum systemic herbicide have been widely applied in agronomic crops and orchards. Furthermore, GLY formulations are currently approved by regulatory bodies and marketed worldwide by many agrochemical companies, such as Bayer, Dow AgroSciences, and Monsanto, in different solution strengths and with various adjuvants.

GLY approval is renewed in the European Union (EU) on 16 December 2017, while its approval expires on 15 December 2022. Therefore, GLY can be used as an active substance in plant protection products (PPPs), until 15 December 2022. GLY has been thoroughly assessed, under an intense debate due to a concern about its effects on the environment and human health, by the Member States, the European Chemicals Agency (ECHA), and the European Food Safety Authority (EFSA) in recent years [6, 7]. An important prerequisite for GLY upcoming renewal as an ingredient in PPPs is that GLY should not adversely affect the environment and human and animal health as delineated by European regulation [8].

2. Glyphosate residues in soil

2.1 Environmental fate of glyphosate

2.1.1 Glyphosate in the soil

Given the widespread use of glyphosate, the investigation of the relationship between glyphosate and soil ecosystem is critical and has great significance for its valid application and environmental safety evaluation. Although herbicides containing glyphosate are not intentionally applied directly to the soil, they may contaminate soils in and around the treated areas, via spray drift during their application and after being washed off from leaf surfaces with rainfall.

The fate of glyphosate in soil is complex and attributed to mineralization, degradation, immobilization, and leaching. Several studies trying to identify and understand the mechanisms that control the fate of chemicals as a source of environmental contamination have been published in previous years, especially in soils and water. Some were conducted with the acid form of glyphosate and others with formulated products, since glyphosate is not introduced into the environment as pure active ingredients but as formulated products containing co-formulant chemicals (adjuvants) and other additives. In a recent review, Mesnage et al. presented an overview of the most common surfactants containing co-formulants in glyphosate-based herbicides and explained whether the presence of such surfactant (e.g., Triton CG-110) has the potential to affect adsorption, leaching, and mineralization of glyphosate in the soil [9].

The fate of glyphosate depends on soil composition, its physicochemical properties (texture, organic matter content, pH), its biological properties (microbial community, climatic conditions), the chemical properties of the specific pesticide, as well as the timing between precipitation and pesticide application [10–13]. A recent study by Muskus et al. showed that temperature, pH, and total organic carbon

(TOC) variations influenced the mineralization kinetics of glyphosate as well as the amount of extractable glyphosate and the extent of bio-NER formation over time in a German soil [14].

Glyphosate degrades at a relatively rapid rate in most soils, with a half-life estimated to be between 7 and 60 days. The relatively rapid degradation of glyphosate has the advantage of limiting its role in polluting the environment, especially soil and water resources. However, its degradation could increase the pollution risk by its metabolites: aminomethylphosphonic acid (AMPA) and/or sarcosine. The degradation of the herbicide molecule as described in the literature (**Figure 1**) can follow two paths: the first is based on the breakdown of the carbon-nitrogen bond and leads to the formation of AMPA (main metabolite of glyphosate) via glyphosate oxidoreductase which is further degraded to carbon dioxide, while the second way is based on the splitting of the carbon-phosphorus (C-P) bond that is mediated by C-P lyase enzyme and results in the formation of sarcosine and glycine [15–20]. However, AMPA also exists in the environment as a photodegradation product of aminopolyphosphonates in water [21].

Glyphosate is a small, amphoteric molecule characterized by three polar functional groups. These are the phosphonomethyl, amine, and carboxymethyl groups arranged in a linear manner. As a result of the presence of those groups in its structure, glyphosate is an ionic compound (log KOW = −3.20), highly polar and soluble in water (10.5 g L^{-1} at 20°C). GPS is a polyprotic acid with four pKa values, 0.7, 2.2, 5.9, and 10.6, 8 meaning that the speciation of the molecule is dependent upon the pH value of the solution. Three pKa values, 0.9, 5.6, and 10.2, characterize AMPA. Over the pH values commonly found in soils, mono- and divalent anions are the predominant species present [6, 22].

Glyphosate is soluble in water, but it also binds onto soil particles under certain conditions, particularly in clays. Numerous laboratory studies have shown that the absorption constant of the molecule in the soil varies between 8 and 377 dm^3/kg. This coefficient value indicates a high absorption in the soil. Glyphosate adsorption to soil, and later release from soil, varies depending on the characteristics and

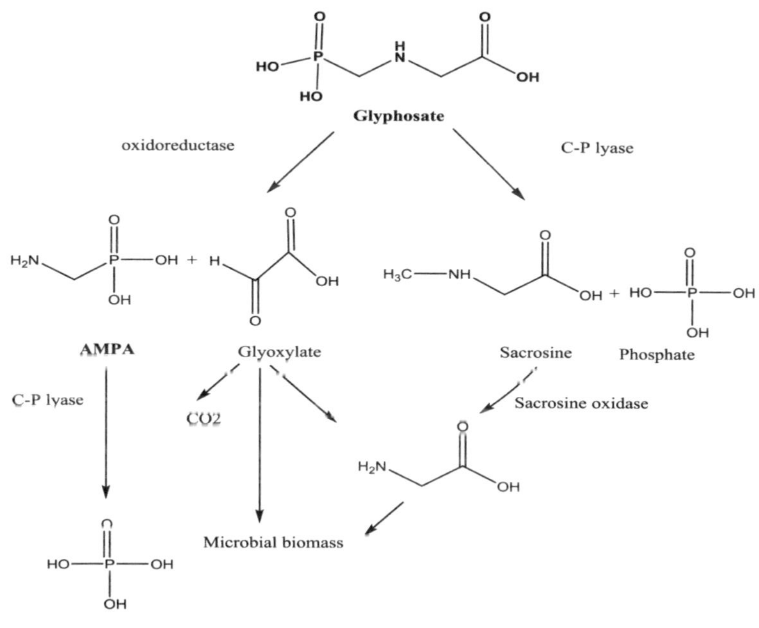

Figure 1.
Main glyphosate biodegradation pathways in the environment [5].

composition of the soil (clay, sand, or gravel), temperature, and soil moisture. So it may quickly wash out of sandy soils or last for more than a year in soils with a high clay content. Even when bound to soil particles, it may dissolve back into soil water later on, for example, in the presence of phosphates. Glyphosate can also form complexes with metal ions, potentially affecting the availability of nutrients in the soil.

The mechanism of glyphosate sorption to soil is similar to that of phosphate fertilizers, the presence of which can reduce glyphosate sorption [23]. Glyphosate compared to most other pesticides strongly absorbs to soil and is not expected to move vertically below the six-inch soil layer, exception made of a colloid-facilitated transport. Its soluble residues are expected to be poorly mobile in the free pore water of soils. The mobility of glyphosate in soil is very low because, as a strong chelating agent through the carboxyl, phosphonate, and amino groups, it creates the complexes that immobilize the mineral micronutrients of the soil (calcium, iron, magnesium, manganese, nickel, zinc, etc.) making them unavailable to plants [11, 24]. Similar to glyphosate, AMPA accumulates in soil and adsorbs in soils with high mineralization rates. Where strong sorption is demonstrated, glyphosate accumulation in soils can be expected. The interaction of pesticide-soil and the diffusion process lead to the formation of non-extractable residues trapped in areas not accessible to water flowing through the soil. The contamination of the environment is therefore considered to be relatively limited.

Nevertheless, this adsorption is not permanent because glyphosate can also be found in lower soil layers. Many studies suggest the possibility of a slow remobilization of these residues, which could explain the low pollution level of groundwater by some pesticides at a long term. Glyphosate does have the potential to contaminate surface waters through erosion, as it adsorbs to soil particles suspended in runoff. Rain events can trigger dissolved glyphosate loss in transport-prone soils [25, 26].

2.2 Glyphosate occurrence in soil

The increase of glyphosate-based herbicides has raised concerns about the occurrence of GLY and AMPA in the environment. Reports of GLY presence in the environment from other parts of the world are numerous. A considerable attention has been given to Argentina [27–30], Canada [31], across the United States [32], Mexico [33], and Portugal [34] as well to Spain [35], New Zealand [36], Austria [37], and French [38].

However, although GLY is the most sold herbicide in Europe, a combined approach on the occurrence and levels of glyphosate residues in European soils and air, in conjunction with analytical methods used for this scope, is still scarce, compared to the magnitude of its use though some research articles and reviews (not only focusing on soil) started to appear (indicatively see [39–41]).

The first large-scale assessment of distribution of GLY and AMPA in soils from agricultural topsoils of the European Union was recently published by Silva, where glyphosate and its metabolite AMPA were tested in 317 EU agricultural topsoils; 21% of the tested EU topsoils contained glyphosate and 42% contained AMPA, while both glyphosate and AMPA displayed a maximum concentration in soil of 2 mg kg^{-1}. Both compounds were present at higher frequencies in northern soils, while eastern and southern regions generally had the most glyphosate- and AMPA-free soils (<0.05 mg kg^{-1}), respectively. In addition, some contaminated soils were observed in areas highly susceptible to water and wind erosion [42]. Therefore, residue threshold values in soils are urgently needed to define potential risks for soil health and off-site effects related to export by wind and water erosion.

2.3 Analytical methods for quantification of GLY and AMPA

In order to detect the presence and quantity of GLY dispersed in the environment, various laboratory analyses are performed on samples taken in situ.

One of the key problems for obtaining reliable results from field samples is the use of the best suitable extraction solution, since sorption and desorption of glyphosate in soils are extremely pH dependent. Some reports showed that humic substances (substances and heterogenic mixtures dispersed and abundant in soils and sediments) adsorb glyphosate strongly due to the hydrogen bonding interactions between the two matrices. Another important aspect is that GLY is a highly polar herbicide, very soluble in water and insoluble in most organic solvents, which does not allow extraction with organic solvents and makes the extraction difficult and the preconcentration step quite lengthy. However, due to the amphoteric character of GLY and AMPA, both anionic and cationic resins have been used for preconcentration and cleanup purposes (commented in the below sections).

2.3.1 Extraction procedure

As already mentioned, GLY has been shown to bind strongly to soils, especially to soils with high amounts of organic matter, iron, and aluminum [43, 44]. There is also evidence that glyphosate binds to clay minerals in a manner similar to inorganic phosphate [44–46]. The strength of the interactions of the phosphonate, carboxyl, and amino groups with iron oxides, silica, alumina, and organic matter depends on factors such as pH, metal cations, phosphate from fertilizers, etc. Therefore, it is hard to detect GLY without a pretreatment method [47].

The choice of the best suitable extraction solution remains a problem that must be addressed accordingly.

Several authors in the past reported different extraction methods of these compounds from soil, mainly using alkaline solutions with different recovery rates [48–51] and most times applicable for one type of soil. In 1980 the FDA's "Pesticide Analytical Manual" (PAM) including a procedure for the analysis of glyphosate residues in soil is published. However low and irreproducible recoveries in soil samples have been reported using this method. Later, Glass in 1983–1984 analyzed soils by alkaline extraction, followed by cleanup using flocculation with $CaCl_2$ and anion exchange [52–54]. Yet, recoveries were still remained poor and ranged from 19 to 55%. Many extractants for soil have been tested in the years that followed with the most commonly used being aqueous bases KOH or NaOH, aqueous NH_4OH or NH_3, or triethylamine. Other extractants include $NaHCO_3$, KH_2PO_4, mixed solutions of KH_2PO_4 and NH_3 or NH_4OH and HPO_4, sodium borate buffers [55–60], or even weak acids such as 10% phosphoric acid buffers [13, 61].

Moreover, it is vital to adjust the concentration of the extraction media in such a way that high recovery rates can be obtained while avoiding matrix problems provoked by excessively aggressive alkaline media, which may enrich the dissolved humic substances in the extraction solution [49]. Humic acids interfere, for example, with the derivatization and suppress the ionization in ESI-MS/MS detectors.

2.3.2 Analytical methods

Although GLY is the most widely used agrochemical in the world, it is also the most cumbersome in its determination in analytical methods, a fact known as the "glyphosate paradox." The challenge to detect GLY using a simple analytical method is an outcome of its ionic character, low volatility and low mass, high polarity and solubility in water, poor solubility in common organic solvents, high boiling points, difficult evaporation, and poor retention on traditional analysis columns. The quantitative and qualitative analyses of GLY (and AMPA) are extremely difficult due to the absence of fluorophores or chromophores in their structure. Furthermore, its determination at the low concentration levels required for residue analysis in different matrices is very difficult. In soil its determination is even more difficult due to the complexity of this matrix and subsequent matrix effects. The derivatization process using different derivatization reagents has been extensively used to overcome some of the above problems [62].

Prior to any attempt, it is important that all analysts to work with a glass that is not silanized to avoid the typical pitfall of GLY analysis. GLY has a profound affinity to glass, and any analytical solution prepared by this way will deviate substantially from its nominal concentration.

Chromatography is the most used and powerful method for the determination of GLY and its main metabolite AMPA, utilizing gas chromatography (GC) and liquid chromatography (LC) after derivatization or directly and capillary electrophoresis (CE). Conventional detectors are difficult to be used (especially for a straightforward analysis) due to the lack of chromophore and fluorophore groups in GLY. Usually, the limits of detection for GLY in soil vary between 0.01 and 0.3 mg/kg.

In all cases, the analytical methodology is practically exclusive for this analyte, since the working conditions cannot be applied to the determination of pesticides different from glyphosate, except for some organophosphorus, such as glufosinate and other polar compounds, and this chemical is difficult to incorporate in the vast majority of multiresidue methods. However, many of the methods published for the determination of GLY are also suitable and report results for the determination of AMPA. The majority of developed analytical methods concerned a single matrix (most often water) and may not be suitable for other matrices. Therefore, the last decade, numerous revised methods have been published on the analysis of glyphosate and AMPA in different matrices such as water, plants, or soils. Many of them just modify several parameters of previously published methods, as the pH of the water in the extraction, cleanup procedure, and derivatization step (volume and/or concentration of the samples or reagents). Other modifications include the use of different separation techniques or detection systems or even new matrices. Fewer new methods have been reported in the past 5 years for more complex matrices such as soil. Very few articles have been published on multimatrix methods.

In **Table 1** numerous analytical methods that have been used for the determination of GLY and AMPA in soil matrices are summarized. Based on the given information, at present LC is the most used method since it is considered the most suitable technique for the detection of phosphonic and amino acid-type herbicides at low concentrations. Hence, the lack of chromophore or fluorophore groups makes it difficult to use conventional detection methods such as ultraviolet (UV) absorption or fluorimetry. LC–MS/MS is currently the method of choice for polar analytes due to its high selectivity and sensitivity.

2.3.3 Gas chromatography - Derivatization

Gas chromatography methods are used after derivatization by simultaneous acylation, esterification, or trialkylsilylation reactions to convert the analytes into volatile compounds [69, 91, 92]. Typically used derivatization reagents are the mixture of trifluoroacetic anhydride (TFAA) and trifluoroethanol (TFE) or N-methyl-N-(tert-butyldimethylsilyl)-trifluoroacetamide (MTBSTFA) containing 1% tertbutyldimethylchlorosilane (TBDMCS) in excess producing sufficiently

Year	Sample preparation (extraction/ cleanup)	Derivatization (pre- or post-column)	Analytical method	LOD/LOQ μg/g	Other information	Reference
1986	0 1 M $(C_2H_5)_3$ N/ SAX cleanup	FDNB	HPLC-UV (405 nm)	0.05 GLY 0.1 AMPA		[63]
1988	0.1 M KH_2PO_4 or 0.2 M KOH	FMOC-CI	HPLC-FLD (λex = 270 nm, λem = 315 nm)	0.5–1.0 GLY	Minutes till some days of analysis time 0.1 M KH_2PO_4 (sandy soils)/0.2 M KOH (high clay soils)	[50]
1989	0.1 M KH_2PO_4	TFAA-TFE	GC-NPD	0.01/0.05	Recoveries 66–75%	[64]
1991, 1999		TsCl	HPLC-UV (240 nm or 280 nm)	3 mg/L GLY 10 mg/L AMPA		[64–66]
1994	0.25 M NH_4OH and 0.1 M KH_2PO_4	HFB/TFAA (1:2)	GC–MS (EI-SIM)	0 01/0.05 GLU, 0.01/0.05 AMPA	Recoveries 84–97%	[67]
1996	0.6 M KOH	FMOC-CI	LC–LC/FLD (263 nm λex = excitation, λem = 317 nm)	0.01/0.05 GLY 0.01/0.05 AMPA	Concerning soil organic matter and clay contents, the LOQ can reach 0.01 μg/g for both analytes for sandy samples, and for soil samples with a high organic matter and clay contents, LOQ is of 0.04 μg/g for glyphosate and 0.1 μg/g for AMPA	[68]
1996	NaOH 0.2 M	iso-PCF	GC-FPD	0.8/8.0 GLY 1.2/12 AMPA	Recoveries 91–106%	[69]
2000	1 M NaOH	TFAA/TFE (2:1)	GC–MS (EI)	0 003/0.006 GLY 0.003/0.006 AMPA	Recoveries 75–78%	[48]
2002			Cyan sensor	0.45 GLY	Phosphonomethyl glycine Inhibits amino acid biosynthesis	[70]
2005	0.6 M KOH	FMOC-CI	LC-ESI-MS/MS	0.005/0.05 GLY 0.005/0.05 AMPA	Recoveries 88–92%	[49]
2007	0.1 N NaOH/ SAX-SPE	FMOC-CI	HPLC-ESI-MS/MS	0.02/0.035 GLY 0.03/0.05 AMPA	Mean recovery values were 70% (7%) for GLU and 63% (3%) for AMPA	[71]

Year	Sample preparation (extraction/ cleanup)	Derivatization (pre- or post-column)	Analytical method	LOD/LOQ µg/g	Other information	Reference
2008	2 M NH_4OH	TFAA/TFE	GC-NPD	0.01 GLY	Recoveries 88–94%	[72]
2009	0.1 M KOH/CAX cleaned up	FMOC-CI	HPLC-FLD	0.025		[55]
2009	2 M NH_4OH	1. CS2 2. Ammonical solution of Cu(II)	UV 435 nm	1.1./3.7 µg/mL GLY	80–87%	[73]
2010	Water		CE/MS (negative with amino capillary)	20 GLY 40 AMPA	Recoveries 76–121% (20/200/2000) for GLY and 83–89% (40/400/4000) for AMPA	[74]
2011	Water	1. $Ca(ClO)_2$ 2.OPA/ME	SIA-FLD (λex =270 nm, λem =315 nm)	0.08/0.25 mmol/L GLY	Pre-column conversion: 1. of glyphosate to glycine by $Ca(ClO)_2$; 2. followed by reaction with OPA/ME in borate buffer (pH 9.5) to produce the fluorescent 1-(2′-hydroxyethylthio)-2-N-alkylisoindole	[75]
2012	0.2 M KOH		SWV using CFME	25/83 µg/L GLY	Recoveries 89–102%	[76]
2013	0.1 M KOH/SPE	FMOC-CI	HPLC-ESI-MS/MS using reversed-phase C18	0.02/0.05 GLY 0.01/0.03 AMPA	Recoveries 79–117%	[77]
2014	0.6 M KOH	FMOC-CI	HPLC-FLD (λex = 267, λem =317 nm) using reversed-phase C18	0.5 GLY 0.5 AMPA		[57]
2014	10% H_3PO_4		SPE-HPLC-ESI-MS/MS (using HILIC)	0.37 GLY 0.61 AMPA	Recoveries 85–126%	[61]
2014	0.01 M	FMOC-CI	SPE-HPLC-FLD (λex =263 nm and λem =317 nm)	0.6/2.0 ng/mL GLY 0.4/1.3 ng/mL AMPA		[78]

Year	Sample preparation (extraction/ cleanup)	Derivatization (pre- or post-column)	Analytical method	LOD/LOQ µg/g	Other information	Reference
2014	Water	HCl/$NaNO_2$	DIPN-GNPs-PGE	0.35 ng/mL GLY	1.0 g soil was suspended in a 30 mL water Recoveries 98.6–102.8%	[79]
2016	Water		IS-FLD using lgG-CDs	0.35 ng/mL GLY	Recoveries 87.4–105.5%	[80]
2015	1 M $Na_2B_4O_7$	FMOC-Cl	HPLC-PDA (206 nm)	0.01/0.1 GLY 0.01/0.1 AMPA	Recoveries 70–76% Confirmation with QTOF MS	[16]
2015	NaOH	FMOC-Cl	HPLC-UV (254 nm)	—		[3]
2015, 2016, 2018	0.6 M KOH	FMOC-CI	HPLCMS/MS/using reversed-phase C18	0.02/0.05 GLY 0.03/0.05 AMPA	Recoveries 77–87%	[81–83]
2016	NaOH	FMOC-CI	SPE-HPLC–MS	0.02 mg/L GLY 0.05 mg/L AMPA		[84]
2018	NH_4Cl	FMOC-CI	UV–Vis (264 nm)	20 GLY		[85]
2018	KOH	FMOC-CI	SPE-HPLC–MS		0.001 GLY 0.001 AMPA	[86]
2018	0.1 M K_2PO_4	FMOC-CI	HPLC-MS (negative ionization)	0.002 GLY 0.005 AMPA	Recovery 80%	[27]
2018	40 mM $Na_2B_4O_7$ or $NaHCO_3$ 0.5 M	2% succinic anhydride IN DMSO	L'ELISA (via microtiter plate reader at 450 nm)	0 8 10–3/0.1 GLY	Recoveries 87.4–97.2% (0.1–10 µg/g) confirmed with HPLC-FLD (by Ibanez)	[87]
2019	0.6 M KOH	FMOC-CI	LC-ESI-MS/MS using reversed-phase C18	GLY 0.01 AMPA	Recoveries from 89.6 to 118.8% for GLY and from 68 to 94.6% for AMPA	[88]
2019	Water	HFBA/FBA	SPE-GC-FPD	0.10/0.37 ng/mL GLY 0.22/0.81 ng/mL AMPA	Recoveries 94–110%	[89]

Year	Sample preparation (extraction/ cleanup)	Derivatization (pre- or post-column)	Analytical method	LOD/LOQ µg/g	Other information	Reference
2019		1. $Ca(ClO)_2$ 2.OPA/ME	SIC-FLD	0.03/0.10 GLY	Pre-column conversion: 1. of glyphosate to glycine by $Ca(ClO)_2$; 2. followed by reaction with OPA/ME in borate buffer (pH 9.5) to produce the fluorescent 1-(2′-hydroxyethylthio)-2-N-alkylisoindole	[90]

Solid-phase extraction = SPE; anion exchange = SAX; cation exchange = CAX; Flame photometric detector = FPD; nitrogen-phosphorus detector = NPD; fluorescence detector = FLD (λex = excitation; λem = emission); capillary electrophoresis = CE; immunosensor = IS; square wave voltammetry = SWV; carbon-fiber microelectrode = CFME; double template imprinted polymer film-gold nanoparticle-modified pencil graphite electrode = DIPN-GNPs-PGE; carbon dot-labeled antibodies = lgG-CDs; sequential injection chromatography = SIC; sequential injection analysis = SIA; linker-assisted enzyme-linked immunosorbent assay = L'ELISA; laser-induced fluorescence detection = LIF; Fluorenylmethylchloroformate = FMOC-Cl; 1-fluoro-2,4-dinitrobenzene = FDNB; trifluoroethanol = TFE; N-methyl-N-(tert-butyldimethylsilyl)-trifluoroacetamide = MTBSTFA; 2,2,3,3,4,4,4-heptafluoro-1-butanol = HFB; trifluoroacetic anhydride = TFAA; heptafluorobutyric anhydride = HFBA; isopropyl chloroformate = iso-PCF; o-phthalaldehyde/2-mercaptoethanol = OPA/ME; 4-toluenesulfonyl chloride = TsCl.

Table 1.
Main characteristics on the methods used for the determination of GLY and its main metabolite in soil matrices.

volatile derivatives for GC analysis. These derivatization approaches can be applied not only to soil (or sediment) but to other commodities as well [91].

GLY as a compound permits its detection (in conjunction with GC) by several detectors such as the flame photometric detector (FPD), flame ionization detector (FID), electron capture detector (ECD), nitrogen-phosphorus detector (NPD),and also the more sensitive and selective mass spectrometer detector (MSD). The quantification of GLY in the soil through NPD has reached an limit of quantification (LOQ) equivalent of 0.02 mg/kg [93, 94]. GC analytical methods are reliable, sensitive, and selective, but the sample preparation is very time-consuming, complicated, and tedious as all ionic groups must be derivatized. In addition, they involve anhydrous conditions, extensive cleanup using solvent partitioning steps, charcoal elimination of pigments, and a large-volume anion of cation exchange.

2.3.3.1 GC-MS

Borjesson and Torstensson using GC-MS managed to sensitively detect GLY and AMPA in soil [48]. One point that should have been addressed was the content of humic acids a problem mentioned above as well. Extraction was tedious, involving extraction under basic conditions, adjustment of the pH to acidic, and then subjected to column purifications to achieve ligand-anion exchanges. To derivatize GLY and AMPA TFE and TFAA were used. By this way the respective ester and acetyl derivative are formed suitable for GC analysis.

Utilizing the S/N approach, they presented one of the lower LOQs of the bibliography for GLY, established at 0.006 mg/kg. The application of the method in soils collected from Swedish embankments after being treated with GLY revealed the gradual degradation of GLY along with the presence of AMPA.

Bergstrom et al. investigated at laboratory level GLY and AMPA in sand and clay soils [13]. More specifically, its degradation was monitored using a GC-MS method, after derivatization with TFE and TFAA. The extraction of the soil was accomplished using an alkaline solution. Selected ion monitoring was utilized to enhance selectivity and optimize sensitivity of the method. The LOQ of the method was established at 0.01 mg/kg. The studied kinetics demonstrated that GLY had a very slow degradation rate in the clay soil. Concerning AMPA, though it is more tenacious than GLY (when derived from GLY), it degrades faster than GLY.

2.3.3.2 GC-FPD

A 2019 study on GLY and AMPA analysis in soil showed that still GC-FPD can function as an adequate tool for such demanding analyses [89]. This research was intrigued by the obstacles observed in derivatization in connection with the by-product interferences in high-performance liquid chromatography (HPLC) analysis (when MS/MS mode is not utilized), which lead to inferior selectivity. Consequently, the soil samples were mixed, turned to powder, and then ultrasonic extracted using water assisted by a solid-phase extraction (SPE). After derivatization, the samples were subjected to chemical analysis. A breakthrough of this process was the three-cross derivatization, and the elaborate investigation of its optimization, aided by the orthogonal experimental design. Such design is fundamental in the selection of the optimum conditions, in this case, reaction temperature, time, and ratio of the derivatizing-coupling reagents.

2.3.3.3 GC-NPD

GC-NPD was utilized by Hu and coworkers to analyze GLY in soil, using GC-MS for verification [72]. Extraction was performed in alkaline environment, followed by acidification in the dry extract. The authors stated that NH_4OH was the most adequate extractant due to lesser extracted interferences than other alkaline agents, though in other works other agents are selected [48]. Derivatization was accomplished by the use of TFE and TFAA, followed by a liquid extraction using methylene chloride. The method verified the degradation of GLY in soil in apple orchards.

2.3.4 Liquid chromatography and derivatization

2.3.4.1 Fundamentals

The availability of derivatization techniques compatible with an aqueous extract or sample and the chromatographic separation makes LC a more attractive pre-column derivatization [91]. Derivatization approach is used to produce fluorescent derivatives and to enhance their retention in hydrophobic stationary phases prior to detection by fluorescence detection (FLD), UV detection, electrochemical detection (ECD), or tandem mass spectrometry (MS/MS). In post-column procedures, the most known reactions are ninhydrin derivatization accompanied by UV detection and fluorogenic labeling with o-phthalaldehyde (OPA) in mercaptoethanol or N,N-dimethyl-2-mercaptoethylamine after oxidation of glyphosate to glycine.

Although GLY and its derivatives show high sensitivity in LC determination, a laborious cleanup procedure such as ion-exchange column chromatography is required which may result in some sample loss and lower reproducibility, or many laboratories do not have the facilities required for this type of pre- or post-column fluorogenic labeling. The use of either hydrophilic/weak exchange or reversed-phase/weak exchange mixed-mode chromatography without any derivatization, followed by diverse detection techniques including tandem mass spectrometry detection, is gaining interest [77, 95]. HPLC methods are highly sensitive with fluorogenic labeling, but they lack specificity and usually require a laborious cleanup procedure such as ion-exchange column chromatography, which may result in some sample loss and lower reproducibility.

2.3.4.2 Pre-column procedures

2.3.4.2.1 FMOC derivatization

Pre-column procedures are a good alternative to post-column ones, and this has gradually come to play an important role in the analysis of glyphosate. The easier, less demanding and more current popular method to analyze these compounds is derivatization with 9-fluorenylmethylchloroformate (FMOC-Cl) followed by HPLC with FLD or MS/MS. A factorial experimental design was applied by a Chilean group in a critical analysis of this derivatization reaction [78]. The design was studied in aqueous soil extracts, unveiling the proper equilibrium between agents for the successful completion of the reaction. For example, excess of FMOC-Cl is required since there are also other active centers (amine-hydroxy groups) with which FMOC-Cl can react. Isotherm data verified the broad applicability of this method.

Back in the 1990s, Sancho et al. established a method for the analysis of GLY in soil samples that involved a pre-column derivatization step with FMOC-Cl and subsequent estimation by coupled-column liquid chromatography with fluorescence detection (LC–LC/FLD) [68]. However, for the determination of glyphosate in soils based on FMOC derivatization analytics, an extraction procedure including an SPE cleanup step has been used in many studies and considered more efficient [37]. In particular, Todorovic et al. extracted soils using sodium tetraborate. Once again, a group devoted substantial time in the extraction of GLY and AMPA from soil due to the complex sorption and desorption in soil which is also pH dependent. The sodium tetraborate performed better in terms of chromatographic efficiency than KOH extraction (more matrix interferences, more humic substances, etc.). The authors after derivatizing GLY and AMPA with FMOC-Cl cleaned up-enriched the extract with a polymeric SPE cartridge. Overall, the method was fit for purpose based on the analytical results on three different types of soils.

Botero-Coy et al. have established a method based on LC–MS/MS, which was successfully applied to soil samples from Colombia and Argentina [77]. This work was an improvement of the previous work in the same domain [49]. In that work, the soil samples were extracted with potassium hydroxide solution and purified with SPE Oasis HLB cartridges. A pre-column derivatization step was also required in this method for which 9-fluorenylmethylchloroformate (FMOC-Cl) was used and the purification method using SPE cartridges was troublesome and expensive. Despite these difficulties, the analysis was conducted in Spanish soils with success. But, when soil samples from the mentioned countries were analyzed by the specific protocol, their high organic content proved an obstacle in the analysis. For this reason the authors introduced a dilution step of the extract assisted by pH adjustment to 9, before the SPE step. For SPE the polymeric reversed-phase Oasis HLB cartridges proved better in retaining-releasing the FMOC derivative than Oasis MAX used with good results.

Internal standard's use compensated possible downsides during sample preparation and corrected matrix effects. An additional tool in this work was the use of high-resolution mass spectrometry exploiting the time-of-flight technology. By this way additional interferences that would affect the analysis were further elucidated using the accurate mass full-acquisition data. It is noteworthy that the authors investigated MS ions used in the MS/MS mode. Interestingly, MRM transition containing the m/z 179 was problematic since it is related to FMOC and lacks specificity. In our work (see below), the specific ion was monitored only in AMPA transition, solving this issue [88]. Overall, the analysis verified the presence of GLY and AMPA in the majority of samples.

Another work in the field of GLY analysis in soil/sludge using FMOC-Cl as a derivatizing agent was presented by Sun and coworkers [96]. In this context, an optimized sample preparation protocol was developed, applying extraction with sodium phosphate and trisodium citrate solutions (aqueous) and a purification step using hexane in acidified soil. The rationale behind the use of trisodium citrate was to counteract the effect of other metal ion complexing agents (such as Mg^{2+}, Ca^{2+}, etc.), in which GLY binds. The method was validated in three types of soils (and sludge samples) verifying that it was fit for purpose. The demonstrated LOQ was determined at 0.04 mg/kg.

A pre-column derivatization was applied by Druart and coworkers, embracing glufosinate also in their portfolio [60]. A detailed study was conducted on the parameters governing the extraction of the analytes from the matrix. Accelerated solvent extraction, ultrasonic extraction, and magnetic stirring agitation were tested to achieve optimum conditions. In the end agitation was selected. The group also optimized derivatization by selecting water as the solvent of the reaction, though the previous study showed that an equivalent mixture of H_2O:ACN would compromise the solubility of both GLY and FMOC-Cl reagents [97]. In the same study, it was demonstrated that a C18 column of 30 cm superseded other columns tested, even a respective NH_2 column broadly used for such separations.

In addition our group has developed a methodology for GLY and AMPA detection in topsoils originating from Greece [88]. The sample preparation was envisaged by previous works (one of our group) [49, 98]. The LC–MS/MS method developed was adequate for the analysis of both active substances, showing that GLY and AMPA were detected in 37 and 45%, respectively, of the samples investigated. A breakthrough of this work was the association of the results with the land use utilizing geographical information system (GIS) databases.

GLY in soil is studied for registering not only its residual prevalence (including AMPA's) but also its degradation dynamics. With this in view, Zhang et al. investigated its dynamics using an HPLC-FD method, utilizing FMOC derivatization [3]. Results of this study showed that the degradation is dependent on the physicochemical parameters of the soil, exemplified by the pH. The behavior of GLY and AMPA was investigated in compost-amended soils by Erban and colleagues [86]. Soil depth was disclosed as a key factor on the concentrations detected. GLY and AMPA though showed a different behavior when moisture and saturated hydraulic conductivity are considered. More specifically, GLY was affected principally by moisture, whereas AMPA was impacted by this conductivity.

2.3.4.2.2 Other fluorophores

Oliveira-Pereira and colleagues, in the context of adsorption studies, determined GLY and AMPA using a low-cost reversed-phase sequential injection chromatography method [90]. More specifically, GLY was converted (pre-column) to glycine (using hypochlorite). Then, by reaction with o-phthaldialdehyde, the respective fluorescent indole was formed. Expectedly, this reaction reduces the polarity of the indole derivative making it adequate for analysis under reversed-phase conditions (e.g., C18).

2.3.5 Direct analysis: a recent cornerstone

Direct analysis of GLY and AMPA, avoiding the derivatization step, is still a challenge for the analysts. In this context, Marek and Koskinen developed a method for the straightforward analysis of GLY and AMPA in soil using for separation a Bio-Rad cation H exchange column coupled to LC–MS/MS [61]. The sample preparation involved mixing of soil with phosphoric acid solutions and sequential extractions advancing from a specific SPE technology. The combined extracts were purified using IC-Chelate cartridges known for their ability to exchange transition metals and divalent cations. A portion of the end extract was reacidified and passed through an IC-RP SPE cartridge to eliminate hydrophobic interferences prior to analysis. This work managed to provide very high recoveries for both substances regardless of the type of soil, which is a clear advantage.

2.3.6 Hydrophilic interaction liquid chromatography (HILIC) and normal phase, a new frontier for GLY and AMPA in soil analysis

Due to the chemical nature of GLY, its analysis can be pursued under normal phase conditions, utilizing the same framework, the golden standard—HILIC. The latter is used in the efficient separation of a plethora of polar compounds, including pesticides. Despite its application for the separation of challenging polar pesticides, including GLY, in a variety of commodities [99], seldom are the reports for GLY analysis in soil. Marek reported a poor chromatographic performance when HILIC conditions (only one HILIC column was used; data were not shown) were used in the determination of GLY in soil and other matrices [61]. Hence, efforts need to be made in this direction, considering the inherent advantages of analyses of polar compounds under these conditions.

2.3.7 Other methods

Capillary electrophoresis methods have been reported in recent years using detection systems such as contactless conductivity, electrochemiluminescence [100], and laser-induced fluorescence [101, 102], as reviewed by Gauglitz et al. [103]. Ion chromatography [104], electrochemical method, surface resonance-enhanced spectrometry, enzyme-linked immunosorbent assay also called ELISA methods [87], spectrophotometry [73, 85], and fluorescent spectrometry [50, 55, 57, 75, 78, 80, 90, 96] were also reported to detect GLY in current literatures. However, the selectivity of ion chromatography was limited. Unlike other pesticides, the application of immunoanalytical techniques for glyphosate determination has been troublesome, although they have made some improvements.

Indicatively, El-Gendy and coworkers studied GLY in Egyptian soil samples using an optimized and sensitive linker-assisted enzyme-linked immunosorbent assay (L'ELISA) [87]. To derivatize GLY succinic anhydride was used. The method was well correlated with an HPLC-FD method that used sodium tetraborate for the extraction.

The advances in cutting-edge technologies can further hyphen such methods with modern mass spectrometers to provide solutions that currently are disregarded or seem problematic.

3. Glyphosate residues in the atmosphere

3.1 General aspects

The environmental pollution instigated by the use of plant protection products, commonly referred to as pesticides, is one of the most serious problems that facing the world due to their potential toxicity, high persistence, and slow degradation. Pesticide fate in the environment is characterized by a number of complex processes occurring in different environmental compartments, such as air, soils, and plants [105]. A wide variety of pesticides has been detected in different environmental media, including water bodies, soil, and the atmosphere. The extended use of pesticides containing persistent active ingredients can lead to raised concentrations due to the accumulation in the environment and long-term exposure to nontarget organisms.

Since the last decades, there has been an increasing global concern over the human health impacts attributed to the environmental pollution and specifically to air pollution. During applications, a noteworthy segment of applied pesticides ranged from 15 to 40% is dispersed in the atmosphere and can travel with long-range atmospheric transport [106]. Thus, the atmosphere has been considered as an important spread vector at local, regional, and global scales. It has been reported in the international literature that air pesticide contamination was observed both in urban and rural areas with concentration levels ranging from some picograms to several nanograms per cubic meter [107]. However, the contamination of air by pesticides is an aspect of atmospheric pollution that remains less documented than that of other environments.

Worry over the transport of pesticides in air started in the 1960s with the detection of persistent and volatile substances such as DDT, dieldrin, and aldrin far from their application sites. The first legislation to consider air as an exposure route was in the United States in 1971. Since then the issue of pesticides in air has been subject to sporadic regulatory concern, especially in Europe [108].

Milestone legislation in Europe concerning pesticides in the atmosphere occurred in 1996 with the Stockholm convention on persistent organic pollutants (POPs). This regulation covers all chemicals, including pesticides, and lays down principles to identify substances for which aerial transport may be noteworthy [108].

Long-range transport in air and water can result in the exposure of remote and particularly vulnerable ecosystems such as the Arctic [109, 110].

Pesticides enter into the atmosphere, and their residues can move away from the application sites resulting in accidental exposure for humans, animals, and plants, close or distant the treated sites. It is well recognized that the exposure and effect assessment of pesticides should not be constrained to the target area, and its close zone because this does not adequately cover possible hazards associated with their use.

The most common routes of pesticide entry into the atmosphere could be the drift during their application, volatilization from the soil, surface water or crop foliage, as well as wind erosion of deposited residues [111–113]. Once they enter in the atmosphere, pesticides are distributed between the gaseous and particulate phases depending on parameters such as:

- Octanol-air partition coefficient
- Vapor pressure
- Henry's law constant
- Water solubility
- Total suspended particulate matter
- Weather conditions [114]

In the atmosphere, pesticides are distributed between particle and vapor phases based on their vapor pressure, the ambient temperature, and the concentration of suspended particulate matter. Taking into account the low volatility of the majority of the most commonly used pesticides; it could be considered that they are often absorbed on the surface of atmospheric particles. In that way they may incur transformation processes resulting in the formation of secondary metabolites which could be even more hazardous than the parent released compounds [106]. Pesticides released into the atmosphere can settle to the ground, be broken down by sunlight and water, or dissipate into the surrounding air.

3.1.1 Transfer processes of pesticides in the air

During and after the application of a pesticide, a considerable portion of the amount applied may enter into the atmosphere through many different routes (the most important will be briefly discussed) and consequently may be transported over shorter and longer distance.

Through spray application of pesticides, a fraction of the spray would exist as pesticides in the gas phase and as small droplets or particles. The latter do not reach their target due to their extremely small size and cannot be captured by drift collectors. This fraction that exists in the gas phase and as aerosol should be taken into account along with drift.

Volatilization is defined as the transfer of pesticide residues into the gas phase after application. Volatilization from treated areas is a constant process and could be the main dissipative route for numerous pesticides [115]. Its extent is governed by the physical and chemical properties of the pesticide such as vapor pressure and Henry's law constant; the application parameters such as the droplet size and the water volume; and finally the climatic conditions during and after application [108, 116]. Volatilization may be swayed by relative humidity, the atmospheric pressure, and the wind velocity [117]. The compound's volatility with medium vapor pressure values is significantly influenced by environmental and application factors, whereas substances with high vapor pressure values present high volatilization which does not depend on other factors. It is broadly established in the literature that vapor pressure can be used to categorize pesticides with a very high or with no volatilization potential. Vapor pressure also rules the partitioning of a semi-volatile constituent between the gas and the airborne particle phases. According to Bidleman substances with a vapor pressure value higher than 10^{-2} Pa are mainly expected in the vapor phase, while those with vapor pressure value lower 10^{-5} Pa solely exist in the particle-adsorbed phase [118]. Pesticides with vapor pressure between 10^{-2} and 10^{-5} Pa values partition between these phases.

A significant amount of pesticides entering into the atmosphere for several days or weeks after pesticide application comprises volatilization from the soil and plant surfaces as well as wind erosion of soil particles containing sorbed pesticides [119, 120]. Many parameters such as the physicochemical properties of the pesticide (vapor pressure, solubility, adsorption coefficient, molecular mass, and chemical nature), the soil properties (water content, soil density, soil organic matter content, clay content/texture, soil pH), the weather conditions (air temperature, solar radiation, rain, air humidity, and wind), and the agricultural practices used (application date and rate and formulation type) may influence the volatilization process [111].

Volatilization from plants is considered up to three times higher than soil volatilization under similar meteorological conditions. The vapor pressure and Henry's law constant are the physicochemical characteristics of the compound that seem to be related with the degree of volatilization. Additionally, application methods and weather conditions may also play an important role in the volatilization process from plants [121].

The Focus Air group has deemed that vapor pressure is the most significant factor affecting volatilization and deemed that active ingredients applied to soil with vapor pressure values higher than 10^{-4} Pa and active ingredients applied to plants with vapor pressure values higher than 10^{-5} Pa have a high possibility to enter in the air and for that reason require a risk assessment evaluation before authorization [108].

Pesticides existing in the aerial phase could be carried by wind and deposited accidentally in untreated areas by dry (gas and particle) and wet (rain and snow) deposition [122].

The atmosphere could be efficiently cleaned of suspended particulate matter to which pesticides might be sorbed by rainfall, and thus gas-phase pesticides can partition directly into a falling raindrop [122].

High pesticide concentrations in the air could be considered seasonal and often associated with local use and thus occur during the spraying months [123]. The physical and chemical properties of each pesticide also play a significant role in determining if a pesticide converts airborne, whether it then exists primarily in the gaseous or particle phase, and how efficiently rainfall removes it from the atmosphere. The period of time that a pesticide is applied, its amount, and the cultivated area play also significant roles in whether a pesticide exists in the atmosphere and at which concentration [113].

3.1.2 Glyphosate occurrence in the air

Glyphosate (N-[phosphonomethyl] glycine), a broad-spectrum, nonselective, and post emergence herbicide, is the most widely used pesticide worldwide.

Although numerous laboratory and field studies have been carried out for the determination of glyphosate and AMPA in the aquatic environment, there are limited studies in field soils. Furthermore, atmospheric concentrations of glyphosate and AMPA are shabbily documented as very few studies have monitored them in the atmosphere [124].

The first report about the atmospheric concentrations of glyphosate and AMPA had been published in 1991 in order to present the results of a study that had been conducted in 1988 in northeastern Finland for measuring the workers' exposure to glyphosate when they used sprayers connected to brush saws. In that study glyphosate was determined from the breathing zone and from urine samples. Based on the results of this study and at the end of the spraying week, two air samples were found to have measurable levels of glyphosate at concentrations 2.8 and 15.7 μg m^{-3}. AMPA had not been detected in any of the air samples [125].

In 2002, Humphries et al. examined the atmospheric samples at three different sites in east-central Alberta. For the purposes of the study, air samples were collected before the application of glyphosate and after its application and for 24 h time period at regular intervals. Glyphosate was not detected in any of the collected air samples at levels above the method LOQ; however, it was detected in few particulate samples [126]. The nonexistence of glyphosate in the polyurethane foam indicates that glyphosate is not released as the vapor forms into the atmosphere but rather is carried by a particulate matter.

In 2004, glyphosate was examined in 59 atmospheric samples in Hauts-de-France Region in France, with a detection occurrence of 14% and a maximum concentration of 0.19 ng m^{-3} [124, 127].

Chang et al. reported that both glyphosate and AMPA had been detected in the ambient air of Iowa, Indiana, and Mississippi during two growing seasons of the years 2007 and 2008. Atmospheric concentrations of glyphosate reached 9.1 and 5.4 ngm^{-3} in Mississippi and Iowa agricultural areas, respectively; however atmospheric concentrations of AMPA touched 0.49 and 0.97 ngm^{-3} in Mississippi and Iowa, correspondingly. It had been concluded that the existence of glyphosate in air is due to spray drift or wind erosion as it is not a volatile compound whereas AMPA presence is due to wind erosion as it is a glyphosate degradation product and it is formed in soil [128]. The authors provided also measurements in rainwater and estimated that 97% of glyphosate existing in the atmosphere could be removed by weekly rainfall greater than 30 mm [129].

Morshed et al. determined the atmospheric concentrations of glyphosate in treated fields in Malaysia during spray applications by a mist blower [129]. The maximum concentration of 42.96 μgm^{-3} was measured for glyphosate, and additionally a first modeling attempt for the estimation of glyphosate emission to the atmosphere at regional level was done; however, there were no measurements to confirm the model output.

In 2014, and specifically from July to November, Sousa et al. performed a study in northeastern Brazil, in the municipality of Limoeiro do Norte-Ceará, in urban and rural areas, for the determination of the atmospheric concentrations of glyphosate. Glyphosate detected at concentrations ranged between 0.313 and 2.939 μg m^3 in all collected atmospheric samples [130].

During the years 2015–2016, glyphosate and AMPA were searched in 142 air samples during a 2-year field campaign in France. Samples were taken from both non-agricultural and agricultural areas, while atmospheric concentrations of glyphosate were detected at an overall frequency of 7%. AMPA was not detected in any sample. The maximum concentration of 1.04 ng m^{-3} was measured for glyphosate in the rural site of Cavaillon. As regards the temporal distribution of glyphosate, it had been pointed out that there was no reproducible detection pattern from 2015 to 2016 [125].

3.1.3 Monitoring studies for pesticides in the air

Generally, a few number of monitoring studies have been conducted for the determination of pesticide residues in atmospheric samples. These studies could not provide consistent results due to the variability in experimental conditions, the lack of consistency in sampling methodologies, the variation in collection time and duration, the analytes selected, the analytical methods used [131], as well as the method detection limits. Most of the studies have been performed at the national level, they are short-term as they lasted from 1 to 2 years, and for that reason, the overall conclusion on the long-term trends and the atmospheric movements of pesticides could not been reached [108].

3.2 Determination of glyphosate

3.2.1 Sampling and extraction procedures

Pesticides existing in the atmosphere are usually at very low concentrations, and thus appropriate sampling and techniques are necessary. The most common sampling techniques used for pesticides in the ambient air could be separated into two categories: the active and the passive or diffuse samplers [132].

3.2.1.1 Active sampling

Active samplers allow the pesticides existing in gaseous and particulate phases to be trapped by pumping air through a filter followed by a solid adsorbent. Thus, pesticides standing in the gas phase are stacked by the solid adsorbent, whereas pesticides in the particulate phase are maintained in the filter.

Pesticides present in the atmosphere could be sampled through low-volume or high-volume samplers. As pesticide residues in the atmosphere are at very low concentrations, high-volume samplers are usually used [121].

For sampling of semi-volatile pesticides, the use of diffusion denuder systems, which consist of a series of coaxial glass tubes coated with an appropriate adsorbent through which the air flows, is proposed [121].

3.2.1.2 Passive sampling

Passive air samplers are devices that collect pesticides from the air without the use of pump, and they are comprised of an accumulating intermediate which has a high retention capacity for the target analytes. Passive samplers are able to gather only the free gaseous phase pesticides, while the length of sampling range from few weeks to several months, considerably larger than the usual time required using the active ones [121].

In 1991, Jauhiainen et al. collected air samples for the determination of glyphosate from the breathing zone through a portable pump onto an absorption liquid [125]. The air samples collected were first evaporated to dryness and then dissolved with trifluoroethanol and trifluoroacetic anhydrite.

Chang et al. used high-volume active samplers for collecting air samples for the determination of glyphosate. The glass fiber filters used were baked at 550°C, cooled to the room temperature, and enfolded in aluminum foil before sampling [128, 130]. The glass fiber filters after sampling were slowly grounded in a polypropylene tube and then extracted with hydrochloric acid (pH 2) and further with a potassium hydroxide solution (pH 11). Cellulose nitrate filters were used under vacuum for filtration [128].

Ravier et al. used also high-volume samplers, and the particulate samples were collected on quartz microfiber filters. The filters after sampling were protected from the light and stored at −20°C [124]. Field air blank samples were also collected for the determination of the background contamination through handling and storage. The extraction of all the samples was performed in polytetrafluoroethylene or polypropylene vessels in order to avoid loss of the studied compounds via wall adsorption. According to Ravier et al., filters were extracted with ultrahigh quality water with the addition of appropriate quantities of Borax (0.05 M) and EDTA solutions. Polyethersulfone membranes were used for sample filtration. FMOC-Cl was used as a derivatization agent.

Morshed et al. performed a study for the determination of glyphosate in the atmosphere by using both active and passive sampling methods. For the purposes of the study, three different air samplers were used. Cellulose filter patches and polyurethane foam were used for passive samplers. Active samplers were also used for sampling and were connected to polyurethane foam plug for the determination of glyphosate existing in the vapor phase and a quartz fiber filter for the particulate phase of airborne glyphosate [129]. Sample extraction for both active and passive extraction methods was performed with borate buffer. FMOC-Cl was used as a derivatizing agent.

High-volume air samplers were used to collect suspended, airborne particulates and trap airborne glyphosate vapors in a study conducted in Alberta's area. A volatile glyphosate was collected on a polyurethane foam plug and particulate glyphosate on a filter paper [126, 133].

Sousa et al. used a glass sample holder in which a polyurethane foam (adsorbent medium) was placed. The particulate material was collected from the glass fiber filters. Glyphosate was determined in the atmosphere after extraction from polyurethane foams with a solution comprising of monobasic potassium phosphate and methanol in ultrapure water while the pH of the solution was maintained at 2 using concentrated phosphoric acid. The samples were concentrated in a C18 solid-phase extraction cartridge.

3.2.2 Analytical method

The chromatographic analysis of glyphosate and AMPA is considered tough in trace analysis. Due to their low molecular weight, low volatility, thermal lability, and excellent water solubility, their extraction and determination are complex.

The main analytical techniques used for the analysis of glyphosate in atmospheric samples are liquid chromatography equipped with diode array or fluorescence detectors and liquid chromatography interfaced with a quadrupole-time-of-flight mass spectrometer or mass spectrometry. However, gas chromatographic technique with ECD has also been used.

In 1991 Jauhiainen et al. reported that a gas chromatographic system equipped with ECD and fused silica has been used for glyphosate determination in air samples. Additionally a triple-quadrupole mass spectrometer equipped with fused silica was used for identification purposes.

In 2011 a liquid chromatographic method for the determination of glyphosate in air samples was reported [129]. The analytical standards (stock and working) were prepared in a 0.025 M sodium borate buffer (pH 9) solution. Prior to HPLC chromatographic analysis, working standards were pre-column derivatized with a derivatizing agent (0.002 M FMOC-Cl). The liquid chromatographic system consisted of a florescence detector and a Hypersil NH2 chromatographic column, while the mobile phase comprised of 50% phosphate buffer (0.05 M potassium phosphate monobasic KH2PO4 adjusted to pH 6.0 with 7 N KOH). The glyphosate retention time was 5.6 min and the total run time was 10 min. The LOD of the method was 0.015 μg ml^{-1}, while the LOQ was 0.05 μg ml^{-1} and determined through the linear calibration curve.

Chang et al. reported another method for the determination of glyphosate and AMPA by using a liquid chromatography tandem mass spectrometer. Both glyphosate and AMPA were derivatized with 9-fluorenylmethylchloroformate before analysis. A gradient elution system comprised of 95% of 5 mM ammonium acetate in HPLC-MS-grade water to 100% HPLC-grade acetonitrile was used. The molecular ion and the fragment ion for glyphosate were 390 and 168. In the case of AMPA the molecular ion and the fragment ions were 332, 110, and 136 [128].

Zhang et al. performed the analyses for the determination of glyphosate in the air samples of workplaces by ion chromatography using a conductivity detector. The limit of detection was found to be 0.003 mg/m^3. The recovery ranged between 94.8 and 97.4% [134].

According to Maria Gizeuda de F. Sousa et al., glyphosate was determined by liquid chromatography equipped with a diode array detector and a C-18 chromatographic column at 195 nm. The mobile phase consisted of 0.006 mM KH2PO4, and the flow rate set at 1.0 mL/min. Under these conditions glyphosate is eluted at 2.97 min, whereas the total analysis time was 7 min. The analytical method LOD was 0.09 μg mL^{-1}, whereas the LOQ was 0.27 μg mL^{-1} [130].

For the determination of glyphosate and its major metabolite AMPA, Ravier et al. used an ultra-performance liquid chromatographic (UPLC) system interfaced with a quadrupole-time-of-flight mass spectrometer and equipped with an electrospray ion source and a C18 UPLC column. The elution system consisted of water with 5 mM ammonium formate and acetonitrile. The analyses are performed in the negative ionization mode. Both the LOD and the LOQ were determined by the calibration curve and were 0.05 and 0.14 ngm^{-3}, respectively, for glyphosate and 0.30 and 0.90 ngm^{-3}, respectively, for AMPA [124].

4. Conclusions

HPLC methods are highly sensitive especially with fluorogenic labeling, but they lack specificity and usually require a laborious cleanup procedure such as ion-exchange column chromatography, which may result in some sample loss and lower reproducibility. At present LC-MS in tandem mode (MS/MS) is considered the most suitable technique for the detection of phosphoric and amino acid-type herbicides at low concentrations. Derivatization is the most common way to analyze GLY and AMPA using LC-ESI-MS/MS systems, a procedure that is described in soil matrix as well.

The maximum concentrations of glyphosate in atmospheric samples correspond to the time of its application. Due to the limited number of monitoring studies for monitoring pesticides and specifically glyphosate in the air, a reliable conclusion about its fate could not be reached.

Conflict of interest

The authors declare no conflict of interest.

Author details

Evagelia Tzanetou*† and Helen Karasali*†
Laboratory of Chemical Control of Pesticides, Department of Pesticides Control and Phytopharmacy, Benaki Phytopathological Institute, Athens, Greece

*Address all correspondence to: ev.tzanetou@bpi.gr and e.karassali@bpi.gr

† Equal contribution of the two authors.

References

[1] Blair A, Fritschi L, McLaughlin J, Sergi CM, Calaf GM, Curieux FL, et al. IARC Monographs Volume 112: Evaluation of Five Organophosphate Insecticides and Herbicides. International Agency for Research on Cancer. Lyon-France: World Health Organization; 2015

[2] Panel of Experts on Pesticide Residues in Food and the Environment and the WHO Core Assessment Group on Pesticide Residues. Report of the Joint Committee on Pesticide Residues. Geneva: WHO/FAO; 2016

[3] Zhang CP, Hu XQ, Luo JY, Wu ZY, Wang L, Li B, et al. Degradation dynamics of glyphosate in different types of Citrus orchard soils in China. Molecules. 2015;**20**(1):1161-1175

[4] Benbrook CM. Trends in glyphosate herbicide use in the United States and globally. Environmental Sciences Europe. 2016;**28**:3. DOI: 10.1186/s12302-016-0070-0

[5] Grandcoin A, Piel S, Baures E. AminoMethylPhosphonic acid (AMPA) in natural waters: Its sources, behavior and environmental fate. Water Research. 2017;**117**:187-197

[6] EFSA-Glyphosate. Conclusion on the peer review of the pesticide risk assessment of the active substance glyphosate. EFSA Journal. 2015;**13**(11):4302

[7] Myers JP, Antoniou MN, Blumberg B, Carroll L, Colborn T, Everett LG, et al. Concerns over use of glyphosate-based herbicides and risks associated with exposures: A consensus statement. Environmental Health. 2016:**19**. DOI: 10.1186/s12940-016-0117-0

[8] European-Regulation-1107/2009. Regulation of the European Parliament and of the Council of 21 October 2009 concerning the placing of plant protection products on the market and repealing Council Directives 79/117/EEC and 91/414/EEC. Official Journal of the European Union; 2009. OJ L 309, 24 November 2009. pp. 1-50

[9] Mesnage R, Benbrook C, Antoniou MN. Insight into the confusion over surfactant co-formulants in glyphosate-based herbicides. Food and Chemical Toxicology. 2019;**128**:137-145

[10] Laitinen P, Siimes K, Eronen L, Ramo S, Welling L, Oinonen S, et al. Fate of the herbicides glyphosate, glufosinate-ammonium, phenmedipham, ethofumesate and metamitron in two Finnish arable soils. Pest Management Science. 2006;**62**(6):473-491

[11] Gimsing AL, Szilas C, Borggaard OK. Sorption of glyphosate and phosphate by variable-charge tropical soils from Tanzania. Geoderma. 2007;**138**(1-2):127-132

[12] Sorensen SR, Schultz A, Jacobsen OS, Aamand J. Sorption, desorption and mineralisation of the herbicides glyphosate and MCPA in samples from two Danish soil and subsurface profiles. Environmental Pollution. 2006;**141**(1):184-194

[13] Bergstrom L, Borjesson E, Stenstrom J. Laboratory and lysimeter studies of glyphosate and aminomethylphosphonic acid in a sand and a clay soil. Journal of Environmental Quality. 2011;**40**(1):98-108

[14] Muskus AM, Krauss M, Miltner A, Hamer U, Nowak KM. Effect of temperature, pH and total organic carbon variations on microbial turnover of (13)CA(3)(15)N-glyphosate in agricultural soil. Science of the Total Environment. 2019;**658**:697-707

[15] Borggaard OK, Gimsing AL. Fate of glyphosate in soil and the possibility of leaching to ground and surface waters: A review. Pest Management Science. 2008;**64**(4):441-456

[16] Bandana B, Sharma N, Joshi R, Gulati A, Sondhia S. Dissipation kinetics of glyphosate in tea and tea-field under northwestern mid-hill conditions of India. Journal of Pesticide Science. 2015;**40**(3-4):82-86

[17] Wang S, Seiwert B, Kastner M, Miltner A, Schaffer A, Reemtsma T, et al. (Bio)degradation of glyphosate in water-sediment microcosms—A stable isotope co-labeling approach. Water Research. 2016;**99**:91-100

[18] Pipke R, Amrhein N. Degradation of the phosphonate herbicide glyphosate by Arthrobacter-Atrocyaneus Atcc-13752. Applied and Environmental Microbiology. 1988;**54**(5):1293-1296

[19] Kishore GM, Jacob GS. Degradation of glyphosate by *Pseudomonas* sp.-Pg2982 via a sarcosine intermediate. The Journal of Biological Chemistry. 1987;**262**(25):12164-12168

[20] Shinabarger DL, Braymer HD. Glyphosate catabolism by *Pseudomonas* sp. strain-Pg2982. Journal of Bacteriology. 1986;**168**(2):702-707

[21] Lesueur C, Pfeffer M, Fuerhacker M. Photodegradation of phosphonates in water. Chemosphere. 2005;**59**(5):685-691

[22] Sidoli P, Baran N, Angulo-Jaramillo R. Glyphosate and AMPA adsorption in soils: Laboratory experiments and pedotransfer rules. Environemental Science and Pollution Research. 2016;**23**(6):5733-5742

[23] Munira S, Farenhorst A, Flaten D, Grant C. Phosphate fertilizer impacts on glyphosate sorption by soil. Chemosphere. 2016;**153**:471-477

[24] Okada E, Costa JL, Bedmar F. Adsorption and mobility of glyphosate in different soils under no-till and conventional tillage. Geoderma. 2016;**263**:78-85

[25] Saunders LE, Pezeshki R. Glyphosate in runoff waters and in the root-zone: A review. Toxics. 2015;**3**(4):462-480

[26] Richards BK, Pacenka S, Meyer MT, Dietze JE, Schatz AL, Teuffer K, et al. Antecedent and post-application rain events trigger glyphosate transport from runoff-prone soils. Environmental Science & Technology Letters. 2018;**5**(5):249-254

[27] Alonso LL, Demetrio PM, Etchegoyen MA, Marino DJ. Glyphosate and atrazine in rainfall and soils in agroproductive areas of the pampas region in Argentina. Science of the Total Environment. 2018;**645**:89-96

[28] Berman MC, Marino DJG, Quiroga MV, Zagarese H. Occurrence and levels of glyphosate and AMPA in shallow lakes from the Pampean and Patagonian regions of Argentina. Chemosphere. 2018;**200**:513-522

[29] Primost JE, Marino DJG, Aparicio VC, Costa JL, Carriquiriborde P. Glyphosate and AMPA, "pseudo-persistent" pollutants under real world agricultural management practices in the Mesopotamic pampas agroecosystem, Argentina. Environmental Pollution. 2017;**229**:771-779

[30] Bonansea RI, Filippi I, Wunderlin DA, Marino DJG, Ame MV. The fate of glyphosate and AMPA in a freshwater endorheic basin: An ecotoxicological risk assessment. Toxics. 2018;**6**:3. DOI: 10.3390/toxics6010003

[31] Van Stempvoort DR, Spoelstra J, Senger ND, Brown SJ, Post R, Struger J. Glyphosate residues in rural

groundwater, Nottawasaga River watershed, Ontario, Canada. Pest Management Science. 2016;**72**(10):1862-1872

[32] Battaglin WA, Meyer MT, Kuivila KM, Dietze JE. Glyphosate and its degradation product AMPA occur frequently and widely in us soils, surface water, groundwater, and precipitation(1). The Journal of the American Water Resources Association. 2014;**50**(2):275-290

[33] Ruiz-Toledo J, Castro R, Rivero-Perez N, Bello-Mendoza R, Sanchez D. Occurrence of glyphosate in water bodies derived from intensive agriculture in a tropical region of southern Mexico. Bulletin of Environmental Contamination and Toxicology. 2014;**93**(3):289-293

[34] Abrantes N, Pereira R, Goncalves F. Occurrence of pesticides in water, sediments, and fish tissues in a Lake surrounded by agricultural lands: Concerning risks to humans and ecological receptors. Water, Air, and Soil Pollution. 2010;**212**(1-4):77-88

[35] Sanchis J, Kantiani L, Llorca M, Rubio F, Ginebreda A, Fraile J, et al. Determination of glyphosate in groundwater samples using an ultrasensitive immunoassay and confirmation by on-line solid-phase extraction followed by liquid chromatography coupled to tandem mass spectrometry. Analytical and Bioanalytical Chemistry. 2012;**402**(7):2335-2345

[36] Stewart M, Olsen G, Hickey CW, Ferreira B, Jelic A, Petrovic M, et al. A survey of emerging contaminants in the estuarine receiving environment around Auckland, New Zealand. Science of the Total Environment. 2014;**468**:202-210

[37] Todorovic GR, Mentler A, Popp M, Hann S, Kollensperger G, Rampazzo N, et al. Determination of glyphosate and AMPA in three representative agricultural Austrian soils with a HPLC-MS/MS method. Soil and Sediment Contamination. 2013;**22**(3):332-350

[38] Zgheib S, Moilleron R, Chebbo G. Priority pollutants in urban stormwater: Part 1—Case of separate storm sewers. Water Research. 2012;**46**(20):6683-6692

[39] Koskinen WC, Marek LJ, Hall KE. Analysis of glyphosate and aminomethylphosphonic acid in water, plant materials and soil. Pest Management Science. 2016;**72**(3):423-432

[40] Grunewald K, Schmidt W, Unger C, Hanschmann G. Behavior of glyphosate and aminomethylphosphonic acid (AMPA) in soils and water of reservoir Radeburg II catchment (Saxony/Germany). Journal of Plant Nutrition and Soil Science. 2001;**164**(1):65-70

[41] Valle AL, Mello FCC, Alves-Balvedi RP, Rodrigues LP, Goulart LR. Glyphosate detection: Methods, needs and challenges. Environmental Chemistry Letters. 2019;**17**(1):291-317

[42] Silva V, Montanarella L, Jones A, Fernandez-Ugalde O, Mol HGJ, Ritsema CJ, et al. Distribution of glyphosate and aminomethylphosphonic acid (AMPA) in agricultural topsoils of the European Union. Science of the Total Environment. 2018;**621**:1352-1359

[43] Sprankle P, Meggitt WF, Penner D. Rapid inactivation of glyphosate in soil. Weed Science. 1975;**23**(3):224-228

[44] Sprankle P, Meggitt WF, Penner D. Adsorption, mobility, and microbial degradation of glyphosate in soil. Weed Science. 1975;**23**(3):229-234

[45] Hance RJ. Adsorption of glyphosate by soils. Pesticide Science. 1976;7(4):363-366

[46] Shoval S, Yariv S. Interaction between roundup (glyphosate) and montmorillonite. 1. Infrared study of the sorption of glyphosate by montmorillonite. Clays and Clay Minerals. 1979;**27**(1):19-28

[47] Stalikas CD, Konidari CN. Analytical methods to determine phosphonic and amino acid group-containing pesticides. Journal of Chromatography A. 2001;**907**(1-2):1-19

[48] Borjesson E, Torstensson L. New methods for determination of glyphosate and (aminomethyl) phosphonic acid in water and soil. Journal of Chromatography A. 2000;**886**(1-2):207-216

[49] Ibanez M, Pozo OJ, Sancho JV, Lopez FJ, Hernandez F. Residue determination of glyphosate, glufosinate and aminomethylphosphonic acid in water and soil samples by liquid chromatography coupled to electrospray tandem mass spectrometry. Journal of Chromatography A. 2005;**1081**(2):145-155

[50] Miles CJ, Moye HA. Extraction of glyphosate herbicide from soil and clay-minerals and determination of residues in soils. Journal of Agricultural and Food Chemistry. 1988;**36**(3):486-491

[51] Aubin AJ, Smith AE. Extraction of [C-14] glyphosate from Saskatchewan soils. Journal of Agricultural and Food Chemistry. 1992;**40**(7):1163-1165

[52] Glass RL. Metal-complex formation by glyphosate. Journal of Agricultural and Food Chemistry. 1984;**32**(6):1249-1253

[53] Glass RL. Liquid-chromatographic determination of glyphosate in soil and water samples. Abstracts of Papers of the American Chemical Society. 1983;**185**(Mar):27

[54] Glass RL. Liquid-chromatographic determination of glyphosate in fortified soil and water samples. Journal of Agricultural and Food Chemistry. 1983;**31**(2):280-282

[55] Laitinen P, Ramo S, Nikunen U, Jauhiainen L, Siimes K, Turtola E. Glyphosate and phosphorus leaching and residues in boreal sandy soil. Plant and Soil. 2009;**323**(1-2):267-283

[56] Al-Rajab AJ, Schiavon M. Degradation of C-14-glyphosate and aminomethylphosphonic acid (AMPA) in three agricultural soils. Journal of Environmental Sciences. 2010;**22**(9):1374-1380

[57] Mardian-Jansar K, Ismail BS. Residue determination and levels of glyphosate in surface water, sediments and soils associated with oil palm plantation in Tasik Chini, Pahang, Malaysia. AIP Conference Proceedings. 1614;**2014**:795-802

[58] Gonzalez-Martinez MA, Brun EM, Puchades R, Maquieira A, Ramsey K, Rubio F. Glyphosate immunosensor. Application for water and soil analysis. Analytical Chemistry. 2005;**77**(13):4219-4227

[59] Huang XJ, Pedersen T, Fischer M, White R, Young TM. Herbicide runoff along highways. 1. Field observations. Environmental Science & Technology. 2004;**38**(12):3263-3271

[60] Druart C, Delhomme O, de Vaufleury A, Ntcho E, Millet M. Optimization of extraction procedure and chromatographic separation of glyphosate, glufosinate and aminomethylphosphonic acid in soil. Analytical and Bioanalytical Chemistry. 2011;**399**(4):1725-1732

[61] Marek LJ, Koskinen WC. Simplified analysis of glyphosate and aminomethylphosphonic acid in water, vegetation and soil by liquid chromatography-tandem mass spectrometry. Pest Management Science. 2014;**70**(7):1158-1164

[62] Arkan T, Molnar-Perl I. The role of derivatization techniques in the analysis of glyphosate and aminomethyl-phosphonic acid by chromatography. Microchemical Journal. 2015;**121**:99-106

[63] Lundgren LN. A new method for the determination of glyphosate and(aminomethyl) phosphonic acid residues in soils. Journal of Agricultural and Food Chemistry. 1986;**34**:535-538. DOI: 10.1021/jf00069a041

[64] Roy DN, Konar SK. Development of an analytical method for the determination of glyphosate and (aminomethyl) phosphonic acid residues in soils by nitrogen-selective gas chromatography. Journal of Agricultural and Food Chemistry. 1989;**37**(2):441-443. DOI: 10.1021/jf00086a038

[65] Kawai S, Uno B, Tomita M. Determination of glyphosate and its major metabolite aminomethylphosphonic acid by high-performance liquid chromatography after derivatization with p-toluenesulphonyl chloride. Journal of Chromatography. 1991;**540**:411-415. DOI: 10.1016/S0021-9673(01)88832-4 411

[66] Forlania G, Mangiagallia A, Nielsena E, Suardib CM. Degradation of the phosphonate herbicide glyphosate in soil: Evidence for a possible involvement of unculturable microorganisms. Soil Biology and Biochemistry. 1999;**31**:991-997. DOI: 10.1016/s0038-0717(99)00010

[67] Alferness PL, Iwata Y. Determination of glyphosate and (Aminomethy1)phosphonic acid in soil, plant and animal matrices, and water by capillary gas chromatography with mass-selective detection. Journal of Agricultural and Food Chemistry. 1994;**42**:2751-2759

[68] Sancho JV, Hidalgo C, Hernandez F, Lopez FJ, Hogendoorn EA, Dijkman E. Rapid determination of glyphosate residues and its main metabolite AMPA in soil samples by liquid chromatography. International Journal of Environmental Science and Technology. 1996;**62**(1):53-63

[69] Kataoka H, Ryu S, Sakiyama N, Makita M. Simple and rapid determination of the herbicides glyphosate and glufosinate in river water, soil and carrot samples by gas chromatography with flame photometric detection. Journal of Chromatography A. 1996;**726**(1-2):253-258

[70] Shao CY, Howe CJ, Porter AJR, Glover LA. Novel cyanobacterial biosensor for detection of herbicides. Applied and Environmental Microbiology. 2002;**68**(10):5026-5033. DOI: 10.1128/aem.68.10.5026-5033.2002

[71] Ghanem A, Bados P, Kerhoas L, Dubroca J, Einhorn J. Glyphosate and AMPA analysis in sewage sludge by LC-ESI-MS/MS after FMOC derivatization on strong anion-exchange resin as solid support. Analytical Chemistry. 2007;**79**(10):3794-3801. DOI: 10.1021/ac062195k

[72] Hu JY, Chen CL, Li JZ. A simple method for the determination of glyphosate residues in soil by capillary gas chromatography with nitrogen phosphorus. Journal of Analytical Chemistry. 2008;**63**(4):371-375

[73] Jan MR, Shah J, Muhammad M, Ara B. Glyphosate herbicide residue determination in samples of environmental importance using spectrophotometric method. Journal of Hazardous Materials. 2009;**169**(1-3):742-745

[74] Iwamuro Y, Iio-Ishimaru R, Chinaka S, Takayama N, Kodama S, Hayakawa K. Analysis of phosphorus-containing amino acid-type herbicides by capillary electrophoresis/mass spectrometry using a chemically modified capillary having amino

groups. Journal of Health Science. 2010;**56**(5):606-612. DOI: 10.1248/jhs.56.606

[75] Colombo SD, Masini JC. A sequential-injection reversed-phase chromatography method for fluorimetric determination of glyphosate and aminomethylphosphonic acid. Analytical Methods. 2014;**6**(2):490-496

[76] Tapsoba I, Paré S, Toé AM, et al. SWV determination of glyphosate in Burkina Faso soils using carbon fiber microelectrode. International Journal of Biological and Chemical Sciences. 2012;**6**:2211-2220

[77] Botero-Coy AM, Ibanez M, Sancho JV, Hernandez F. Improvements in the analytical methodology for the residue determination of the herbicide glyphosate in soils by liquid chromatography coupled to mass spectrometry. Journal of Chromatography A. 2013;**1292**:132-141

[78] Baez ME, Fuentes E, Espina MJ, Espinoza J. Determination of glyphosate and aminomethylphosphonic acid in aqueous soil matrices: A critical analysis of the 9-fluorenylmethyl chloroformate derivatization reaction and application to adsorption studies. Journal of Separation Science. 2014;**37**(21):3125-3132

[79] Prasad BB, Jauhari D, Tiwari MP. Doubly imprinted polymer nanofilm-modified electrochemical sensor for ultra-trace simultaneous analysis of glyphosate and glufosinate. Biosensors and Bioelectronics. 2014;**59**:81-88. DOI: 10.1016/j.bios.2014.03.019

[80] Wang D, Lin B, Cao Y, Guo M, Yu Y. A highly selective and sensitive fluorescence detection method of glyphosate based on an immune reaction strategy of carbon dot labeled antibody and antigen magnetic beads. Journal of Agricultural and Food Chemistry. 2016;**64**(30):6042-6050. DOI: 10.1021/acs.jafc.6b01088

[81] Yang X, Wang F, Bento CPM, Xue S, Gai L, van Dam R, et al. Short-term transport of glyphosate with erosion in Chinese loess soil—A flume experiment. Science of the Total Environment. 2015;**512-513**:406-414. DOI: 10.1016/j.scitotenv.2015.01.071

[82] Bento CPM, Yang X, Gort G, Xue S, van Dam R, Zomer P, et al. Persistence of glyphosate and aminomethylphosphonic acid in loess soil under different combinations of temperature, soil moisture and light/darkness. Science of the Total Environment. 2016;**572**:301-311. DOI: 10.1016/j.scitotenv.2016.07.215

[83] Bento CPM, van der Hoeven S, Yang X, Riksen MMJPM, Mol HGJ, Ritsema CJ, et al. Dynamics of glyphosate and AMPA in the soil surface layer of glyphosate-resistant crop cultivations in the loess Pampas of Argentina. Environmental Pollution. 2019;**244**:323-331. DOI: 10.1016/j.envpol.2018.10.046

[84] Larsbo M, Sandin M, Jarvis N, Etana A, Kreuger J. Surface runoff of pesticides from a clay loam field in Sweden. Journal of Environmental Quality. 2016;**45**(4):1367. DOI: 10.2134/jeq2015.10.0528

[85] Felton DE, Ederer M, Steffens T, Hartzell PL, Waynant KV. UV-Vis spectrophotometric analysis and quantification of glyphosate for an interdisciplinary undergraduate laboratory. Journal of Chemical Education. 2018;**95**(1):136-140

[86] Erban T, Stehlik M, Sopko B, Markovic M, Seifrtova M, Halesova T, et al. The different behaviors of glyphosate and AMPA in compost-amended soil. Chemosphere. 2018;**207**:78-83

[87] El-Gendy K, Mosallam E, Ahmed N, Aly N. Determination of glyphosate residues in Egyptian soil samples. Analytical Biochemistry. 2018;**557**:1-6

[88] Karasali H, Pavlidis G, Marousopoulou A. Investigation of the presence of glyphosate and its major metabolite AMPA in Greek soils. Environmental Science and Pollution Research International. 2019;**26**(36):36308-36321

[89] Zhang W, Feng Y, Ma L, An J, Zhang H, Cao M, et al. A method for determining glyphosate and its metabolite aminomethyl phosphonic acid by gas chromatography-flame photometric detection. Journal of Chromatography A. 2019;**1589**:116-121

[90] Pereira EAO, Melo VF, Abate G, Masini JC. Determination of glyphosate and aminomethylphosphonic acid by sequential-injection reversed-phase chromatography: Method improvements and application in adsorption studies. Analytical and Bioanalytical Chemistry. 2019;**411**(11):2317-2326

[91] Saito T, Miura N, Namera A, Oikawa H, Miyazaki S, Nakamoto A, et al. Mixed-mode C-C-18 monolithic spin-column extraction and GC-MS for simultaneous assay of organophosphorus compounds, glyphosate, and glufosinate in human serum and urine. Forensic Toxicology. 2012;**30**(1):1-10

[92] Deyrup CL, Chang SM, Weintraub RA, Moye HA. Simultaneous esterification and acylation of pesticides for analysis by gas-chromatography. 1. Derivatization of glyphosate and (aminomethyl)phosphonic acid with fluorinated alcohols-perfluorinated anhydrides. Journal of Agricultural and Food Chemistry. 1985;**33**(5):944-947

[93] Pei MQ, Lai J. Qualitative and quantitative analysis of glyphosate. Chinese Journal Guangdong Police Science Technology. 2004;**1**:14-15

[94] Ding J, Guo H, Liu W-W, Zhang W-W, Wang J-W. Current progress on the detection of glyphosate in environmental samples. Journal of Applied Biomedicine. 2015;**3**(6):88-95

[95] Yoshioka N, Asano M, Kuse A, Mitsuhashi T, Nagasaki Y, Ueno Y. Rapid determination of glyphosate, glufosinate, bialaphos, and their major metabolites in serum by liquid chromatography-tandem mass spectrometry using hydrophilic interaction chromatography. Journal of Chromatography A. 2011;**1218**(23):3675-3680

[96] Sun LS, Kong DY, Gu WD, Guo XY, Tao WQ, Shan ZJ, et al. Determination of glyphosate in soil/sludge by high performance liquid chromatography. Journal of Chromatography A. 2017;**1502**:8-13

[97] Nedelkoska TV, Low GKC. High-performance liquid chromatographic determination of glyphosate in water and plant material after pre-column derivatisation with 9-fluorenylmethyl chloroformate. Analytica Chimica Acta. 2004;**511**(1):145-153

[98] Karanasios E, Karasali H, Marousopoulou A, Akrivou A, Markellou E. Monitoring of glyphosate and AMPA in soil samples from two olive cultivation areas in Greece: Aspects related to spray operators activities. Environmental Monitoring and Assessment. 2018;**190**(6):361

[99] Vass A, Robles-Molina J, Perez-Ortega P, Gilbert-Lopez B, Dernovics M, Molina-Diaz A, et al. Study of different HILIC, mixed-mode, and other aqueous normal-phase approaches for the liquid chromatography/mass spectrometry-based determination

of challenging polar pesticides. Analytical and Bioanalytical Chemistry. 2016;**408**(18):4857-4869

[100] Hsu CC, Whang CW. Microscale solid phase extraction of glyphosate and aminomethylphosphonic acid in water and guava fruit extract using alumina-coated iron oxide nanoparticles followed by capillary electrophoresis and electrochemiluminescence detection. Journal of Chromatography A. 2009;**1216**(49):8575-8580

[101] Jiang J, Lucy CA. Determination of glyphosate using off-line ion exchange preconcentration and capillary electrophoresis-laser induced fluorescence detection. Talanta. 2007;**72**(1):113-118

[102] Cao L, Deng T, Liang S, Tan X, Meng J. Determination of herbicides and its metabolite in soil and water samples by capillary electrophoresis-laser induced fluorescence detection using microwave-assisted derivatization. Analytical Sciences. 2014;**30**(7):759-766

[103] Gauglitz G, Wimmer B, Melzer T, Huhn C. Glyphosate analysis using sensors and electromigration separation techniques as alternatives to gas or liquid chromatography. Analytical and Bioanalytical Chemistry. 2018;**410**(3):725-746

[104] Mallat E, Barcelo D. Analysis and degradation study of glyphosate and of aminomethylphosphonic acid in natural waters by means of polymeric and ion-exchange solid-phase extraction columns followed by ion chromatography-post-column derivatization with fluorescence detection. Journal of Chromatography A. 1998;**823**(1-2):129-136

[105] Özkara A, Akyil D, Konuk M. Pesticides, environmental pollution, and health. Environmental Health Risk - Hazardous Factors to Living Species. 2016. Chapter 1. pp. 4-28. DOI: 10.5772/63094

[106] Socorro J, Durand A, Temime-Roussel B, Gligorovski S, Wortham H, Quivet E. The persistence of pesticides in atmospheric particulate phase: An emerging air quality issue. Scientific Reports. 2016;**6**:33456. DOI: 10.1038/srep33456

[107] Désert M, Ravier S, Gille G, Quinapallo A, Armengaud A, Pochet G, et al. Spatial and temporal distribution of current-use pesticides in ambient air of Provence-Alpes-Côte-d'Azur region and Corsica, France. Atmospheric Environment. 2018;**192**(439):241-256. DOI: 10.1016/j.atmosenv.2018.08.054

[108] FOCUS. Pesticides in air: Considerations for exposure assessment. Report of the FOCUS Working Group on Pesticides in Air, EC Document Reference SANCO/10553/2006 Rev 2 June 2008. pp. 327

[109] Matthies M, Klasmeier J, Beyer A, Ehling C. Assessing persistence and long-range transport potential of current-use pesticides. Environmental Science & Technology. 2009;**43**(24):9223-9229. DOI: 10.1021/es900773u

[110] Sarigiannis DA, Kontoroupis P, Solomou ES, Nikolaki S, Karabelas AJ. Inventory of pesticide emissions into the air in Europe. Atmospheric Environment. 2013;**75**:6-14. DOI: 10.1016/j.atmosenv.2013.04.003

[111] Van den Berg F, Kubiak R, Benjey WG, Majewski MS, Yates SR, Reeves GL, et al. Emission of pesticides into the air. Water, Air, and Soil Pollution. 1999;**115**(1/4):195-218. DOI: 10.1023/a:1005234329622

[112] Bedos C, Cellier P, Calvet R, Barriuso E, Gabrielle B. Mass transfer of pesticides into the atmosphere by volatilization from soils and plants:

Overview. Agronomie. 2002;**22**:21-33. DOI: 10.1051/agro:2001003

[113] Voutsas E, Vavva C, Magoulas K, Tassios D. Estimation of the volatilization of organic compounds from soil surfaces. Chemosphere. 2005;**58**(6):751-758. DOI: 10.1016/j.chemosphere.2004.09.057

[114] Sauret N, Wortham H, Strekowski R, Herckès P, Nieto LI. Comparison of annual dry and wet deposition fluxes of selected pesticides in Strasbourg, France. Environmental Pollution. 2009;**157**(1):303-312. DOI: 10.1016/j.envpol.2008.06.034

[115] Prueger JH, Gish TJ, McConnell LL, Mckee LG, Hatfield JL, Kustas WP. Solar radiation, relative humidity, and soil water effects on metolachlor volatilization. Environmental Science & Technology. 2005;**39**(14):5219-5226. DOI: 10.1021/es048341q

[116] Wienhold BJ, Gish TJ. Effect of formulation and tillage practice on volatilization of atrazine and Alachlor. Journal of Environmental Quality. 1994;**23**(2):292. DOI: 10.2134/jeq1994.00472425002300020011x

[117] Harper LA, White AW, Bruce RR, Thomas AW, Leonard AA. Soil and microclimate effects on trifluralin volatilisation. Journal of Environmental Quality. 1976;**5**:236-242

[118] Bidleman TF. Atmospheric processes: Wet and dry deposition of organic compounds are controlled by their vapor particle partitioning. Environmental Science & Technology. 1988;**22**:361-367

[119] Glotfelty DE, Leech MM, Jersey J, Taylor AW. Volatilization and wind erosion of soil surface applied atrazine, simazine, alachlor, and toxaphene. Journal of Agricultural and Food Chemistry. 1989;**37**(2):546-551. DOI: 10.1021/jf00086a059

[120] Majewski MS. Micrometeorological methods for measuring the post-application volatilization losses of pesticides. Water, Air, and Soil Pollution. 1999;**115**:83-113

[121] Yusà V, Coscollà C, Mellouki W, Pastor A, de la Guardia M. Sampling and analysis of pesticides in ambient air. Journal of Chromatography A. 2009;**1216**(15):2972-2983. DOI: 10.1016/j.chroma.2009.02.019

[122] Slinn WGN. Some approximations for the wet and dry removal of particles and gases from the atmosphere. Water, Air, and Soil Pollution. 1977;**7**(4):513-543. DOI: 10.1007/bf00285550

[123] Kuang Z, McConnell LL, Torrents A, Meritt D, Tobash S. Atmospheric deposition of pesticides to an agricultural watershed of the Chesapeake Bay. Journal of Environmental Quality. 2003;**32**(5):1611. DOI: 10.2134/jeq2003.1611

[124] Ravier S, Désert M, Gille G, Armengaud A, Wortham H, et al. Monitoring of Glyphosate, Glufosinate-ammonium and (Aminomethyl) phosphonic acid in ambient air of Provence-Alpes-Côte-d'Azur Region, France. Atmospheric Environment. 2019;**204**:102-109. DOI: 10.1016/j.atmosenv.2019.02.023hal-02059173

[125] Jauhiainen A, Räsänen K, Sarantila R, Nuutinen J, Kangas J. Occupational exposure of forest workers to glyphosate during brush saw spraying work. American Industrial Hygiene Association Journal. 1991;**52**(2):61-64. DOI: 10.1080/15298669191364334

[126] Humphries D, Byrtus G, Anderson AM. Glyphosate residues. In: Alberta's Atmospheric Deposition, Soils and Surface Waters. Vegreville, Alberta: Water Research Users Group Alberta Environment; 2005

[127] Prouvost H, Declercq C. Exposition de la population aux pesticides dans la region Nord-Pas-de-Calais: Apports du programme PHYTO AIR. 2005:**78**. Project report. Available from: http://www.orsnpdc.fr/wp-content/uploads/2015/02/05-5.pdf [Assessed: 20 April 2020]

[128] Chang F, Simcik MF, Capel PD. Occurrence and fate of the herbicide glyphosate and its degradate aminomethylphosphonic acid in the atmosphere. Environmental Toxicology and Chemistry. 2011;**30**(3):548-555. DOI: 10.1002/etc.431

[129] Morshed MM, Dzolkhifli O, Rosli BM, Samsuri BAW. Determination of glyphosate through passive and active sampling methods in a treated field atmosphere. African Journal of Agricultural Research. 2011;**6**(17):4010-4018. DOI: 10.5897/AJAR11.533

[130] De F. Sousa MG, da Silva AC, dos Santos Araújo R, Rigotto RM. Evaluation of the atmospheric contamination level for the use of herbicide glyphosate in the northeast region of Brazil. Environmental Monitoring and Assessment. 2019;**191**:604. DOI: 10.1007/s10661-019-7764-x

[131] Van Dijk HFG, Guicherit R. Atmospheric dispersion of current-use pesticides- a review of the evidence from monitoring studies. Water, Air, & Soil Pollution. 1999;**115**:21-70

[132] Esteve-Turrillas FA, Pastor A, Yusà V, de la Guardia M. Using semi-permeable membrane devices as passive samplers. TrAC Trends in Analytical Chemistry. 2007;**26**(7):703-712. DOI: 10.1016/j.trac.2007.05.006

[133] Boethling RS, Howard PH, Meylan WM. Finding and estimating chemical property data for environmental assessment. Environmental Toxicology and Chemistry. 2004;**23**(10):2290. DOI: 10.1897/03-532

[134] Zhang RN, Liu HL, Huo ZL. Determination of glyphosate in air of workplaces by ion chromatography. Zhonghua Lao Dong Wei Sheng Zhi Ye Bing Za Zhi. 2013;**31**(10):779-782

Chapter 9

Herbicide Resistance in Brazil: Status, Impacts, and Future Challenges

Ricardo Alcántara-de la Cruz,
Guilherme Moraes de Oliveira, Leonardo Bianco de Carvalho and Maria Fátima das Graças Fernandes da Silva

Abstract

Brazil is a large producer and exporter of crops in global terms. Weeds may be responsible for ~14% of crop losses, depending on the crop system. Herbicides occupy 58% of the Brazilian pesticide market; however, the continuous use of these products and the high selection pressure have led to the emergence of weeds resistant to herbicides. Today, there are 51 weed species reported as being resistant to herbicides in Brazil, of which 17 involves cross and multiple-resistance. Acetolactate synthase (ALS), acetyl coenzyme A carboxylase (ACCase) and 5-enolpiruvylshikimate-3-phosphate synthase (EPSPs) inhibitors are the herbicidal groups with the most resistance cases. Soybean, corn, rice, wheat and cotton present 30, 12, 10, 9 and 8 cases, respectively, occurring mainly in herbicide-resistant crop fields from the Southern and Central West regions of the country. To better understand the dimensions of herbicide resistance, in this chapter, we will explore the size of agricultural activity in Brazil, the pesticide market and the use of herbicides in the main crops. In addition, the agronomic, scientific-technical and economic aspects that have contributed, directly or indirectly, to the selection of resistant weeds will be discussed in order to have an overview of the economic impact of herbicide resistance management.

Keywords: Brazilian pesticide market, glyphosate-resistant crops, herbicide resistance mechanisms, integrated weed management, management cost of weed

1. Introduction

Brazil is one of the leading manufacturers and exporters of food, fibers and energy, being one of the largest producers of coffee, maize, grapes, oil plants, oranges (fruit and juice), soybeans, sugarcane and meat [1]. These agricultural commodities have a crucial role in the development and agribusiness of the country, being the focus of Brazilian production and exports [2]. Agricultural pests limit global food security by reducing crop yields [3, 4]. The crop losses caused by pest can be over 80% if they are not controlled. Even when pests are controlled, crop yield losses range from 23 to 38% [4]. To reduce these losses, synthetic pesticides have become the main pest management tool globally [5].

Brazil has the fourth largest cultivation area worldwide, after India, China, and the United States; however, to ensure its agricultural productivity, Brazil has become in the largest pesticide market since 2011 [6]. The pesticides consumption increased 300% from 1991 to 2010 in this country [5]. Although weeds are responsible for ~14% of crop losses, depending on the agricultural system and crop situation [4], herbicides represent ~58% of the Brazilian pesticide market, where only the herbicide glyphosate occupied 35% of pesticide sales [7]. This increase has been largely due to the cultivation of herbicide resistant (HR) crops, mainly those resistant to glyphosate (GR) [8]. In addition, the loss of the glyphosate patent by Monsanto in 2000, and consequently a reduction of its price, was decisive for its widespread use as the main tool to control weeds in GR crops as well as another agricultural systems [9]. The almost exclusive reliance of glyphosate to control weeds, but not only, selected for glyphosate resistant weeds forcing to test/use alternative herbicides to control them.

Know the cause of the herbicide resistance, i.e., characterize the resistance mechanisms that govern it, is important for the proper choice of management methods [10]. However, of the 51 cases of herbicide resistance recorded for Brazil [11], only in few cases such resistance mechanisms have been studied [12–16]. Of the 17 cases of multiple or cross-resistance reported in Brazil, 14 occurred in the last 10 years. The most worrying case is *Conyza sumatrensis*, which was found as being resistant to the 5-enolpyruvylshikimate-3-phosphate synthase (EPSPs) and photosystem I and II (PSI and PSII), protoporphyrinogen oxidase (PPO) inhibitors and synthetic auxins in a GR-soybean field from Assis Chateaubriand, Paraná [11]. Considering these data, the trend is that cases of herbicide resistance, mainly of the multiple resistance, continue increasing in the coming years in Brazil, if little effort is devoted to understanding the cause of herbicide resistance.

In this chapter, we will describe the current overview of the situation of resistance to herbicides in Brazil, discuss the agronomic, scientific, technical and economic factors that have contributed, directly or indirectly, to increase cases of herbicide resistance, as well as the future trends of these agronomic issues according to the weed management measures that are currently being implemented in the country.

2. Pesticide use in Brazil

Brazil, with 77.8 million ha (8.9% of the national territory) in 2018 and with the goal of obtaining 85.7 million ha in 2029 [17], is one of the largest agricultural powers in the world. The area planted in Brazil represents only 3.4% of the global planted area, while countries like India, United States, China and Russia contribute with 9.68, 9,06, 8.96 and 8.38%, respectively [18]. However, Brazil is the main consumer of pesticides since 2011 (20% of the global market in 2017) [6, 7, 19].

Pesticide consumption has almost doubled from 300.5 thousand tons of pesticide active ingredient in 2009 to 549.3 in 2018 in Brazil (**Figure 1A**). According to the pesticide trade reports of the Instituto Brasileiro do Meio Ambiente e dos Recursos Naturais Renováveis (IBAMA) [7], the use profile of pesticides has maintained a growth and similar trend in use in the last 10 years (2009–2018), where the sale of insecticides/acaricides and fungicides accounted for 28.9% of the national market, and the other classes of pesticides (nematicides, bactericides, adjuvants, growth regulators, etc.) occupied only 12.9%. However, the most striking is that herbicides are the products that dominate the national pesticide market with 58.2%, i.e., of every 10 kg of pesticides sold, 5.8 kg were herbicides. Additionally, the herbicide market has been dominated by five active ingredients (glyphosate, 2,4-D, atrazine, paraquat, and diuron), but glyphosate accounts for 36% of the national market (**Figure 1B**). Sales of glyphosate grew from 118.5 thousand tons of active ingredient in 2009 to 195.0 in 2018, i.e., increased 65% (**Figure 1A**).

As already noted, herbicides were the main pesticide class used in Brazil between 2009 and 2018, with oscillations from 52.4% (2011) to 62.5% (2012). The top 10 active ingredients used in this period were: 2,4-D, atrazine, paraquat, diuron, clomazone, tebuthiuron, picloram, trifluralin, MSMA, with some peaks in specific years of clethodim, hexazinone, and triclopyr, but the main herbicide has been glyphosate, consuming more than 50% of the herbicide market in the country (**Figure 2A**). In percentage terms, glyphosate consumption decreased 15% from 2009 (73%) to 2018 (58%) in favor of the use of other herbicide active ingredients that increased sales such as atrazine, 2,4-D and paraquat. The last active ingredient, with an average of 2.6% in the period 2009–2018, presented a regular increase in its sales going from occupying 1.2% of the herbicide market in 2009 to 3.9% in 2018. Already 2,4-D and atrazine have presented a variable preference on the

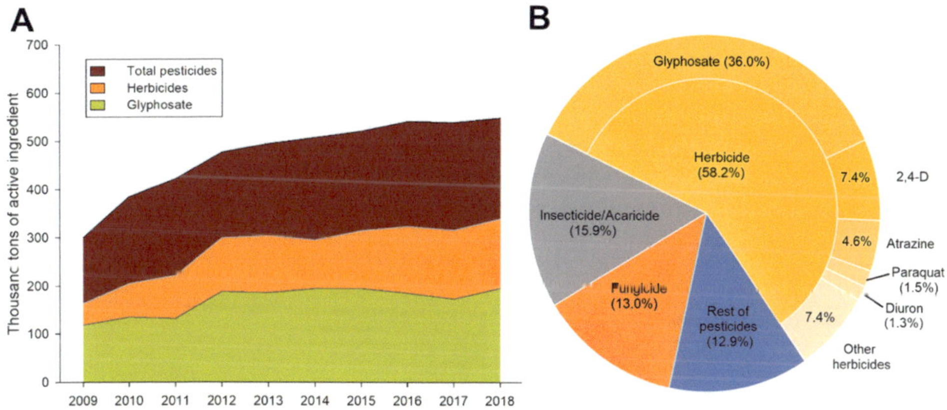

Figure 1.
(A) Commercialization of pesticides (tons of active ingredient × 1000) from 2009 to 2018 in Brazil. (B) Pesticide market share (%) according to their biological activity. Charts were constructed from the pesticide trade reports of the IBAMA [7].

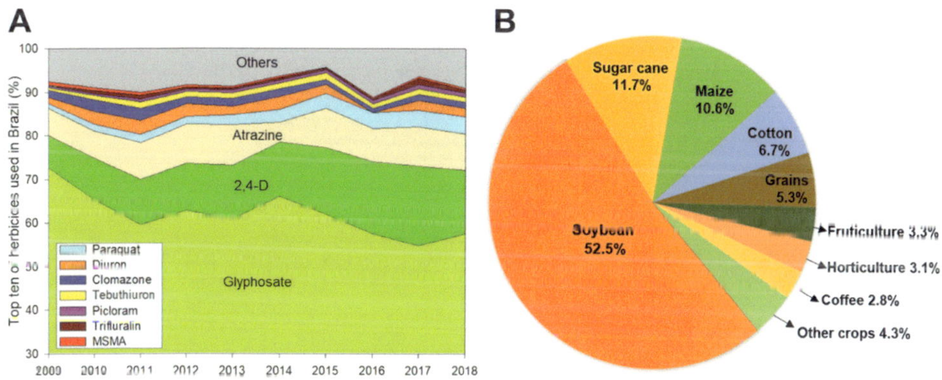

Figure 2.
(A) Percent evolution of the herbicide market in Brazil from 2009 to 2018. (B) Percentage of pesticides occupied in the main production systems of the country. Charts were constructed from the pesticide trade reports of the IBAMA [7].

part of the farmers. For example, 2,4-D (average of the period 12.7%) occupied 7.4% of the herbicide market in 2009, however, in 2017 it reached 18.2%, while atrazine (7.8%) represented the 4.7% in 2014 reached its highest peak in 2013 with 9.4% (**Figure 2A**). A large part of pesticides used in Brazil (81%) is destined to the production of four crops. Soybean is the main consumer being responsible for 52.2% of sales, followed by sugarcane (11.7%), maize (10.6%), and cotton (6.7%) (**Figure 2B**) [20].

The increase in the use of pesticides is related to the evolution of agricultural production, mainly to the increase of agricultural areas destined to monoculture of transgenic crops, i.e., crop varieties that carry traits of resistance to herbicides (HR), insects and diseases, mainly the events that stack glyphosate resistance (GR) traits [21]. According to the Instituto Brasileiro de Geografia e Estatística (IBGE) [22], more than 45% of Brazil's cultivated area is occupied by soybean followed by maize (22%) and sugarcane (14%) (**Figure 3A**), which contributed 62% of the value of agricultural production in 2017 [1]. Between 2009 and 2018, soybean, maize, and cotton showed increases in cultivated area of 60, 17, and 41% [21]. However, the highest growth was observed in relation to the area destined for the cultivation of GR crops. For example, in 2008 there were 14.1 million hectares (64.8%) of GR soybean, but in 2018 the area destined for GR soybeans had more than doubled, occupying 33.4 million hectares (95.8%) (**Figure 3B**). The total area destined for the cultivation of maize showed a lower growth, but the area cultivated with GR varieties tripled in the same period from 4.4 million hectares (31.8%) in 2009 to 14.7 (89%) in 2018 (**Figure 3C**). The total cultivated area of cotton had highs and lows in this period, where the area devoted to the cultivation of GR varieties remained constant between 2012 and 2017 with ~0.75 million hectares.

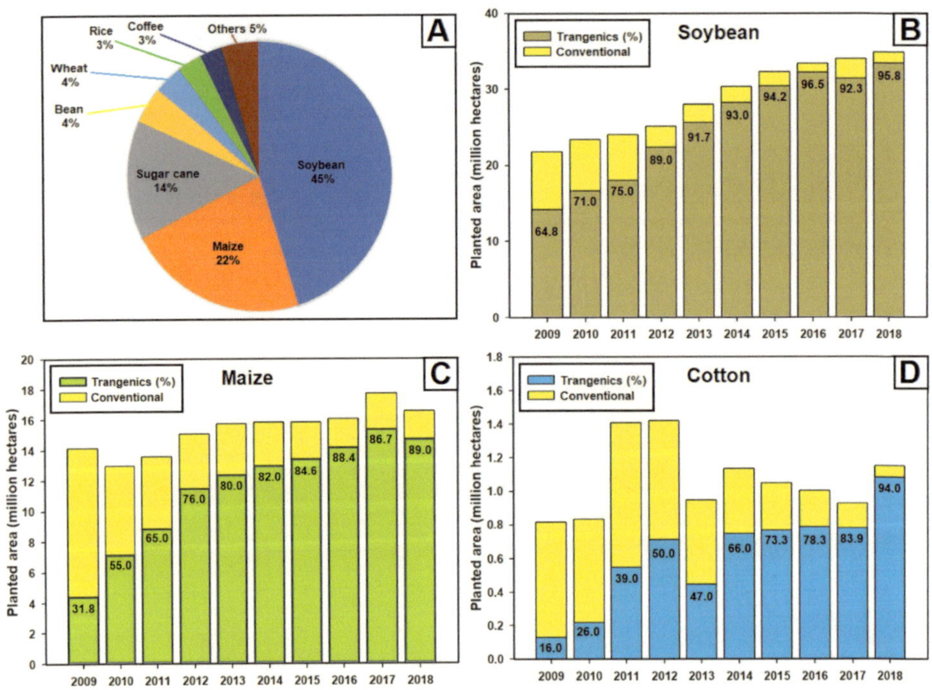

Figure 3.
(A) Percentage of planted area by type of crop in 2015, and total area (million ha) and percentage occupied by transgenic varieties resistant to herbicides of soybean (B), maize (C), and cotton (D). Charts were constructed from the municipality productivity reports of the IBGE [22] and the Conselho de Informações sobre Biotecnologia [21].

However, the area of conventional varieties was reduced from 50 to 16% in the same period. Today, 94% of the area devoted to cotton production is occupied by GR varieties (**Figure 3D**).

The increase in area cultivated with GR varieties has impacted the pesticide market, since more than 70% of pesticides are used in the cultivated area with these crops. However, pesticide statistics do not provide information on how defenses are used in individual crops; therefore, it is not possible to conclude how pesticide use has changed as a result of large-scale adoption of GR varieties [23]. However, this scenario, specifically the herbicide market, reflects the great concern of farmers about the interference of weeds in the agricultural production, but also, how the use and high dependence of these products have had a direct impact on the selection and emergence of weeds resistant to herbicides.

3. History and status of herbicide resistance

The rapid acceptance of GR crops, but not only, the addition of new productive areas and the increasing difficulty in obtaining labor in the fields, has established herbicides as the main control tool, even in integrated systems of weed management. This almost exclusive dependence on herbicides for weed management has contributed to the selection of herbicide resistant weeds with higher frequency. Herbicide resistance is the inherited ability of a plant to survive following application of the commercially used dose of the herbicide recommended for its control [24]. Currently 262 weeds (152 dicots and 110 monocots) have presented 512 unique cases (species x site of action) of herbicide resistance worldwide in 93 crops in 70 countries [11]. In Brazil, there are 51 weed species resistant to herbicides confirmed.

The Brazilian situation of weed resistant to herbicides, mainly to the acetoacetate synthase (ALS) and acetyl coenzyme A carboxylase (ACCase) inhibitors, in conventional soybean cultivation in the mid-2000s was already considered unsustainable due to control difficulties, high cost and low efficiency of the available herbicides to control weed resistant species. The solution to this problem was the introduction of GR soybean varieties [23, 25]. Therefore, to understand the current status of herbicide resistance, it is important to note that GR crops were officially approved in 2005 in Brazil, although GR soybean was irregularly introduced and cultivated in Rio Grande do Sul since 2000. Therefore, the chronological appearance of herbicide resistant weeds is divided into two periods: the pre-glyphosate era preceding 2005 when the use of herbicides was more diversified, and the post-glyphosate era, beginning after approval of GR crops involving an almost exclusive use of glyphosate. In the pre-glyphosate era, from 1993 to 2004, 16 cases were reported, of which only one case presented multiple resistance to two sites of action. In the post-glyphosate era, 35 cases have been reported, of which 16 are cases of multiple resistance. The weed genera with the most resistance cases are *Amaranthus* (7), *Conyza* (8), and *Lolium* (5) (**Figure 4**).

The main groups of herbicides with resistance are the ALS, ACCase, EPSPs, and PSII inhibitors with 30, 9, 16, and 7 cases, respectively (**Figure 5A**). The crop systems with more frequency of herbicide resistance were soybean (30), maize (12), rice (10), wheat (9), and cotton (8) (**Figure 5B**). The Southern, comprising the states of Paraná, Santa Catarina and Rio Grande do Sul, and the Central-West (only in Mato Grosso and Mato Grosso do Sul) regions present 82% of the cases, being Paraná the state where more cases of herbicide resistance were reported (**Figure 5C**). Most of these cases were found in GR crop fields and occurred after 2005, i.e., in the post-glyphosate era, evidencing the drastic changes that GR crop technology caused in weed management.

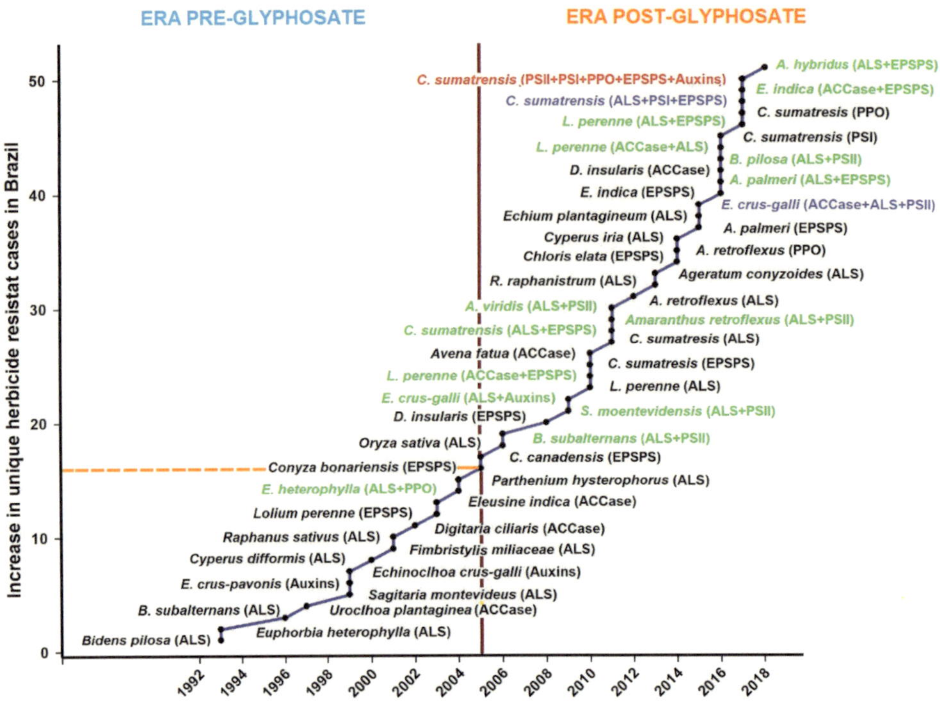

Figure 4.
History of reports of herbicide-resistant weeds in Brazil. Vertical bar indicates the official introduction of transgenic crops resistant to glyphosate. Chart was constructed from the information available in the International Survey of Herbicide Resistant Weeds database [11].

3.1 Resistance to ALS inhibitors

The first cases of resistance to ALS inhibitors were *Euphorbia heterophylla* and *Bidens pilosa* reported in 1993 in soybean areas in the states of Mato Grosso do Sul and Rio Grande do Sul, which showed cross-resistance to sulfonylureas and imidazolinones [26]. After, resistant biotypes of *B. subalternans* (1996) [27], *Parthenium hysterophorus* (2004) [28], *Conyza sumatrensis* (2011) [29] and *Ageratum conyzoides* (2013) were found in Paraná. The latter species was also reported in cotton in Mato Grosso [11]. However, the greatest resistance challenges to ALS inhibitors are found in irrigated rice cultivation. The species reported with ALS resistance in this culture are: *Sagittaria montevidensis* (1999) [30], *Echinochloa* sp. (1999) [31], *Cyperus difformis* (2000) [32], *Fimbristylis miliaceae* (2001), *Oryza sativa* (2006), and *Cyperus iria* (2014) [33] in Rio Grande do Sul and Santa Catarina.

Cases of resistance in rice cultivation are associated with the rapid adoption of Clearfield® technology (crops tolerant to imidazolinones, a chemical group of ALS inhibitors), which were introduced in 2002 in areas of southern Brazil [34]. Although the emergence of new resistant species after the adoption of Clearfield® cultivars did not increase significantly, the dispersion of weed populations resistant to ALS inhibitors, mainly of red rice, was favored by genetic flow of cultivated rice to red rice, representing a great agricultural, economic, and social restriction in the use of Clearfield® technology [35].

Other specific, but not least, cases of resistance to ALS inhibitors are *Raphanus sativus* (2001), *Lolium multiflorum* (2010), and *R. raphanistrum* (2013), found in wheat and barley in Rio Grande do Sul and Paraná; and *Amaranthus retroflexus* (2012) in cotton in the states of Mato Grosso, Mato Grosso do Sul and Goiás [11, 36, 37].

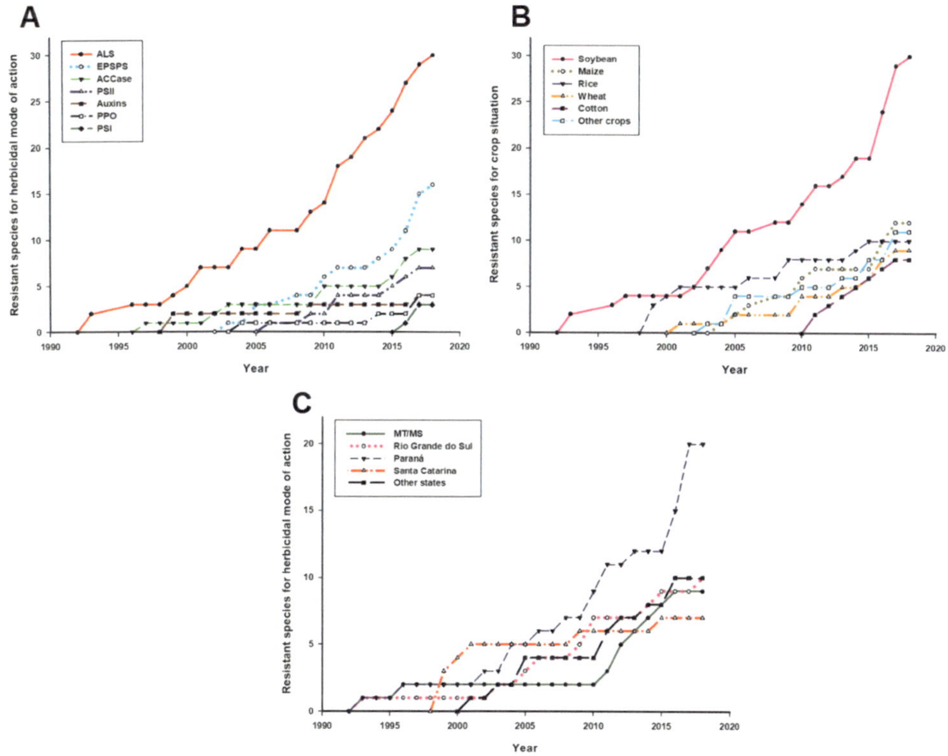

Figure 5.
History of reports of herbicide-resistant weeds in Brazil per mode of action of herbicide (A), crop situation (B), and state of first record (C). MT/MS are the abbreviation of the states Mato Grosso and Mato Grosso do Sul. Charts were constructed from the information available in the International Survey of Herbicide Resistant Weeds database [11].

3.2 Resistance to EPSPs inhibitors

Currently, nine weed species have been reported with glyphosate resistance in Brazil, some of these species have multiple resistance to other modes of action [11]. *Lolium multiflorum* (2003) was the first species identified with glyphosate resistance in orchards and vineyards from Rio Grande do Sul [38]. After, *Conyza bonariensis* (2005), *C. canadensis* (2005) [39], *C. sumatrensis* (2010) [40], *Digitaria insularis* (2008) [12], *Chloris elata* (2014) [13], *Amaranthus palmeri* (2015) [14], *Eleusine indica* [15], and *A. hybridus* (2018) [11] were identified with this resistance mainly in maize and soybean, and wheat fields, but also in citrus and coffee orchards in the states of Mato Grosso, Paraná, Rio Grande do Sul and São Paulo.

With the exception of *L. multiflorum*, the selection of glyphosate resistance in these species is related to the use of GR cultivars, which has also influenced their dispersion throughout the country. Resistant populations of *L. multiflorm* have gone from infesting apple orchards and vineyards to invading GR-soybean fields in the southern states of Brazil [41]. The species of the genus *Conyza*, which have a high invasive potential due to the large seed production, the rapid and high germination capacity, cause great damage to agriculture, and due to their poor interspecific differentiation, it can be an exchange of resistant alleles between species [42]. However, *D. insularis* has been, among glyphosate resistant species, one of the main problems to be faced; therefore, greater efforts have been made to characterize the factors involved in its resistance, dispersal and management [12, 43–48]. Molecular studies showed that the first glyphosate resistant *D. insularis* populations found in the country (Guairá—Paraná) came from Paraguay and were dispersed to other states of Brazil, partly due to their biology and perennial capacity, but mainly due to anthropogenic activities, such as the lack of cleanliness of agricultural implements, but also events of independent selection [47, 48].

Other weeds that pose a major challenge to Brazilian agriculture are species of the genus *Amaranthus*, as they are often reported with glyphosate resistance in GR fields in the United States and Argentina [49]. In addition, *Amaranthus* sp. can hybridize interspecifically facilitating dispersion of resistance alleles [50]. In Brazil, *A. palmeri* was reported to have glyphosate resistance in 2015 [11], when its multiple resistance to the ALS inhibitors was also corroborated [14]. However, the Instituto Mato-Grossense do Algodão had records of the occurrence of glyphosate resistant populations of this species since 2012 in the municipalities of Ipiranga do Norte and Tepurah, Mato Grosso, that was imported from Argentina in cotton harvesting machines in 2011 [51]. Recently, multiple resistance of *A. hybridus* to glyphosate and ALS inhibitors was also confirmed in Rio Grande do Sul in soybeans [11, 52]. With respect to the latter case, there is great concern because it is feared that it has also been introduced from Argentina, where populations of *A. hybridus* with this resistance profile carry mutations in the genes encoding the target enzymes [53]. In the case of glyphosate resistance, it is a triple mutation that confers high levels of resistance and that had not previously been observed in any other species [10, 54]. In addition, in Argentina there are also populations of the species with multiple resistance to 2,4-D and dicamba [55]. Therefore, if it is confirmed that the resistant populations of *A. hybridus* found in Brazil were introduced from Argentina, the scenario faced by Brazilian farmers in the coming years in relation to weed management will be very difficult.

3.3 Resistance to ACCase inhibitors

Urochloa plantaginea (1997) [56], *Digitaria ciliaris* (2002) [57], *Eleusine indica* (2003) [58], *Avena fatua* (2010) [11], and *D. insularis* [16] were reported with resistance to ACCase inhibitors, mainly in non-transgenic soybean fields. These findings demonstrate the importance of these herbicides for the control of grasses in soybean fields, due to the low availability of selective herbicides that effectively control these weeds in pre-emergence conditions, allied to the difficulties of using graminicides, since these products have high retention in the organic matter [29].

3.4 Resistance to other mechanisms of action

The majority of herbicide resistance cases reported in Brazil are included in the three groups of herbicides described above, following the global trend. However, cases of resistance to other modes of action have also been found. In 1999, *Echinochloa crus-pavonis* and *E. crus-galli* were reported with resistance to synthetic auxins, specifically quinclorac, in rice fields of Itajai, Santa Catarina [59]. *Amaranthus retroflexus* (2014) and *C. sumatrensis* (2017) were reported with resistance to PPO inhibitors [11]. The first showed fomesafen resistance and it was found in GR-soybean and -cotton fields of Mato Grosso; and *C. sumatrensis* presented resistance to saflufenacil in soybean fields in the western region of Paraná in the municipalities of Palotina and Assis Chateaubriand [11]. This last species had already been confirmed to be resistance to chlorimuron-ethyl (ALS inhibitor) in 2011 [60] and paraquat (PSI inhibitor) in 2016 [61] within the same region.

3.5 Cross- and multiple-resistance

Cross resistance is expressed when a weed resistant biotype shown resistance against two or more herbicides with the same mode of action, and multiple resistance occurs when a weed resistant to a given herbicide manifests resistance to two or more different modes of action. Most cases of resistance to ALS inhibitors have cross resistance, that is, weeds resistant to imidazolinones often have a degree of resistance to sulfunylureas and vice versa [29]. *Eleusine indica* resistant to sethoxydim (cyclohexanediones) showed resistance to the ariloxifenoxipropionatos (FOPs) [58], and quinclorac resistant *E. crus-galli* showed cross resistant to others synthetic auxins [11]. Weeds with cross resistance represent a great challenge for Brazilian agricultural sustainability; however, weeds with multiple resistance are more challenging by reducing chemical alternatives for their control.

The occurrence of multiple resistance has increased significantly in recent years, and most of the reported cases occurred in the post-glyphosate era. The first case of multiple resistance was *E. heterophylla*, which was found in fields of maize and soybeans in 2004 and showed resistant to triclopyr and fomesafen (ALS + PPO) [62]. In 2009, *E. crus-galli* was found with resistance to synthetic auxins and ALS inhibitors in rice fields in Rio Grande do Sul [34]. Biotypes of *B. subalternans* (2006) and *B. pilosa* (2016) were found to be resistant to atrazine (PSII inhibitors) and ALS inhibitors in soybean and maize fields from Paraná [63]. Among the cases that involves glyphosate resistance are *C. sumatrensis* (2014), *A. palmeri* (2015) and *A. hybridus* (2018) as dicots, that also shown resistance to the ALS inhibitors and were found in soybean fields [11, 14, 60], and *L. multiflorum* (2010), *D. insularis* (2016), and *E. indica* (2016) as monocots with resistance to the ACCase inhibitors. However, the most worrying case is *Conyza sumatrensis* reported in 2017, which was found as being resistant to EPSPs, PSI, PSII, PPO and synthetic auxins in a GR-soybean field from Assis Chateaubriand-PR [11].

This brief account shows the global scenario of the current situation of herbicide resistance in Brazil; however, it is far from reality, because only the first occurrence of a unique case (species x site of action) is reported, while in countries like the United States and Australia, there are multiple reports for the same unique case of herbicide resistant occurring in different regions. For example, the case of *A. palmeri* resistant to glyphosate have more than 30 reports along of the United States [11]. To have an idea of the real problem in Brazil, we have as an example the study conducted by Lopez-Ovejero et al. [45], who determined the frequency and dispersion patterns of glyphosate resistant *D. insularis* revealing the existence of 1299 (of 2596) populations with different resistance levels to this herbicide distributed only in the areas of soybean production. In the scientific-academic environment it is commonly said that it is more difficult to find a population susceptible to the glyphosate of *C. sumatrensis* or *D. insularis* than a resistant one. In addition, from the botanical point of view, more species of the Amaranthaceae, Asteraceae, Cyperaceae, and Poaceae families have high potential to select for resistant to the inhibitors of ALS, ACCase, EPSPs, PPO, and synthetic auxin herbicides in the coming years [64].

4. Herbicide resistance: the problem and the cause

Genetic factors such as genetic variability (mutations localized in a single locus), heredity patterns (dominance of genes enable rapid dispersion), type of pollination (cross-pollination allows for greater genetic recombination and recessive alleles are more easily established in autogenous species), flow gene (transfer resistance characteristics to a susceptible population) and number of resistance genes involved; and bioecological factors such as short life cycle, high seed yield, low dormancy, multiple generations per year, mechanism of propagule dispersion, extreme susceptibility to herbicides, population size, and low biodiversity are key factors in the selection of herbicide resistant weed populations [65]. However, in this section only the agronomic, economic, and even scientific-technical factors that may have contributed to the increase in herbicide resistance in Brazil will be discussed.

4.1 Agronomic factors

Among the agronomic factors that favored the rapid selection of resistance are the characteristics of the herbicide used and the cultural practices. Some herbicide chemical groups have a higher risk of selecting for resistance, especially those with a single mechanism of action or detoxification way (high specificity). High dose applications provide greater selection pressure for resistant weed individuals. The greater persistence of a herbicide also favors the selection for resistance, since the period of exposure is longer, therefore, the ideal is that the herbicide only has effect in the critical period of competence. Reduced crop rotation (monoculture), lack of alternative herbicides, nonuse of herbicide mixtures or sequential applications, nonremoval of weeds from field that escaped herbicide control, and poor inclusion of nonchemical methods are major cultural practices that can lead to emergence of herbicide resistance [66].

In Brazil, a large part of crop production systems is intensive, and today effective weed management without herbicides is inconceivable in the short term in these systems [67]. In addition, a large part of the agricultural areas is occupied with HR crops, resistant to glyphosate or imidazolines, as described in Section 2. The adoption and the use of these technologies caused great changes in weed management, which in most cases, implied the substitution of different herbicidal molecules, that were traditionally used before the insertion of HR crops, by the almost exclusive herbicide associated with said technology in question, at least in the first years after its adoption [23, 68]. For example, in the United States, glyphosate applications replaced a large part of previously used herbicides in GR crops [69].

In Brazil, during the first years after the adoption of GR crops, glyphosate was used in various steps of the production process (chemical fallow (pre-planting), weed management (single or sequential), and desiccation) in doses ranging from 2 to 8 L ha^{-1}, and in some cases, those doses exceeded 10 L ha^{-1} per application [69]. In other cases, many GR soybean farmers delayed the management of weeds that germinated before planting in order to control them with post-emergent applications of glyphosate made on the crop when the competition between the soybean and weeds had already begun [23, 69].

The almost exclusive use of glyphosate quickly showed deficiencies in weed control [23]. Species such as *A. palmeri*, *Conyza* sp., *C. elata*, *D. insularis*, and *E. indica* selected for resistance to this herbicide, forcing farmers to use other herbicides in areas cultivated with GR crops [70]. Herbicides such as 2,4-D, ACCase inhibitors, and ALS were retaken for weed control during pre-sowing (chemical fallow) and crop development, and glufosinate, diuron, and paraquat for desiccation. Currently, glyphosate is applied in isolation only 14% of the time [71]. At the same time, the relative amount of glyphosate used per hectare decreased. For example, 118.5 tons of glyphosate were sold in 2009 and there were 18.6 million ha of GR crops (14.1 soybean +4.4 maize +0.13 cotton), and by 2018, there were 49.2 million ha of GR crops (33.4 soybean +14.7 maize +1.15 cotton) and 195.1 tons of glyphosate were sold. Considering that only these three crops consume 70% of pesticides market of Brazil, in 2009, 4.46 kg of glyphosate ha^{-1} $year^{-1}$ were used, while in 2018, that amount was 2.78 kg of glyphosate ha^{-1} $year^{-1}$, i.e., there was a reduction of at least 26% (**Figure 1A** and **2A**). On the other hand, sales of herbicides such as 2,4-D, paraquat, atrazine, increased between 2009 and 2018 (**Figure 2A**). However, the increase in the use of herbicides with different mode of action, applied in mixture or in sequence with glyphosate, has contributed to the emergence of weeds with multiple resistance.

Paraquat and diuron are considered as bodyguard of glyphosate and are essential tools for Brazilian farmers to hamper the spread of glyphosate resistant weeds [72]. However, the use of paraquat is only authorized until 2020 by the Agência Nacional de Vigilância Sanitaria (ANVISA) after several studies demonstrated that this herbicide can cause Parkinson's and irreversible damage to the genome [73]. Therefore, this legal determination will represent a new challenge in relation to the management of glyphosate resistant weeds, not only in GR fields.

Currently, soybean farmers are anxious and have high expectations with the introduction of new varieties of transgenic soybeans of the technologies Enlist E3™ (2,4-D + glyphosate + glufosinate) and Intacta 2 Xtend® (dicamba + glyphosate) that, have stacked traits of resistance to lepidoptera and up to three herbicides and, will be available in the Brazilian market for commercial use as of the 2020/2021 and 2021/2022 cycles, respectively [74, 75]. On the other hand, the use of synthetic auxins has also been questioned. According to the Instituto Brasileiro do Vinho (Ibravin), the 2,4-D drift used to control weeds in pre-planting of GR soybeans caused damage estimated in R$ 100 million only in 2018 in vineyards of Rio Grande do Sul [76]. Moreover, the use of auxinic herbicides needs to be done with caution, especially in periods with lower temperatures, since any problems related to the application technology, which allows the contact of *Digitaria sp.* plants with low doses of these herbicides, will promote the re-growth of these weeds, which will hinder its control and favor the dispersion of this species [77].

Crop rotation is a consolidated weed management strategy in most of the grain-producing agricultural regions of Brazil. However, it often involves the use of the same technology, i.e., GR soybean is replaced by GR maize and vice versa. This situation is due to the high competitiveness of global commodity markets, which have led farmers to specialize in the production of one or few closely related crops, avoiding the implementation of more complex crop rotations (grain by vegetables). The efficiency obtained by the specialization, which allows the use of the same seeder, combine and marketing infrastructure, has led to the widespread adoption of monocultures [67]. This limited crop rotation (grains by grains), has impacted on the use of herbicides because the number of applications is doubled per agricultural year (3–5 applications per agricultural summer or winter cycle), since second crop requires similar agricultural tasks to the first crop. This practice has increased the herbicide selection pressure on weed populations, but also have provoked the occurrence of voluntary plants from the previous crop, which are difficult to control because they have a similar herbicide resistance profile as the current crop, reducing the crop yield [78, 79].

Direct sowing systems in conjunction with other cultural weed control practices such as cover crops and crop rotations reduce weed population densities [67]. Brazil is one of the few countries that have widely adopted the direct sowing system. This production system reduce the annual weed density compared to conventional agricultural systems [67]; however, the appearance of biannual (*Conyza* sp.) or perennial (*C. elata, D. insularis*, and *E. indica*) weeds have been favored [29, 40, 66], which under continuous selection pressure of glyphosate selected for resistance; therefore, weed management strategies more complex are required.

4.2 Scientific and technical aspects

The Brazilian technical and scientific community specialized on weed science is very large, and their efforts to prevent, monitor, identify, and establish management programs of herbicide resistant weeds, as well as to alert farmers about the occurrence of new cases in order to reduce their dispersion are also very large. Symposiums, congresses, and multiple regional, national, and even international extension activities are frequently held to improve the sustainability of the main production systems, bringing together farmers, researchers, agricultural companies (machinery, pesticides, seeds, etc.), politicians, and agronomy students. In this way we can affirm that the Brazilian agricultural community knows in depth the negative impacts of herbicide resistance. The main efforts of the scientific community are addressed in combating the "problem," as evidenced by the greater amount of research papers proposing alternative management strategies of herbicide resistant weeds. For example, in 2019, of the 150 articles published in *Planta Daninha* (scientific journal published by the Brazilian Society of Weed Science), 22 papers addressed issues related to alternative methods of weed management (chemical, nonchemical, or combined measures), four articles reported the occurrence of new herbicide resistant cases, and only two articles fully or partially characterized the mechanisms of resistance involved (Material S1), i.e., studied the "cause" of herbicide resistance. *Planta Daninha* is not the only scientific journal where Brazilian researchers publish their results, but if it is the main one; therefore, these data reflect the trend in which they invest their main efforts to combat herbicide resistance.

Knowing the "problem" is one thing, but knowing the "cause" is another. Herbicide resistance would be equivalent to referring to a headache. In both cases, the "problem" is known, but the "cause" is unknown. We often underestimate headaches (which can be caused by muscle tension, stress, anxiety, head trauma, etc.) by resorting to self-medication or requesting medication from the pharmacist on duty, who asks a series of questions and recommends some type of analgesic. The pharmacist has not identified the "cause" of the problem, but his recommendation could totally or temporarily relieve the headache and, at the same time, we avoid the consultation with a specialist doctor. Similarly, agricultural field technicians have a deep understanding of the negative impact of herbicide resistant weeds and often recommend different management alternatives; however, they do not know the "cause" of herbicide resistance. Implementing herbicide resistance management measures without knowing the cause of it, by characterizing the resistance mechanisms that govern it would be equivalent to self-medicating. In many cases, herbicide resistant weeds are satisfactory controlled initially, but often the problem worsens over time, resulting in cases of cross and/or multiple resistance. This analogy allows us to infer, that in many cases, Brazilian technicians, and even weed scientist, have acted more as pharmacists than as doctors. This scenario can be added that many field professionals (agronomists and sales agents of pesticides) act without professional ethics prescribing pesticides in a superhuman rhythm [80]. In Brazil, farmers need of prescriptions to purchase these products; however, a professional is often an employer of cooperatives or reseller pesticide offices, so he needs to sell supplies to guarantee his employment [80]. A study carried out by the Agência de Defesa Agropecuária do Paraná (ADAPAR) showed that at least 30% (600 of 2000) of the field professionals signed daily between 7 and 17 prescriptions between 2015 and 2017, i.e., they signed 1–2 prescription for every hour of work. According to ADAPAR, this fact is technically inviable due to the long distance between properties, because to sing a prescription, the field professional must visit the crop fields [80].

In theory, Brazilian weed scientists know the cause of herbicide resistance, i.e., they are familiar with the possible physiological, biochemical, and molecular mechanisms that can confer it. However, studies aimed at characterizing these mechanisms are scarce, often conducted only with the first population(s) that confirmed the occurrence of a given case of herbicide resistance. In addition, the resistance mechanisms characterized in a herbicide resistant population have been adopted in a generalized manner by technicians and other weed researchers, assuming that new occurrences of a case of herbicide resistance already reported (species x herbicide) will have the same mechanisms observed in the first resistant populations. This conclusion may be partially correct when a new occurrence is found in agricultural areas near where the first occurrence was found (dispersion) [48]. However, resistance within the same area and between geographically distant areas may be due to independent herbicide resistance selection events [81], so the resistance mechanisms involved may be different. When multiple studies on the characterization of resistance mechanisms have been carried out on the same weed, the results have been interpreted in a controversial way by the scientific community. For example, in the case of glyphosate resistant *D. insularis* there is no consensus of the mechanism that governs such resistance [82]. In the first populations of glyphosate-resistant *D. insularis*, collected in the state of São Paulo in 2009, the reduced absorption, translocation and metabolism of the herbicide, and a gene mutation (at the Pro-106 EPSPs gene position referred to as Pro-182) were the mechanisms conferring that resistance [12]. After, differences in absorption, but not in the translocation of glyphosate and the occurrence of mutations were observed in other populations [83, 84], while collected populations in different regions of the State of São Paulo presented mutations and enhanced activity of the EPSPs [85]. In the most recent study, including glyphosate resistant *D. insularis* populations collected in different states, it was not possible to characterize the mechanism (s) involved in the resistance [86]. These results show that resistance to a given herbicide can be governed by different mechanisms, acting in isolation or together, in the same species. In addition, these divergent results observed in the different studies show that each new occurrence must be evaluated individually, therefore, this information should not be used to generate a consensus on the mechanisms involved in the resistance of a given case (herbicide x species).

Knowing the mechanisms that govern herbicide resistance is fundamental to plan a proper management strategy, since in some cases, a specific mutation (target site mechanism) does not represent that a herbicide with the same mode of action to which resistance was observed can be used. For example, the Ala-122-Thr mutation in the ALS gene confers high resistance to imidazolinones, but does not confer resistance to sulfonylureas [87]. In the case of herbicide metabolism (non-target-site mechanism) regulated by the cytochrome P-450 enzyme complex, which can confer multiple resistance up to six or possibly more groups of herbicides [88], the use of the same herbicide to which resistance was reported it may be possible using a cytochrome P-450 inhibitor such as malathion or phorate before applying the herbicide in question [89, 90]. Paraquat is an alternative for the management of glyphosate resistant weeds only if resistance is not governed by vacuolar sequestration, since although they have different mechanisms of action, this non-target-site mechanism confers resistance to both herbicides [91]. It is important to note that the management of herbicide resistance is not as simple as described here, since it often involves the participation of different resistance mechanisms, but the timely and appropriate characterization of them could reduce the use of complex mixtures of herbicides in high doses, reducing the impact environmental [92].

Some Brazilian weed research groups have partnerships with weed scientists from the United States and Spain and other countries, who collaborate actively in studies that characterize the mechanisms of the most important cases of herbicide resistance of the country [12–16, 93]. However, these efforts are still insufficient and the previous information allows us to infer that in most cases, the management of herbicide resistance in Brazil has been faced in an inverse way, i.e., first, efforts are made to test and propose different alternatives to solve the "problem" and, in some cases, then try to characterize the "cause." This would be equivalent to placing the shoes first and then the socks. Therefore, to face the problem of herbicide resistance, more efforts must be made to characterize the resistance mechanisms involved in each case, and only later, evaluate different alternative management strategies that are technical and economically viable.

4.3 Economic factors

In this aspect there are two great scenarios. On the one hand, there are the pesticide manufacturers that are valued according to the volume of sales of their products; therefore, they invest their efforts in "conquering" more farmers every day so that they use their products and thus have greater presence in the market and consequently greater prestige. On the other hand, there are farmers who in turn want to obtain the highest profit margin with the least investment, often, in the short term.

As highlighted in the previous section, Brazilian scientists focus their efforts on developing herbicide-resistant weed management strategies, mainly through the applications of herbicides with different modes of action applied in tank mix or in sequence [71]. These investigations are often funded by pesticide manufacturers. Although the conclusions are not biased, objectively reflecting which treatments are the best alternative to control certain weed resistant herbicide(s) in a particular production system, and the researchers also do not recommend the use of commercial formulations of a specific manufacturer, obviously the intention of the financing pesticide manufacturer is to increase the sales of its products and technologies.

Weed researchers evaluating alternative management programs often find at least one efficient control option, both for the level of control achieved (> 80%) and for the period that a treatment maintains the level of control, i.e., there are solutions to the "problem," and Brazilian weed scientists never have stopped looking for new herbicide management alternatives. However, if research is abundant in this regard in the country, why do cases of herbicide resistance continue to increase? The answer to this question is possibly related to the fact that in most of these studies the costs (herbicides + cost of operations + worker's payment) of the resistance management programs evaluated are not considered. In addition, the yield (kg ha^{-1}) that a given management program can guarantee to the farmer is rarely determined. A specific case that addresses these two aspects (cost vs. yield) is the study developed by Piasecki et al. [94], who evaluated 16 treatments, of which 11 did not show differences in soybean yield (3600–3750 kg ha^{-1}), but there were differences in the costs of each treatment, since they were composed by 3 or 4 herbicides. In that study, the highest yield of soybean (3888 kg ha^{-1}) was achieved with the treatment consisting of glyphosate + chlorimuron-ethyl +2,4-D + saflufenacil (T13: 1080 + 25 + 670 + 50 g ia ha^{-1}), which had a cost of R$ 180.00 ha^{-1}. However, the best relative economic return was obtained with the treatment of glyphosate + chlorimuron-ethyl +2,4-D (T12: 1080 + 22.5 + 670 g ia ha^{-1}), which presented a yield of 3749 kg ha^{-1} and cost only R$ 85.00 ha^{-1}. This study did not include the costs related to the application operations, but contrasting the cost of a management program with the crop yield can be an additional tool for the farmer, so that he can estimate his profit margin and decide whether or not to adopt given weed management program.

This situation is also reflected in the type of HR crop technology used by farmers. For example, Liberty Link® technology (glufosinate resistant crops) is available in Brazil since 2016/2017 cycle [95]; however, its use is low compared to GR crops, since glufosinate is, in average, three times more expensive than glyphosate. Total glufosinate sales exceeded 1000 tons $year^{-1}$ in 2017 (1137 tons) and 2018 (1450 tons), but they are still very far from glyphosate sales (173,150 and 195,056 tons in 2017 and 2018, respectively) [7]. This shown that farmers often prefer to continue living with glyphosate resistance than to adopt a new but more expensive technologies, i.e., the adoption of an HR technology is motivated by the cost–benefit ratio by saving costs devoted to pest control guaranteeing high yields [2]. Therefore, the success of Enlist E3™ and Intacta 2 Xtend® technologies, which will be available in the Brazilian market from 2020/2021 and 2021/2022 crop cycles, respectively, will depend on their final cost; meanwhile, farmers will continue to be reluctant to adopt integrated management measures for herbicide-resistant weed control or new HR technologies [96, 97]. The Brazilian scientific community has the task of demonstrating to the farmers that, although the implementation of an integrated weed management program is complex and expensive initially, in the long term it is profitable and environmentally sustainable [98].

5. Economic impacts of herbicide resistance

The economic impact of herbicide resistance management is related to the need to use alternative herbicides with different modes of action, yield losses caused by competition, but mainly to the weed species resistant to being controlled [25]. The cost of alternative herbicides varies according to the choice of farmer, as there is often more than one herbicide option available. Yield losses caused by competition vary according to weed and crop competitive ability, number of plants per area, vegetative stage of crops and weeds, soil fertility, and water availability, among other factors. Therefore, estimating the real economic impact of herbicide resistance on Brazilian agricultural activity is difficult.

Embrapa's Herbology Research Group (GherbE) has been continuously monitoring herbicide resistant weeds in grain production systems in Brazil since 2010, through questionnaires and consultations with technical assistance, farmers, and other researchers; seed collection from areas suspected of resistance with subsequent tests for resistance in a greenhouse; field experiments; and visits to areas with suspected resistance. Resistance monitoring by GherbE researchers was made possible through the joint implementation of the projects "Identification and characterization of glyphosate resistant weeds in Brazil" and "Integrated management of herbicide resistant weeds in soybean production systems" [99]. Relevant information is now available showing the potential economic impact of glyphosate resistant weeds on soybean production and the most representative results are summarized here [25, 99, 100].

The average cost of nonresistant weed control in 2017, restricted to two postemergence glyphosate applications and one for desiccation, was estimated in R\$ 120.00 ha^{-1}. In a scenario of glyphosate resistant *L. multiflorum* infestation, in addition to glyphosate, it is necessary to add a graminicide (ACCase inhibitor), increasing the average cost to R\$ 177.65 ha^{-1}. If the infestation is of *Conyza* sp., the use of a latifolicide such as 2,4-D increases the average cost to R\$ 170.50 ha^{-1}. In areas infested with *D. insularis*, a weed more difficult to control than *L. multiflorum*, requires the use of graminicides in both postemergence and desiccation, and may be interspersed with contact herbicides such as paraquat and glufosinate, increasing the average management cost of this species up to R\$ 318.35 ha^{-1}. However, in mixed infestation scenarios, herbicide resistance management is complicated because herbicide options are reduced. For example, infestations of *Conyza* sp. and *L. multiflorum* require selective herbicides for cultivation during soybean vegetative phase, with flumioxazin and trifluralin being the main options, while for desiccation 2,4-D and paraquat are required for control of *Conyza* sp. and *L. multiflorum*, respectively. The average control cost in this scenario may reach R\$ 197.55 ha^{-1}. If the infestation is of *Conyza* sp. and *D. insularis*, the control cost can be up to R\$ 386.65 ha^{-1}, i.e., R\$ 266.65 ha^{-1} more expensive compared to one scenario without resistance. These estimates do not consider the possible occurrence of multiple resistance of *L. multiflorum* and *D. insularis* to graminicides or *Conyza* sp.

to latifolicides, scenarios in which the cost of management is more expensive and restricted in relation to the alternative herbicide options available [25].

According to GherbE monitoring, 59% of soybean area (20.1 out of 34.0 million ha) had infestations of glyphosate-resistant populations of *Conyza* sp., *D. insularis* and/or *L. multiflorum* in 2017. *Lolium multiflorum* affected 4.2 million ha in the southern states, and this weed occurred simultaneously with *Conyza* sp. in 3.4 million ha. The areas infested by *Conyza* sp. and *D. insularis* were estimated at 7.7 and 8.2 million ha, respectively, of which 2.7 million ha correspond to mixed infestations of these two species [100]. On the GherbE website it can visualize distribution maps of these weeds in the different agricultural regions of Brazil [99]. Analyzing the infested area and the control cost according to the infesting weed species, the average cost of resistance management was R$ 4,918,820,000.00 in 2017 [25]. If a conservative 5% yield loss by weed competition is added, the total cost of herbicide resistance in Brazil exceeds R$ 9 billion annually in soybean cultivation alone [25].

6. Future trends, challenges, and conclusions

Brazil is a consolidated agricultural power; however, the large size of its agricultural activity, especially the intensive production, makes it highly dependent on pesticides for the management of phytosanitary issues, which has led to the emergence of pests resistant to these products.

Much of the Brazilian agricultural activity (68.4%) is focused on the production of grains (52.5% soybean, 10.6% maize, and 5.3% other grains) by cultivating herbicide-resistant crop varieties. The introduction, rapid adoption, and high dependence on these technologies and their associated herbicides (58% of the national pesticide market) caused major changes in weed management practices, contributing to the selection of herbicide resistance weeds.

Today, herbicide resistance is a fait accompli in Brazil; however, the problem is not rooted in the cultivation of herbicide resistant crops but in the inappropriate use of these technologies as a whole, mainly related to off-season applications and herbicide overdose. Clearly, weed management practices must be constantly changed to prevent or delay the emergence of resistant plants in an area. However, the high specialization of farmers to grow, manage and market one or few crops with similar agricultural tasks limits the implementation of alternatives weed management measures as well as reduce more complex crop rotations (i.e., grains by vegetables instead of grains by grains), since transferring their production system to other crops requires investments in professional training, infrastructure, new agricultural implements as well as in the creation of new marketing networks; otherwise, farmers have no guaranteed economic return. In addition, farmers prefer to continue living with the herbicide resistance, and they are reluctant to adopt integrated weed management measures or new herbicide resistant crop technologies for herbicide resistance control if their profit margins are not severely compromised.

The management of the herbicide resistance may represent an increase ranging from 100 to 350% ha^{-1} of the costs devoted for weed control in relation to fields with no resistance. However, the dimensions of this phytosanitary issue is incalculable, as five glyphosate resistant weeds (*C. bonariensis, C. canadensis, C. sumatrensis, D. insularis*, and *L. multiflorum*), occurring only in soybean, infested ≥25% of the total planted area (20.1 out of 77.8 million ha) of Brazil, and caused R$ 9 billion of losses in 2017. To know the true economic impact of herbicide resistance, the areas of other crops infested by these glyphosate resistant weeds, as well as areas affected by the other 46 cases of herbicide resistance (species x herbicide x crop situation) reported in Brazil should also be considered.

Brazilian technical and scientific community specialized on weed science continually made great efforts to prevent, monitor, identify as well as discuss and establish new weed-resistant weed management strategies. However, in most cases, herbicide resistance has been fought in an inverted way, i.e., it has been tested/implemented for solutions to the problem without determining the cause; therefore, if little effort continues to be devoted to characterize the resistance mechanism involved in each case of herbicide resistance before implementing weed management strategies, new occurrences of herbicide resistance weeds, mainly with cross- and multiple-resistance, will continue to appear in the coming years in Brazil.

Besides inherent biological factors of weeds to select herbicide resistance, agronomic, economic and scientific-technical factors have, directly or indirectly, contributed to increasing cases of herbicide resistance. These factors are generally linked to each other but they often are analyzed separately. Therefore, in order to achieve sustainable weed management, future studies aimed at addressing herbicide resistance problems by evaluating different weed management programs should consider these factors, as well as practical and economic aspects for their large-scale implementation.

The Brazilian weed science community have the great challenge of demonstrate to farmers that the implementation of integrated weed management programs may be expensive initially, but in the long term it is profitable and environmentally sustainable.

Acknowledgements

RAC, GMO, and MFGFS thank the "Fundação de Amparo à Pesquisa do Estado de São Paulo—FAPESP" for the financial support (main-grant: 2014/50918-7, sub-grants: 2018/15910-6 and 2019/15527-0).

Author details

Ricardo Alcántara-de la Cruz[1*], Guilherme Moraes de Oliveira[2], Leonardo Bianco de Carvalho[3] and Maria Fátima das Graças Fernandes da Silva[1]

1 Department of Chemistry, Federal University of São Carlos, São Carlos, Brazil

2 Center of Biological and Health Sciences, Federal University of São Carlos, São Carlos, Brazil

3 School of Agricultural and Veterinary Sciences, São Paulo State University (FCAV/UNESP), Jaboticabal, Brazil

*Address all correspondence to: ricardo.cruz@ufscar.br

References

[1] Buainain AM, Lanna R, Navarro Z. Agricultural Development in Brazil: The Rise of a Global Agro-Food Power. 1st ed. Abingdon: Routledge-Taylor & Francis; 2019. p. 272. DOI: 10.4324/9781351029742

[2] Kamali FP, Meuwissen MPM, MPM d B, et al. Evaluation of the environmental, economic, and social performance of soybean farming systems in southern Brazil. Journal of Cleaner Production. 2017;**142**:385-394. DOI: 10.1016/j.jclepro.2016.03.135

[3] Oliveira CM, Auad AM, Mendes SM, Frizzas MR. Crop losses and the economic impact of insect pests on Braizilian agriculture. Crop Protection. 2014;**56**:50-54. DOI: 10.1016/j.cropro.2013.10.022

[4] Oerke EC. Crop losses to pests. The Journal of Agricultural Science. 2006;**144**:31-43. DOI: 10.1017/s0021859605005708

[5] Zhang W, Jiang F, Ou J. Global pesticide consumption and pollution: With China as a focus. Proceedings of the International Academy of Ecology and Environmental Sciences. 2011;**1**:125-144

[6] Brazilian Agrochemicals Market Continues Growth in 2014 [Internet]. 2015. Available from: https://www.kleffmann.com/en/kleffmann-group/news--press/press-releases/20150818_brazilian_agrochemicals_market_continues/ [Accessed: 12 December 2019]

[7] IBAMA - Relatórios de comercialização de agrotóxicos. [Internet]. 2019. Available from: https://ibama.gov.br/agrotoxicos/relatorios-de-comercializacao-de-agrotoxicos#historicodecomercializacao [Accessed: 12 December 2019]

[8] Bonny S. Genetically modified herbicide-tolerant crops, weeds, and herbicides: Overview and impact. Environmental Management. 2015;**57**:31-48. DOI: 10.1007/s00267-015-0589-7

[9] Duke SO. The history and current status of glyphosate. Pest Management Science. 2018;**74**:1027-1034. DOI: 10.1002/ps.4652

[10] García MJ, Palma-Bautista C, Rojano-Delgado AM, et al. The triple amino acid substitution TAP-IVS in the *EPSPS* gene confers high glyphosate resistance to the superweed *Amaranthus hybridus*. International Journal of Molecular Sciences. 2019;**20**:e2396. DOI: 10.3390/ijms20102396

[11] Heap I. International Survey of Herbicide Resistant Weeds. [Internet]. 2019. Available from: http://www.weedscience.org/ [Accessed: 11 January 2020]

[12] Carvalho LB, Alves P, Gonzalez-Torralva F, et al. Pool of resistance mechanisms to glyphosate in *Digitaria insularis*. Journal of Agricultural and Food Chemistry. 2012;**60**:615-622. DOI: 10.1021/jf204089d

[13] Brunharo CA, Patterson EL, Carrijo DR, et al. Confirmation and mechanism of glyphosate resistance in tall windmill grass (*Chloris elata*) from Brazil. Pest Management Science. 2016;**72**:1758-1764. DOI: 10.1002/ps.4205

[14] Küpper A, Borgato EA, Patterson EL, et al. Multiple resistance to glyphosate and acetolactate synthase inhibitors in palmer amaranth (*Amaranthus palmeri*) identified in Brazil. Weed Science. 2017;**65**:317-326. DOI: 10.1017/wsc.2017.1

[15] Takano HK, Mendes RR, Scoz LB, et al. Proline-106 EPSPS mutation imparting glyphosate resistance in goosegrass (*Eleusine indica*)

emerges in South America. Weed Science. 2019;**67**:48-56. DOI: 10.1017/wsc.2018.71

[16] Takano HK, Melo MSC, Ovejero RFL, et al. Trp2027Cys mutation evolves in *Digitaria insularis* with cross-resistance to ACCase inhibitors. Pesticide Biochemistry and Physiology. 2019 [in press]. DOI: 10.1016/j.pestbp.2019.12.011

[17] Ministério da Agricultura, Pecuária e Abastecimento – MAPA. Em dez anos, área plantada será ampliada em 10,3 milhões de hectares no Brasil [Internet]. 2019. Available from: http://www.agricultura.gov.br/noticias/em-dez-anos-area-plantada-no-brasil-sera-ampliada-em-10-3-milhoes-de-hectares [Accessed: 2 December 2019]

[18] Empresa Brasileira de Pesquisa Agropecuária – EMBRAPA. NASA confirma dados da EMBRAPA sobre área plantada no Brasil [Internet]. 2017. Available from: https://www.embrapa.br/busca-de-noticias/-/noticia/30972114/nasa-confirma-dados-da-embrapa-sobre-area-plantada-no-brasil [Accessed: 11 January 2020]

[19] Vasconcelos Y. Agrotóxicos na berlinda [Internet]. 2018. Available from: https://revistapesquisa.fapesp.br/2018/09/18/agrotoxicos-na-berlinda/ [Accessed: 11 January 2020]

[20] Pignati WA, Lima FANDS, SSD L, et al. Spatial distribution of pesticide use in Brazil: A strategy for health surveillance. Ciência & Saúde Coletiva. 2017;**22**:1281-3293. DOI: 10.1590/1413-812320172210.17742017

[21] Conselho de Informações sobre Biotecnologia. 20 anos de transgênicos: benefícios ambientais, econômicos e sociais no Brasil [Internet]. 2018. Available from: https://croplife-prd.s3-sa-east-1.amazonaws.com/2019/10/Vinte-anos-transgenicos.pdf [Accessed: 16 December 2019]

[22] Instituto Brasileiro de Geografia e Estatística (IGBE). Produção agrícola municipal [Internet]. 2019. Available from: https://www.ibge.gov.br/estatisticas/economicas/agricultura-e-pecuaria/9117-producao-agricola-municipal-culturas-temporarias-e-permanentes.html?t=o-que-e [Accessed: 2 December 2019]

[23] Meyer DE, Cederberg C. Pesticide Use and Glyphosate- Resistant Weeds—A Case Study of Brazilian Soybean Production. Swedish Inst Food Biotechnol. SIK-Rapport Nr 809. 2010

[24] Vencill WK, Nichols RL, Webster TM, et al. Herbicide resistance: Toward an understanding of resistance development and the impact of herbicide-resistant crops. Weed Science. 2012;**60**:2-30. DOI: 10.1614/WS-D-11-00206.1

[25] Adegas FS, Vargas L, Gazziero DLP, et al. Impacto econômico da resistência de planta daninhas a herbicidas no Brasil. EMBRAPA-Circular Técnica. 2017;**132**:1-11. DOI: 10.13140/RG.2.2.34334.43840

[26] Monquero PA, Christoffoleti PJ, Carrer H. Biology, management and biochemical/genetic characterization of weed biotypes resistant to acetolactate synthase inhibitor herbicides. Scientia Agricola. 2003;**60**:495-503. DOI: 10.1590/S0103-90162003000300013

[27] Monquero PA, Christoffoleti PJ, Dias CTS. Weed resistance to ALS–inhibiting herbicides in soybean (*Glycine max*) crop. Planta Daninha. 2000;**8**:419-425. DOI: 10.1590/S0100-83582000000300005

[28] Gazziero DLP, Brighenti AM, Voll E. Ragweed parthenium (*Parthenium hysterophorus*) cross-resistance to acetolactate synthase inhibiting herbicides. Planta Daninha. 2006;**24**:157-162. DOI: 10.1590/S0100-83582006000100020

[29] Oliveira RS Jr, Constantin J, Inoue MH. Biologia e manejo de plantas daninhas. 1st ed. Omnipax: Curitiba; 2011. p. 348

[30] Concenço G, Noldin JA, Lopes NF, Comiotto A. Aspects of *Sagittaria montevidensis* resistance to the ALS-inhibiting herbicide pirazosulfuron-ethyl. Planta Daninha. 2007;**25**:187-194. DOI: 10.1590/S0100-83582007000100021

[31] Merotto A Jr, Fleck NG, Reia B, Andres A. Resistência de *Echinochloa* sp. à quinclorac. In: Resumos do 22° Congresso Brasileiro da Ciência das Plantas Daninhas. Londrina, PR: SBCPD/Embrapa Clima Temperado; 2000. p. 513

[32] Galon L, Panozzo LE, Noldin JA, et al. Herbicide resistance of *Cyperus difformis* to ALS-inhibitors in paddy rice of Santa Catarina. Planta Daninha. 2008;**26**:419-427. DOI: 10.1590/S0100-83582008000200019

[33] Chiapinotto DM, Schaedler CE, Fernandes JPS, Andres A, Lamego FP. Cross-resistance of rice flatsedge to ALS-inhibiting herbicides. Planta Daninha. 2008;**26**:419-427. DOI: 10.1590/S0100-83582008000200019

[34] Andres A, Theisen G, Concenco G, Galon L. Weed resistance to herbicides in rice fields in southern Brazil. In: Price AJ, Kelton JA, editors. Herbicides - Current Research and Case Studies in Use. 1st ed. Rijeka: IntechOpen; 2013. pp. 3-25. DOI: 10.5772/55947

[35] Merotto A Jr, Goulart ICGR, Nunes AL, et al. Evolutionary and social consequences of introgression of nontransgenic herbicide resistance from rice to weedy rice in Brazil. Evolutionary Applications. 2016;**9**:837-846. DOI: 10.1111/eva.12387

[36] Costa LO, Rizzardi MA. Resistance of *Raphanus raphanistrum* to the herbicide metsulfuron-methyl. Planta Daninha. 2014;**32**:181-187. DOI: 10.1590/S0100-83582014000100020

[37] Francischini AC, Constantin J, Oliveira RS Jr, et al. Resistance of *Amaranthus retroflexus* to acetolactate synthase inhibitor herbicides in Brazil. Planta Daninha. 2014;**32**:437-446. DOI: 10.1590/S0100-83582014000200022

[38] Roman ES, Vargas L, Rizzardi MA, Mattei RW. Resistance of Italian ryegrass (*Lolium multiflorum*) to glyphosate. Planta Daninha. 2004;**22**:301-306. DOI: 10.1590/S0100-83582004000200018

[39] Moreira MS, Nicolai M, Carvalho SJP, Christoffoleti PJ. Glyphosate-resistance in *Conyza canadensis* and *C. bonariensis*. Planta Daninha. 2007;**25**:157-164. DOI: 10.1590/S0100-83582007000100017

[40] Santos G, Oliveira RS Jr, Constantin J. *Conyza sumatrensis*: A new weed species resistant to glyphosate in the Americas. Weed Biology and Management. 2014;**14**:106-114. DOI: 10.1111/wbm.12037

[41] Ferreira EA, Concenço G, Silva AA, et al. Competitive potential of ryegras (*Lolium multiflorum*) biotypes. Planta Daninha. 2008;**26**:261-269. DOI: 10.1590/S0100-83582008000200002

[42] Constantin J, Oliveira RS Jr, Oliveira Neto AM. Buva: Fundamentos e recomendações para manejo. 1st ed. Curitiba: Omnipax; 2013. p. 104. DOI: 10.7436/2013.bfrm.0

[43] Barroso AAM, Galeano E, Albrecht AJP, Reis FC, Victoria-Filho R. Does sourgrass leaf anatomy influence glyphosate resistance? Comunicata Scientiae. 2015;**6**:445-453. DOI: 10.14295/cs.v6i4.1124

[44] Silveira HM, Langaro AC, Alcántara-de la Cruz RA, Sediyama T, Silva AA. Glyphosate efficacy on sourgrass biotypes with suspected

resistance collected in GR-crop fields. Acta Scientiarum Agronomy. 2018;**40**:e35120. DOI: 10.4025/actasciagron.v40i1.35120

[45] Lopez Ovejero RF, Takano HK, Nicolai M. et al. Frequency and dispersal of glyphosate-resistant sourgrass (*Digitaria insularis*) populations across Brazilian agricultural production areas. Weed Science. 2017;**65**:285-294. DOI: 10.1017/wsc.2016.31

[46] Martins JF, Barroso AAM, Alves PLCA. Effects of environmental factors on seed germination and emergence of glyphosate resistant and susceptible sourgrass. Planta Daninha. 2017;**35**:e017164499. DOI: 10.1590/s0100-83582017350100039

[47] Martins JF, Barroso AAM, Carvalho LB, et al. Plant growth and genetic polymorphism in glyphosate-resistant sourgrass (*Digitaria insularis* L. Fedde). Australian Journal of Crop Science. 2016;**10**:1466-1473. DOI: 10.21475/ajcs.2016.10.10.p7761

[48] Takano HK, Oliveira RS Jr, Constantin J, et al. Spread of glyphosate-resistant sourgrass (*Digitaria insularis*): Independent selections or merely propagule dissemination? Weed Biology and Management. 2018;**1**:50-59. DOI: 10.1111/wbm.12143

[49] Molin WT, Wright AA, VanGessel MJ, et al. Survey of the genomic landscape surrounding the 5-enolpyruvylshikimate-3-phosphate synthase (*EPSPS*) gene in glyphosate-resistant *Amaranthus palmeri* from geographically distant populations in the USA. Pest Management Science. 2018;**74**:1109-1117. DOI: 10.1002/ps.4659

[50] Nandula VK, Wright AA, Bond JA, et al. EPSPS amplification in glyphosate-resistant spiny amaranth (*Amaranthus spinosus*): A case of gene transfer via interspecific hybridization from glyphosate-resistant palmer amaranth (*Amaranthus palmeri*). Pest Management Science. 2014;**70**:1902-1909. DOI: 10.1002/ps.3754

[51] Bogalin DC, Silveira OR, Gazziero DLP, et al. *Amaranthus plameri*: Programa de erradicação da espécie no Estado de Mato Grosso. In: Meschede DK, Pisa-Gazziero DL, editors. A era glyphosate: Agricultura, meio ambiente e homen. 1st ed. Midiograf: Londrina; 2016. pp. 321-336

[52] HRAC-BR - Associação Brasileira de Ação à Resistência de Plantas Daninhas aos Herbicidas. Comunicado de Alerta [Internet]. 2019. Available from: https://b73f4c7b-d632-4353-826f-b62eca2c370a.filesusr.com/ugd/48f515_ae60b1830ba94f899339ef8e3f3b58b9.pdf [Accessed: 28 December 2019]

[53] Larran AS, Lorenzetti L, Tuesca D, Perotti VE, Permingeat HR. Molecular mechanisms endowing cross-resistance to ALS-inhibiting herbicides in *Amaranthus hybridus* from Argentina. Plant Molecular Biology Reporter. 2018;**36**:907-912. DOI: 10.1007/s11105-018-1122-y

[54] Perotti VE, Larran AS, Palmieri VE, et al. A novel triple amino acid substitution in the EPSPS found in a high-level glyphosate-resistant *Amaranthus hybridus* population from Argentina. Pest Management Science. 2019;**75**:1242-1251. DOI: 10.1002/ps.5303

[55] Dellaferrera I, Cortés E, Panigo E, De Prado R. First report of *Amaranthus hybridus* with multiple resistance to 2,4-D, dicamba, and glyphosate. Agronomy. 2018;**8**:140. DOI: 10.3390/agronomy8080140

[56] Gazziero DLP, Christoffoleti PJ, Brighenti AM, Prete CEC, Voll E. Alexandergrass (*Brachiaria plantaginea*) resistance to ACCase inhibitor

herbicides. Planta Daninha. 2000;**18**:169-180. DOI: 10.1590/S0100-83582000000100017

[57] López-Ovejero RF, Carvalho SJP, Nicolai M, Christoffoleti PJ. Resistance of crab-grass (*Digitaria ciliaris*) populations to acetyl-Co-a carboxylase-inhibiting herbicides. Planta Daninha. 2015;**23**:543-549. DOI: 10.1590/S0100-83582005000300019

[58] Vidal RA, Portes ES, Lamego FP, Trezzi MM. *Eleusine indica* resistance to ACCase inhibitors. Planta Daninha. 2006;**24**:163-171. DOI: 10.1590/S0100-83582006000100021

[59] Lopez-Martinez N, Salva AP, Finch RP, Marshall G, Prado RD. Molecular markers indicate intraspecific variation in the control of *Echinochloa* spp. with quinclorac. Weed Science. 1999;**47**:310-315. DOI: 10.1017/S0043174500091827

[60] Santos FM, Vargas L, Christoffoleti PJ, Agostinetto D, Mariani FI, Dal Magro TV. Differential susceptibility of biotypes of *Conyza sumatrensis* to chlorimuron-ethyl herbicide. Planta Daninha. 2014;**32**:427-435. DOI: 10.1590/S0100-83582014000200021

[61] Zobiole LHS, Pereira VGS, Albrecht AJP, et al. Paraquat resistance of sumatran fleabane (*Conyza sumatrensis*). Planta Daninha. 2019;**37**:e019183264. DOI: 10.1590/s0100-83582019370100018

[62] Trezzi MM, Felippi CL, Mattei D, et al. Multiple resistance of acetolactate synthase and protoporphyrinogen oxidase inhibitors in *Euphorbia heterophylla* biotypes. Journal of Environmental Science and Health. Part. B. 2005;**40**:101-109. DOI: 10.1081/pfc-200034254

[63] Takano HK, Oliveira RS Jr, Constantin J, et al. Multiple resistance to atrazine and imazethapyr in hairy beggarticks (Bidens pilosa). Ciencia e Agrotecnologia. 2016;**40**:547-554. DOI: 10.1590/1413-70542016405022316

[64] Andres A, Concenço G, Schreiber F, et al. Predictions for weed resistance to herbicides in Brazil: A botanical approach. In: Pacanoski Z, editor. Herbicide Resistance in Weeds and Crops. 1st ed. Rijeka: IntechOpen; 2017. pp. 135-157. DOI: 10.5772/intechopen.68336

[65] Moss S. Herbicide resistance weeds. In: Hatcher PE, Froud-Williams RJ, editors. Weed Research: Expanding Horizons. 1st ed. Chichester: Wiley; 2017. pp. 181-214. DOI: 10.1002/9781119380702.ch7

[66] Christoffoleti PJ, Nicolai M. Aspectos de resistência de plantas daninhas a herbicidas. 4th ed. ESALQ: Piracicaba; 2016. p. 262

[67] Lamichhane JR, Devos Y, Beckie HJ, et al. Integrated weed management systems with herbicide-tolerant crops in the European Union: Lessons learnt from home and abroad. Critical Reviews in Biotechnology. 2017;**37**:459-475. DOI: 10.1080/07388551.2016.1180588

[68] Green JM, Owen MDK. Herbicide-resistant crops: Utilities and limitations for herbicide-resistant weed management. Journal of Agricultural and Food Chemistry. 2011;**59**:5819-5829. DOI: 10.1021/jf101286h

[69] Cerdeira AL, Gazziero DLP, Duke SO, et al. Review of potential environmental impacts of transgenic glyphosate-resistant soybean in Brazil. Journal of Environmental Science and Health. Part. B. 2007;**42**:539-549. DOI: 10.1080/03601230701391542

[70] Correia NM, Durigan JC. Weed control in glyphosate tolerant soybean crop. Bragantia. 2010;**69**:319-327. DOI: 10.1590/S0006-87052010000200009

[71] Gazziero DLP. Mixtures of pesticides in tank in Brazilian farms. Planta Daninha. 2015;**33**:83-92. DOI: 10.1590/S0100-83582015000100010

[72] Paraquat Information Center. Rainfast Paraquat Quickly Controls a Broad Spectrum of Weeds and Fights Glyphosate Resistance [Internet]. 2020. Available from: https://paraquat.com/en/use/agronomy/glyphosate-resistant-weeds [Accessed: 01 June 2020]

[73] Ministério da Agricultura, Pecuária e Abastecimento – MAPA. Anvisa autoriza uso do paraquat por mais três anos [Internet]. 2017. Available from: http://www.agricultura.gov.br/noticias/anvisa-autoriza-uso-do-paraquat-por-mais-tres-anos [Accessed: 01 June 2020]

[74] Corteva Agriscience. [Internet]. 2019. Available from: https://www.corteva.com.br/produtos-e-servicos/tecnologias/sistema-enlist.html [Accessed: 05 January 2020]

[75] Bayer Brasil [Internet]. 2018. Available from: https://www.bayer.com.br/midia/sala-de-imprensa/crop-science/releases/intacta-2-xtend-sera-apresentada-em-campo-pela-primeira-vez-durante-o-evento-gigantes-da-soja.php [Accessed: 05 January 2020]

[76] Instituto Brasileiro do Vinho - IBRAVIN. Ministério Público recebe documento com prejuízos causados pelo 2,4-D à vitivinicultura gaúcha [Internet]. 2019. Available from: https://www.ibravin.org.br/Noticia/ministerio-publico-recebe-documento-com-prejuizos-causados-pelo-2-4-d-a-vitivinicultura-gaucha/417 [Accessed: 05 January 2020]

[77] Gazola T, Dias MF, Dias RC, Carbonari CA, Velini ED. Effects of 2,4-d herbicide on species of the *Digitaria* genus. Planta Daninha. 2019;**37**:e019220694. DOI: 10.1590/s0100-83582019370100131

[78] Marca V, Procópio SO, Silva AG, Volf M. Chemical control of glyphosate-resistant volunteer maize. Revista Brasileira de Herbicidas. 2015;**14**:103-110. DOI: 10.7824/rbh.v14i2.408

[79] Costa LL, Almeida DP, Timossi PC, et al. Control of soybean RR voluntary in culture of corn with different doses and mixtures of herbicides. Revista Brasileira de Herbicidas. 2019;**18**:1-8. DOI: 10.7824/rbh.v18i2.655

[80] The Intercept-Brasil. Agrônomos do Paraná estão sendo investigados por receitar agrotóxicos num ritmo sobrehumano [Internet]. 2019. Available from: https://theintercept.com/2019/05/15/agronomos-prescrevem-agrotoxicos-ritmo-sobrehumano/ [Accessed: 05 January 2020]

[81] Alcántara-de la Cruz R, Romano Y, Osuna-Ruíz MD, et al. Genetic relationships between tropical sprangletop (*Leptochloa virgata*) populations from Mexico: Understanding glyphosate resistance spread. Weed Science. 2016;**64**:579-587. DOI: 10.1614/WS-D-15-00183.1

[82] Carvalho SJP, Nicolai M. Problemática do capim-amargoso (*Digitaria insularis* (l.) Fedde) em áreas agrícolas brasileiras [Internet]. 2019. Available from: https://b73f4c7b-d632-4353-826f-b62eca2c370a.filesusr.com/ugd/48f515_226326ff3a3e4f7f97392cbd6481ff21.pdf [Accessed: 12 December 2019]

[83] Melo MSC et al. Absorção e translocação de glyphosate em biótipos de *Digitaria insularis* resistente e suscetível ao glyphosate. In: 28° Congresso brasileiro da ciência das plantas daninhas. Anais da SBCPD, Campo Grande; 2012

[84] Melo MSC et al. Investigação do mecanismo de resistência ao glyphosate na espécie *Digitaria insularis* através de possível mutação na EPSPs. In: 29° Congresso brasileiro da ciência das

plantas daninhas. Anais da SBCPD, Gramado; 2014

[85] Galeano E, Barroso A, Vasconcelos TS, et al. EPSPS variability, gene expression, and enzymatic activity in glyphosate-resistant biotypes of *Digitaria insularis*. Genetics and Molecular Research. 2016;**15**:gmr.15038730. DOI: 10.4238/gmr.15038730

[86] Melo MSC, Rocha LJFN, Brunharo CACG, et al. Sourgrass resistance mechanism to the herbicide glyphosate. Planta Daninha. 2019;**37**:e019185746. DOI: 10.1590/S0100-83582019370100033

[87] Li D, Barclay I, Jose K, Stefanova K, Appels R. A mutation at the Ala122 position of acetohydroxyacid synthase (AHAS) located on chromosome 6D of wheat: Improved resistance to imidazolinone and a faster assay for marker assisted selection. Molecular Breeding. 2008;**22**:217-225. DOI: 10.1007/s11032-008-9168-4

[88] Shergill LS, Bish MD, Jugulam M, Bradley KW. Molecular and physiological characterization of six-way resistance in an *Amaranthus tuberculatus* var. *rudis* biotype from Missouri. Pest Management Science. 2018;**74**:2688-2698. DOI: 10.1002/ps.5082

[89] Oliveira MC, Gaines TA, Dayan FE, et al. Reversing resistance to tembotrione in an *Amaranthus tuberculatus* (var. *rudis*) population from Nebraska, USA with cytochrome P450 inhibitors. Pest Management Science. 2018;**74**:2296-2305. DOI: 10.1002/ps.4697

[90] Busi R, Gaines TA, Powles S. Phorate can reverse P450 metabolism-based herbicide resistance in *Lolium rigidum*. Pest Management Science. 2017;**73**:410-417. DOI: 10.1002/ps.4441

[91] Moretti ML, Hanson BD. Reduced translocation is involved in resistance to glyphosate and paraquat in *Conyza bonariensis* and *Conyza canadensis* from California. Weed Research. 2017;**57**:25-34. DOI: 10.1111/wre.12230

[92] Perotti VE, Larran AS, Palmieri VE, Martinatto AK, Permingeat HR. Herbicide resistant weeds: A call to integrate conventional agricultural practices, molecular biology knowledge and new technologies. Plant Science. 2020;**290**:110255. DOI: 10.1016/j.plantsci.2019.110255

[93] Rojano-Delgado AM, Portugal JM, Palma-Bautista C, et al. Target site as the main mechanism of resistance to imazamox in a *Euphorbia heterophylla* biotype. Scientific Reports. 2019;**9**:15423. DOI: 10.1038/s41598-019-51682-z

[94] Piasecki C, Bilibio MI, Cechin J, et al. Technical effectiveness and economic return of the glyphosate-resistant hairy fleabane management in soybean pre-emergence. Revista Brasileira de Herbicidas. 2017;**16**:20-29. DOI: 10.7824/rbh.v16i1.510

[95] Bayer Brasil [Internet]. 2016. Available from: https://www.bayer.com.br/midia/sala-de-imprensa/crop-science/releases/bayer-lanca-tecnologia-libertylink-para-sojicultores-brasileiros.php [Accessed: 05 January 2020]

[96] Ervin DE, Breshears EH, Frisvold GB, et al. Farmer attitudes toward cooperative approaches to herbicide resistance management: A common pool ecosystem service challenge. Ecological Economics. 2019;**157**:237-245. DOI: 10.1016/j.ecolecon.2018.11.023

[97] Moss S. Integrated Weed Management (IWM): Why are farmers reluctant to adopt non-chemical alternatives to herbicides? Pest Management Science. 2019, 2019;**75**:1205-1211. DOI: 10.1002/ps.5267

[98] Livingston M, Fernandez-Cornejo J, Frisvold GB. Economic returns to herbicide resistance management in the short and long run: The role of neighbor effects. Weed Science. 2016;**64**:595-608. DOI: 10.1614/WS-D-15-00047.1

[99] EMBRAPA Trigo. Grupo de Herbologia da Embrapa – GherbE. Resistência a Herbicidas. [Internet]. 2019. Available from: https://www.embrapa.br/trigo/infraestrutura/plantas-daninhas/gherbe/resistencia [Accessed: 01 June 2020]

[100] Vargas L, Adegas F, Gazziero D, Agostinetto D, Silva WT. Resistência de plantas daninhas a herbicidas no Brasil: histórico, distribuição, impacto econômico, manejo e prevenção. In: Meschede DK, Pisa-Gazziero DL, editors. A Era Glyphosate: Agricultura, meio ambiente e homen. 1st ed. Midiograf: Londrina; 2016. pp. 219-239

Chapter 10

Variable Rate Application of Herbicides for Weed Management in Pre- and Postemergence

Alessandro da Costa Lima and Kassio Ferreira Mendes

Abstract

With the advent of precision agriculture, it was possible to integrate several technologies to develop the variable rate application (VRA). The use of VRA allows savings in the use of herbicides, better weed control, lower environmental impact and, indirectly, increased crop productivity. There are VRA techniques based on maps and sensors for herbicide application in preemergence (PRE) and postemergence (POST). The adoption of the type of system will depend on the investment capacity of the producer, skilled workforce available, and the modality of application. Although it still has some limitations, VRA has been widespread and has been occupying more and more space in chemical management, the tendency in the medium- and long term is that there is a gradual replacement of the conventional method of application. Given the benefits provided by VRA along with the engagement of companies and researchers, there will be constant evolution and improvement of this technology, cheapening the costs of implementation and providing its adoption by an increasing number of producers. Thus, the objective of this chapter was to address an overview of the use of herbicides in VRA for weed management in PRE and POST.

Keywords: VRA, precision agriculture, chemical control, automation

1. Introduction

The growing demand for food and the limitation of territorial expansion of agricultural areas direct agriculture toward an increasing intensification with the rational use of resources and maximization of production [1]. For 2050, the world population is estimated at 9 billion people; this represents a need for an increase in food production around 70 to 100% that can be achieved if more efficient cultivation techniques are adopted with fewer impacts on the environment [2]. For this to be possible, it is necessary to have knowledge and control of the variables that interfere in the costs of production and productivity of crops. In this sense, precision agriculture is a tool that makes it possible to meet these needs.

Precision agriculture comprises a set of technologies that combines sensors, information systems, improved machinery, and informed management to optimize production, considering variability and uncertainties in agricultural systems [3].

This modern agriculture starts from the concept that an area of production is not homogeneous, that is, it has great variation. Thus, it is not appropriate to use agricultural inputs and management techniques equally for areas that have different characteristics. The aggregated knowledge throughout history helps to scientifically explain the variability observed and offers paths to localized management with more technique and rigor [4].

This new approach mainly benefits from the emergence and convergence of various technologies, including the global positioning system (GPS), geographic information system (GIS), microcomputers, control automation, remote sensing, mobile computing, advanced information processing, and telecommunications [5]. With these technologies, it is possible to analyze spatial variability, through data collection, information management, application of inputs at varying rate, and, finally, the economic and environmental evaluation of the results [6].

Precision agriculture allowed to perform not only the mapping of the physicochemical properties of the soil, application of fertilizers in a localized way, pest monitoring, harvesting and post-harvest operations, among others [3] but also the mapping and control of weeds, with localized sprays through mapping equipment or real-time systems and thus rationalize the use of pesticides and also minimize damage to the environment. Thus, the objective of this chapter was to address an overview of the use of herbicides in variable rate application (VRA) for weed management in PRE and POST.

2. Variable rate application (VRA) of herbicides

Weed control with herbicides makes up much of the production costs of a crop. In conventional agriculture, herbicide doses are recommended for large areas, without considering many aspects of spatial and temporal variation. When the use of herbicides is made at a fixed rate, economic losses occur directly and indirectly, both due to the above—what is necessary for herbicides and for possible control failures that decrease productivity. In addition, environmental contamination may occur by leaching herbicides into groundwater and rivers. To fix these problems, it is necessary to use the precision agriculture tools and implement a VRA system [7].

VRA refers to the application of herbicides based on area, location, and soil conditions, among other characteristics. Important characteristics such as the variation in infestation and weed density in the application of herbicides in POST and in the sorption capacity that the soil exerts in the application of herbicides in PRE are considered in this system. This allows us to control weeds more efficiently and reduce environmental risks, as there are no applications of underdoses or overdoses. This technology works by integrating a variable rate control system with the sprayer for herbicide application [8, 9].

VRA systems can be different in many ways, but have components in common; the basic system deployment consists of five components that are represented in **Figure 1**: GPS receiver for location and orientation of the machinery at the time of application, a computer that will perform the data processing, a software capable of relating the data collected in the area and determine the dose to be applied, in addition to controllers that will be responsible for changing the flow and pressure of the spray syrup [7].

The application at a varied rate can be fundamentally based on maps or sensors (**Table 1**). Such methodologies require specific resources that differ greatly from each other.

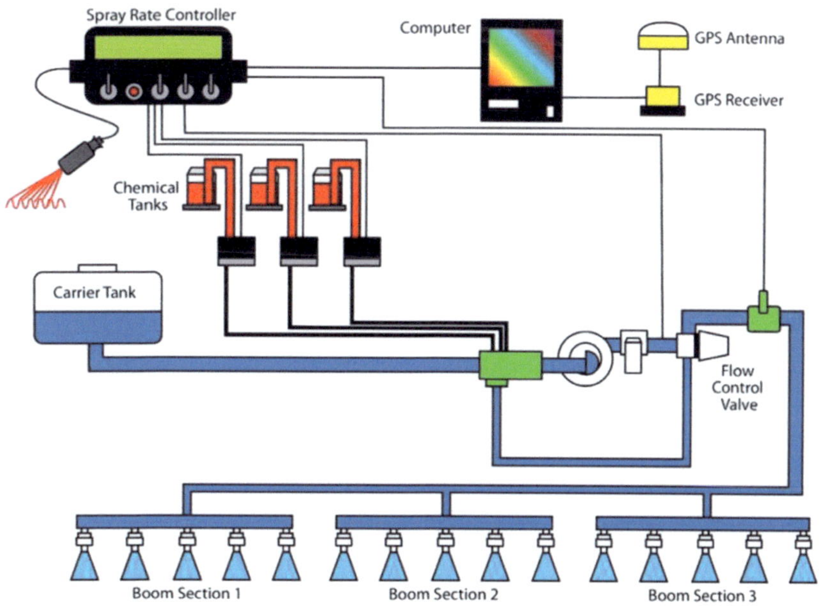

Figure 1.
Main components of a variable rate spraying system (spray rate controller, computer and software, GPS receiver, and control valve). Source: adapted from Grisso et al. [7].

Parameter	Map based	Sensor based
Methodology	Grid sampling—lab analyses—site-specific maps and the use of variable rate applicator	Real-time sensors—feedback control measures and the use of variable rate applicator
GPS/DGPS	Very much required	Not necessary
Laboratory analysis (plant and soil)	Required	Not required
Mapping	Required	May not required
Time consumption	More	Less
Limitations	Cost of soil testis and analysis limit the usage	Lack of sufficient sensors for getting crop and soil information
Operation	Difficult	Easy
Skills	Required	Required
Sampling unit	2 to 3 acres	Individual spot
Relevance	Popular in developing countries	Popular in developed countries

Source: Ahmad and Mahdi [10].

Table 1.
Comparison of the application in varied rate based on maps and sensors.

2.1 Map-based variable rate application (VRA)

Application maps of specific areas are generated by analyzing previous georeferenced samples of soil or plants of the area to be managed. Due to the need to collect many samples to create a representative map of the area, the costs of analysis tend to increase with this method and need more time to get ready. The map-based system is highly dependent on GPS and differential global positioning system (DGPS), as it is necessary to cross-reference the coordinates of the samples collected with the coordinate occupied by the machinery at the time of application. Thus, the operational difficulty of map-based systems is greater.

Although it has some disadvantages referring to operating costs and complexity, the map method is very efficient when used correctly and with accurate equipment. **Figure 2** shows a mapping of weed distribution in a given area and correlated with the required amount of herbicide needed to control weeds according to their density. The result of this crossing of information is a varied rate application map. In the area, there were infestations ranging from 0 to >30 plants m^{-2}; so, it is not necessary to apply the same dose at all levels of infestations [11]. Areas with higher infestation will receive more herbicide than areas with low infestation. In the specific case, the volume of syrup varies from 100 to 250 L ha^{-1}, which corresponds to a variation of 150%. If the volume of syrup was kept constant, there would certainly be herbicide wasting due to excess or lack in certain places. In the example of **Figure 2**, the VRA allowed uniform yield of the crop that was implanted, reduced environmental impacts, and provided savings of 29% in the amount of herbicide.

2.2 Sensor-based variable rate application (VRA)

Data collection of weed presence and processing in sensor-based VRA are made fractions of seconds before herbicide application, avoiding the need to generate a previous map of the area. Sensor-based systems have the ability to vary application rate without any mapping or prior data collection. Sensors measure in real time the desired properties while they are in motion. The measurements made by the system are processed immediately and sent to the controller who will perform the application at a varied rate.

The use of sensors does not necessarily require the use of a positioning system, map generation, or extensive data analysis before making the VRA. Thus, it is an easier-to-use system, consumes less time, and has greater accuracy when compared to the map-based method. Its current limitation is related to the state of

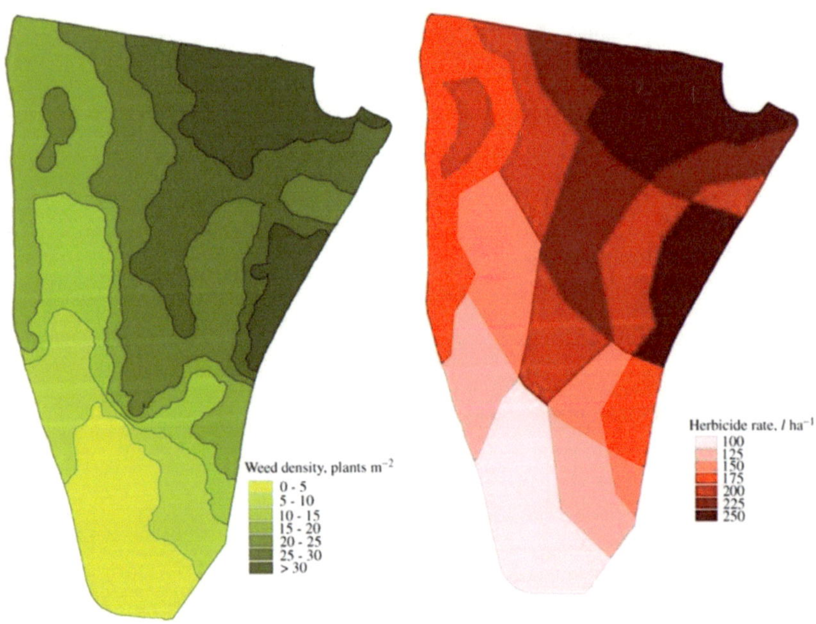

Figure 2.
Weed density map (left) and variable rate application (VRA) of herbicide (right). Source: Carrara et al. [11].

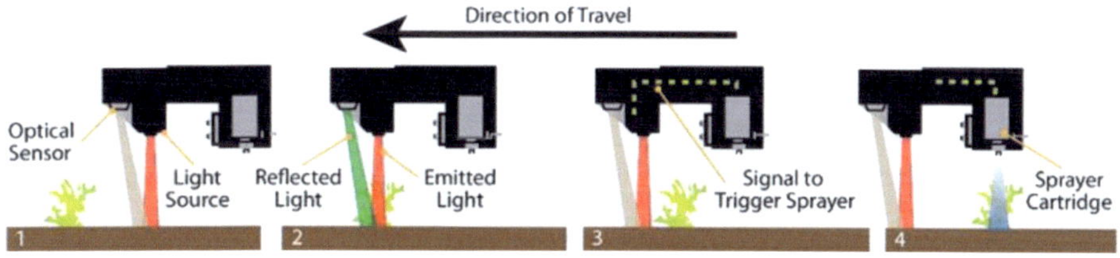

Figure 3.
Acting of an optical sensor in the control of spray nozzles. Source: Grisso et al. [7].

the development of sensors and algorithms with sufficient accuracy to collect and process more detailed information of plants and soil.

In **Figure 3**, there is an example of this type of method, where an optical sensor along with an infrared light source is implanted in the machinery spray bar. This set will be responsible for identifying weeds in the field by reflecting the green color of the leaves and indicating to the controller which sites will be necessary to carry out herbicide application.

3. Variable rate application (VRA) in preemergence (PRE)

The objective of an herbicide application in preemergency is to manage weeds that have not yet germinated, and the herbicide application is made directly in the soil so that as soon as the seeds/propagules germinate, they can absorb the herbicide. But for this to occur, the herbicide must be bioavailable in the soil solution. The application of herbicides in PRE follows different destinations due to the herbicide-soil interactions regulated by physical, chemical, and biological processes [12].

The efficiency of chemical control is associated with several factors that will determine whether the herbicides will be in the soil solution, thus being absorbed by the vegetables; leached, including groundwater; transported by the process of erosion or runoff; and volatilized [13]. In addition, they can be sorbed by soil colloids, thus becoming unavailable to plants.

The variability of soil properties can cause a differential sorption of herbicides, which, in turn, reflects on the different availability of the herbicide in the soil solution, and may generate variation in weed control [14, 15], especially in large cultivated areas where herbicide application is made in a single dose. Thus, the VRA for herbicides in PRE should obtain the main data related to herbicide retention and availability in the soil solution in order to have the correct deposition of the product.

Herbicide sorption is dependent on the interaction of the molecules of the product with the soil, and the process is influenced by the management and climate, mainly soil temperature and humidity. The main physicochemical characteristics of the soil that affect herbicide sorption are organic matter (OM), texture, pH, and cation exchange capacity (CEC). Regarding the herbicide physicochemical characteristics, water solubility (S_w), acid/base dissociation constant (pK_a/pK_b), octanol-water coefficient (K_{ow}) half-life degradation time (DT_{50}), and mainly sorption/desorption coefficient (K_d) [10].

Each herbicide will have a type of behavior in different soil classes. Therefore, to perform VRA in PRE, a previous study of sorption and desorption of the herbicide molecule in the soil type of interest is necessary for the VRA to be efficient. Currently, the technique for sorption and desorption studies of herbicides most used and mentioned in the literature is liquid or gas chromatography. The chromatographic technique can identify individual compounds quantitatively and qualitatively even at small concentrations, being very useful to identify herbicide concentrations in a solution. However, sorption and desorption studies can also be performed with radioisotopes (^{14}C and ^{3}H), in addition to bioassay with plant species sensitive to herbicide [16–18].

Data on soil characteristics are difficult to obtain with sensors in the field; so, most methods for applying herbicides in PRE are based on the generation of maps from laboratory analyses of soil samples. From soil information and herbicide sorption and desorption, a map is interpolated with application information at varying rate [10].

A study of sorption and desorption of the herbicide cyanazine was carried out in different soils (**Table 2**). From this study, the herbicide application was recommended based on soil texture and OM content. Herbicide doses increase as clay and OM contents increases.

Thus, for the application of PRE, herbicide is necessary to analyze the soil's physicochemical properties to interpolate the VRA map. **Figure 4** contains the VRA map in which the different colors represent doses of herbicide to be applied. In this study [15], the use of VRA in PRE decreased the total amount of herbicide by 13%. In addition to the herbicide economy, it should be considered that other benefits are obtained such as better efficiency in weed control, which can help in an increase in productivity, in addition to reducing environmental risks.

Laboratory analyses of soil characteristics are very efficient and accurate. The major disadvantage is the high costs of soil analysis, compromising its use for very large areas. An alternative to map the soil characteristics responsible for herbicide retention and availability without the need for labor collection and analysis is the use of electrical conductivity sensors in the field. The mapping of electrical conductivity with the aid of GPS is a simple tool, which is used to estimate soil texture, in addition to other properties [19]. This quantification considers the clay and ion contents in the soil, resulting in significant correlations [20].

An example of a sensor used to measure electrical conductivity is the VARIS 3100 platform (**Figure 5**). The operation of the equipment consists in the emission of an electric current by two intermediate discs, while two internal discs and two external discs detect the potential difference, which occurs in the electromagnetic field generated in the soil resulting from the applied electric current [21]. The spacing between the discs is calculated so that values of electrical conductivity are measured at depths of 0–0.30 m and 0–0.90 m. Data obtained in the field can be

Soil texture	**Soil organic matter content (%)**					
	<1.0	**1.0**	**2.0**	**3.0**	**4.0**	**≤5.0**
Sand	0.60	0.75	1.25	1.50	1.75	2.00
Sandy loam	0.75	1.25	1.50	1.75	2.00	2.25
Loam, silty loam, silt	1.25	1.50	1.75	2.00	2.25	2.50
Sand clay loam, clay loam, and silty clay loam	1.50	1.75	2.00	2.25	2.50	2.75
Sandy clay, silty clay, and clay	1.75	2.00	2.25	2.50	3.75	3.00
Peat or muck			Not recommended			

Source: Mohammadzamani et al. [15].

Table 2.
Recommendation of doses of cyanazine (L ha^{-1}) according to the texture and organic matter content of the soil.

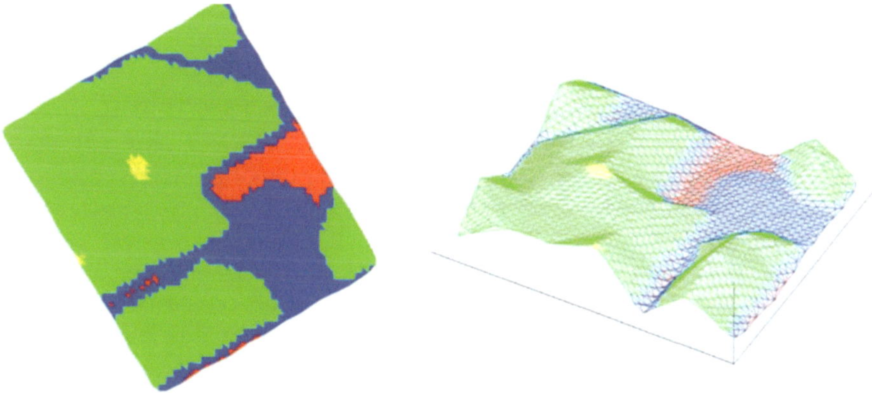

(I) (II)

Color	Soil organic matter content range (%)	Area (m^2)	Area ratio (%)	Herbicide application rate (L ha^{-1})
■	1.25-1.55	408	6.40	1.4
■	1.56-1.85	1616	25.1	1.7
■	1.86-3.35	4374	67.9	2.9
■	3.36-3.65	40	0.60	3.5

Figure 4.
Two-dimensional (I) and three-dimensional (II) maps for variable rate application (VRA) of cyanazine. Source: Mohammadzamani et al. [15].

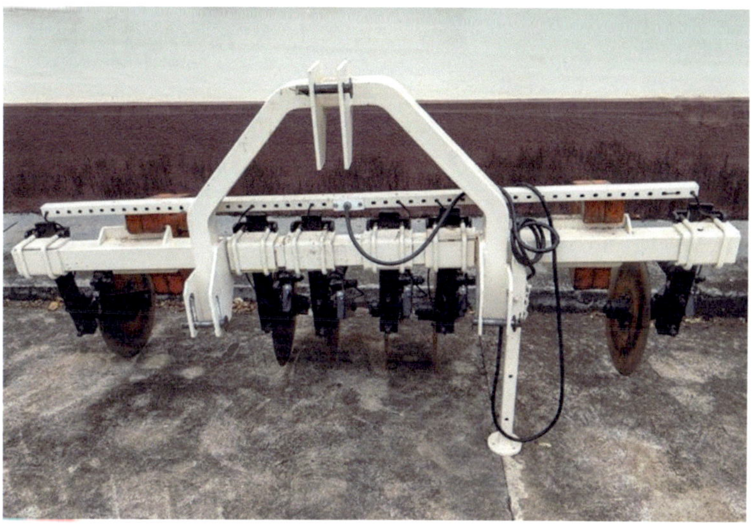

Figure 5.
Veris Platform® 3100 to measure the electrical conductivity of the soil. ESALQ/USP, Piracicaba, SP, Brazil.

visualized, recorded, and exported, since the sensor has a data logger. Data collection occurs with moving equipment, coupled to a tractor and the whole process can be georeferenced by a global navigation satellite system (GNSS) receiver. According to the manufacturer's instructions, two tests must be performed to confirm the correct calibration of the equipment. After data collection, the electrical conductivity is correlated with the clay content for the generation of a textural map.

Studies show that the electrical conductivity measured by contact sensor adequately reflects the variation in clay contents of the studied soil, being efficient to generate soil texture maps, including in no-tillage areas [21]. **Figure 6** shows a conductivity map elaborated with the data collected in VARIS 3100; the lowest conductivity values correlated with lower clay contents. However, for high clay contents, the model was less efficient. Thus, the mapping of electrical conductivity can be a useful tool in the design of more homogeneous areas, which present more similar soil conditions.

Considering that other factors such as moisture, salt concentration, and total carbon remain in the same conditions, soils with higher clay contents conduct more electricity than those with sandier texture. However, these factors may vary and affect the correlation between electrical conductivity and soil texture. Therefore, as the electrical conductivity method does not quantify the CEC and soil OM contents, the use of the same may have reduced efficiency in some situations.

There are companies on the market that provide the VRA service for herbicides in PRE, one of which is APagri which has the HTV® method which consists of a process developed and patented for the application of herbicides in PRE at the varied rate based on maps (**Figure 7**), that considers the clay, OM, and CEC content of the soil [22]. The objective is to adjust the dose according to the soil ability to retain each type of herbicide so that the final concentration in the soil solution is equal regardless of the position in space.

Due to technological limitations, there is still no VRA available on the market for PRE herbicides based on sensors that read, process, and apply the herbicide without the need for the generation of maps. One of the great challenges of this market is precisely to eliminate this stage, in view of the costs of generating the maps.

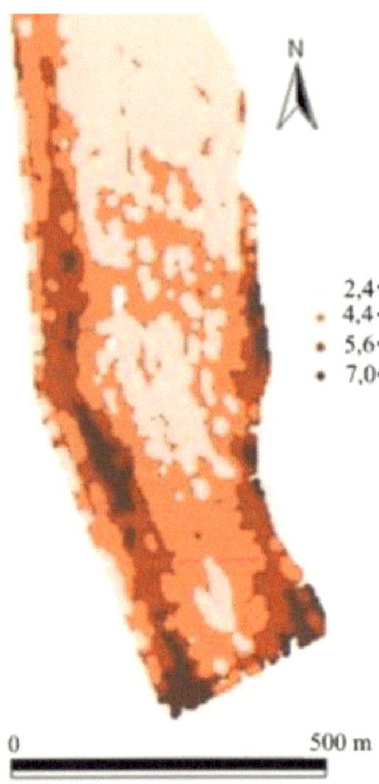

Figure 6.
Interpolated map of electrical conductivity measured with mobile contact measurement equipment. Source: Machado et al. [21].

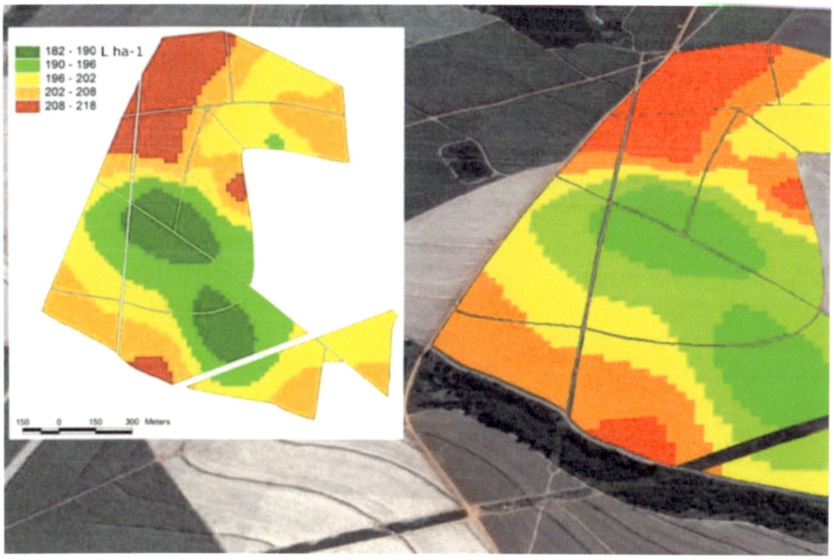

Figure 7.
Variable rate application (VRA) map drawn up with the system HTV®. Source: APagri [22].

4. Variable rate application (VRA) in postemergence (POST)

The purpose of a POST application is to control weeds that have already emerged in the field. Thus, the target of the application is the aerial part of the plant species. For the VRA to be used in POST, it is necessary that the system has information about the weed population in the area. This information can be collected by the map-based and sensor-based systems. Therefore, both methods can be used VRA in POST.

4.1 Map based: weed mapping

The literature mentions several methodologies for weed mapping, where each one has its specificity. Some have processing algorithms to differentiate monocot and eudicot plants [23]. Others use machine learning with deep neural network to identify weeds [24, 25]. However, all have the principle based on the quantitative and qualitative identification of the infested area, generation of the recommendation map, and integration with the VRA system.

Remote sensing is generally considered one of the most important technologies for precision agriculture. This technology can be used in weed mapping. Remote sensing can monitor many crops and vegetation parameters through images at various wavelengths. Images can be acquired by satellites, manned aircrafts, or unmanned aerial vehicles (UAV). However, satellite imagery is often not the best option because of the low spatial resolution of images acquired and the restrictions of the temporal resolutions as satellites are not always available to capture the necessary images [26]. Considering the use of manned aircrafts, usually it results in high costs, and many times, it is not possible to carry out multiple flights to obtain more than a few crop images. UAVs' ability to fly at a low altitude results in ultra-high spatial resolution images of the crops (i.e., a few centimeters). This significantly improves the performance of the monitoring systems. Furthermore, UAV-based monitoring systems have high temporal resolution as they can be used at the user's will. This enhances the flexibility of the image acquisition process [27]. In addition, UAVs are a lot simpler to use and also cheaper than manned aircrafts. Moreover, they are more efficient than the ground systems as they can cover a large field in a short amount of time and in a non-destructive way, which is very important. UAVs can gather images and derive data from the whole field that can be used to generate a precise weed cover map depicting the spots where the herbicide are needed in different rates [28].

A variety of different types of sensors can be used in an agricultural UAV depending on the different vegetation parameters that should be monitored. The main sensors used that meet the limitations mentioned above are: visible light sensors, red, green, and blue (RGB) color model, multispectral sensors, hyperspectral sensors, and thermal sensors. RGB are relatively low cost compared to the other types and can acquire high resolution images, are easy to use and operate, and are lightweight [29]. In addition, the information acquired requires simple processing. However, they are inadequate for analyzing a lot of vegetation parameters that require spectral information in the non-visible spectrum. Thus, commonly are used with the other types of sensors.

Multispectral or hyperspectral imaging sensors can acquire information about the vegetation's spectral absorption and reflection on several bands. Spectral information can be significantly helpful in assessing a lot of biological and physical characteristics of the plants. This information is important to determinate which weed species are in the field [30]. Multispectral and hyperspectral sensors are frequently used, despite their higher costs. However, a drawback of these sensors arises from the fact that it is required to apply more complex preprocessing methods in order to extract useful information from the captured images. The preprocessing procedure of spectral images often contains the radiometric calibration, geometric correction, image fusion, and image enhancement. The main difference between multispectral and hyperspectral sensors is the number of bands (or channels) that each sensor can capture and the width of the bands. Multispectral sensors capture 5–12 channels, while hyperspectral images can usually capture hundreds or thousands of bands, but in a narrower bandwidth. Multispectral sensors are used much more frequently than hyperspectral sensors due to their lower cost, but hyperspectral technology appears to have a lot of potential and is considered the future trend for crop phenotyping research. Thermal infrared sensors capture information about the temperature of the objects and generate images displaying them based on this information and not their visible properties. This type of sensors is used for very specific applications (irrigation management). As a result, they are not frequently used in remotely piloted aircraft applications of UAV systems that focus on monitoring other characteristics of the crops [26–28].

UAVs can acquire information for various features of the cultivated field by using specialized sensors. However, as mentioned above, there is still no standardized workflow or well-established techniques to analyze and visualize the information acquired. The most commonly used image processing methods to analyze UAV imagery for weed mapping are photogrammetry and machine learning. Photogrammetry regards the accurate reconstruction of a scene or an object from several overlapping pictures. Photogrammetric techniques are very commonly used in all types of applications as they are also required to create vegetation indices maps. However, photogrammetric techniques are in most cases used to compliment other types of data processing methods [29].

Machine learning is used to process the data acquired, for prediction and/or identification purposes, with great results in many domains. Machine learning techniques are often applied in precision agriculture to exploit the information from the large amount of data acquired by the UAVs. Machine learning is able to estimate some parameters regarding the crop growth rate, detect diseases, or even identify/discriminate objects in the images [30].

The most promising technique for weed mapping is machine learning, especially those based on object-based image analysis (OBIA). Weed detection with UAVs based on object-based image analysis appears to be at an advanced stage and can be used for specific weed management.

In an example of weed mapping performed on corn, an UAV coupled with a six-band multispectral camera (visible and near infrared range) was used to map the area (**Figure 8**).

After mapping, an OBIA procedure processes the data and generates a classification of weed, crop, and bare soil (**Figure 9**).

The identification and delimitation of the weeds allows generating maps showing the infestation level (**Figure 10**). The information of this map can be integrated into VRA system and used for POST herbicide application. In this study, weed-free areas corresponded to 23% and areas with low infestation (<5% of weeds) to 47% of the total, indicating a high potential to reduce herbicide application [31].

Figure 8.
Unmanned aerial vehicle (UAV) used for weed mapping. Source: Peña [31].

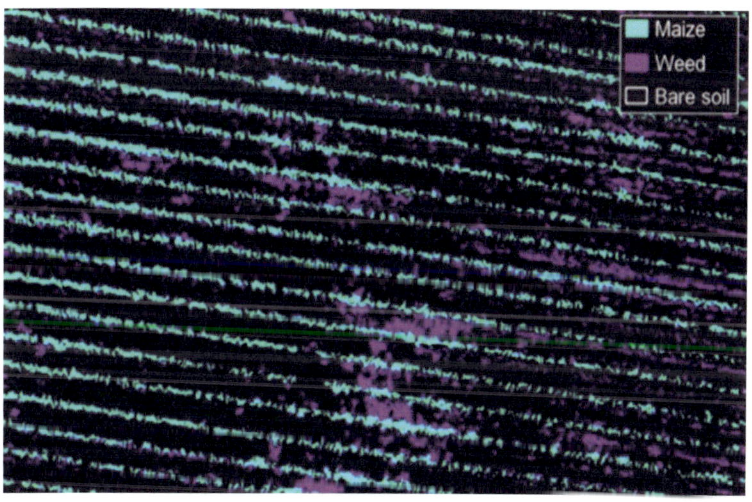

Figure 9.
Partial view of the outputs of the object-based image analysis (OBIA) procedure: classified image with crop, weeds, and bare soil. Source: Peña [31].

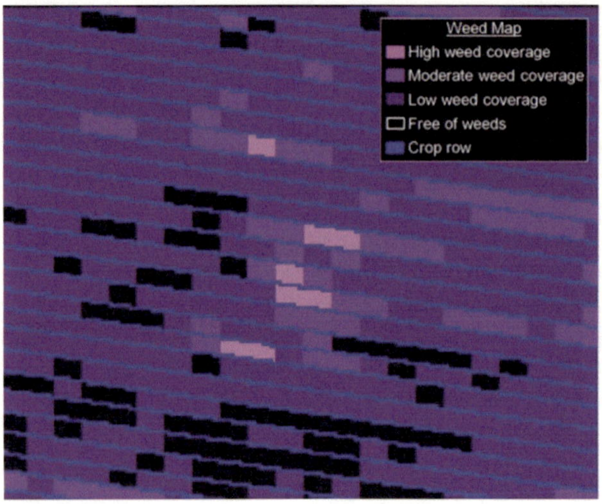

Figure 10.
Partial view of the outputs of the OBIA procedure: weed coverage map showing three levels of infestation (low, moderate, and high), crop rows, and weed-free zones. Source: Peña [31].

When data collection and map generation is done for POST herbicide application, the whole process must be done as quickly as possible because in a few days, the weed dynamics can be changed and infestation levels can increase, making the recommendation map obsolete.

4.2 Sensor based: real time

When applying POST herbicides using a real-time based sensor method, there is no need of a prior area mapping. Spraying is based on sensors attached to the sprayer responsible for detecting weeds and applying the herbicide dose. In **Figure 11**, there is a basic model for this application type.

In real-time-based sensor method, the optical sensor collects data that are immediately processed by the computer, where the locations and doses to be applied are determined. This information is sent as a command to a nozzle controller.

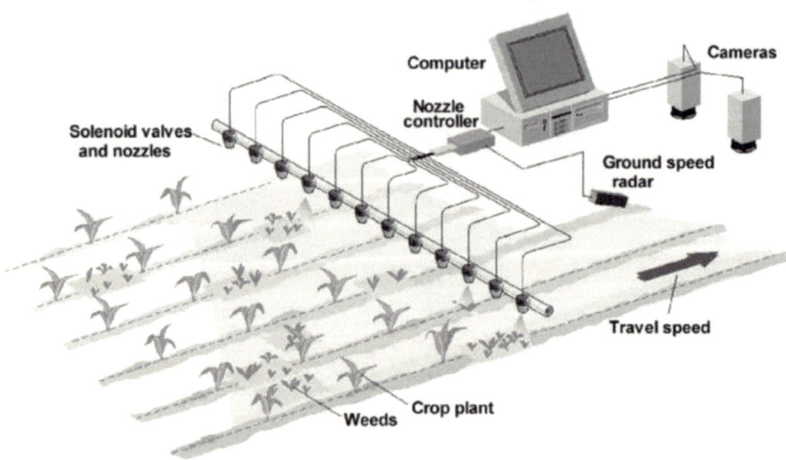

Figure 11.
Sensor-based VRA model for POST herbicide application. The system includes a multiple-camera vision system, a ground speed sensor, and nozzle controller. Source: Tian [32].

In the spray boom, each nozzle is opened or closed by a solenoid valve connected to the controller, so that the nozzle controller can vary the flow applied or the total opening and closing of each nozzle. The presence of a GPS system is not essential for the operation of system, but it does provide guidance to the machinery operator and is useful for recording sprayed areas. The database can be used to improve weed control in the following years, especially for perennial species that reproduce vegetatively, in view of their stability in spatial distribution [33].

Depending on the model, the system components can vary in several characteristics. Optical sensors can be multispectral or infrared. The software can be composed from algorithms that can only identify green plants to deep neural networks that have the ability to learn to differentiate weed species. The controller can only open or close a spray nozzle or it can even coordinate the herbicide mixture and control the alternating flow of dozens of nozzles. The variations are huge, and the more research evolves, the greater the accuracy and reliability of the VRA [7, 32, 33].

Commercially, some companies have consolidated in recent years with VRA systems for application in POST with sensor methods based on real time. Among the most widespread are Weed-it and WeedSeeker.

4.2.1 WEED-IT

WEED-IT is a high-performance localized spraying system, formed by chlorophyll detection sensors and extremely fast valves to guarantee application only where necessary (**Figure 12**). The system is based on the principle of chlorophyll fluorescence: a light source in the set of sensors emits a constant beam of infrared light that is absorbed by the plants chlorophyll and re-emits near infrared light (NIR). This emission is detected by the sensors by performing 40,000 readings per second and capture even the lowest chlorophyll fluorescence emissions activating the nozzle set only on the identified weeds, applying only what is necessary, according to the size of the plant (**Figure 13**) [34].

The system can be installed in self-propelled and trailed sprayers, operating at speeds of up to 25 km h^{-1}. In the spray bar, each sensor is responsible for covering 1 m in width and independently activating up to five nozzles with an opening time of 1 ms. Its valves have a system for modulating the width of the energy pulses that generate extremely rapid interruptions in the spray nozzle outlet; the greater the number of interruptions, the lower the applied dose (**Figure 14**) [34].

In curves or maneuvers, the speed on the outside of the bar is greater than the inside; the system is able to correct the flow along the bar to apply equal amounts of herbicide even in curves or with speed variations (**Figure 15**).

Figure 12.
WEED-IT performing application with weed detection by infrared sensors. Source: SmartSensing [34].

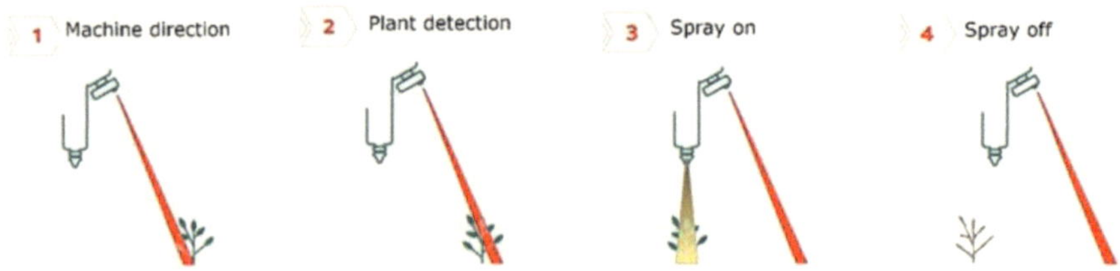

Figure 13.
WEED-IT operating system. Source: SmartSensing [34].

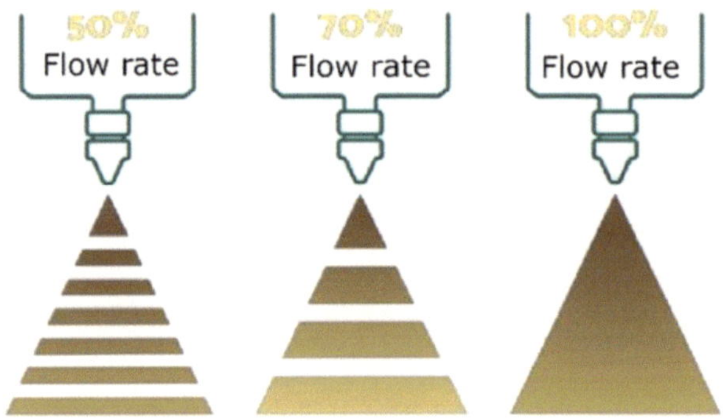

Figure 14.
WEED-IT valve system modulation. Source: SmartSensing [34].

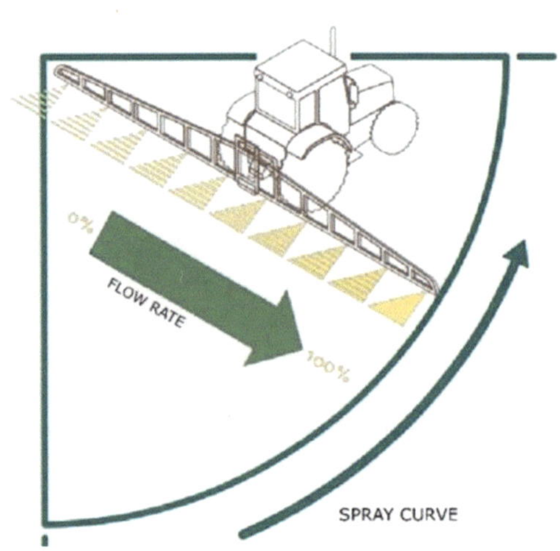

Figure 15.
Differential flow compensation system. Source: SmartSensing [34].

The system has an important limitation. As the sensor is based only on the chlorophyll fluorescence, the system is not able to differentiate the crop and weeds, both are interpreted as living plants. Therefore, it is necessary to be careful with the application of nonselective herbicides in POST, as the crop will certainly be sprayed together with weeds.

4.2.2 WeedSeeker

The WeedSeeker is another widely used commercial system that has the same WEED-IT operate principle, where a sensor emits red and near infrared light and a photodiode detects the intensity of the reflected light (**Figure 16**). Afterward, the reading is converted into a command to apply or not the herbicide (**Figure 17**) [35].

The system can be operated at speeds of 20 km h^{-1} installed in trailed and self-propelled sprayers. Nozzles are opened by solenoid valves connected to a central controller. The sensor spacing is 38 cm, and each sensor controls one spray nozzle. Although WEED-IT and WeedSeeker have many similarities, some aspects differentiate the two systems. The WeedSeeker requires a prior calibration of the sensors in order for the system to operate correctly, while the WEED-IT does not require any calibration [35].

As both systems have own light source, they can perform applications at night. Both are highly efficient systems that fulfill your proposals well. There are few studies that compare two systems. In a study focused on methods of comparing commercial precision spraying technology, the authors compared the efficiency and precision of WEED-IT and WeedSeeker and however, this comparison was only undertaken with a 0.16 ha^{-1}. In this way, WEED-IT can be more efficient for identifying newly emerged plants [35].

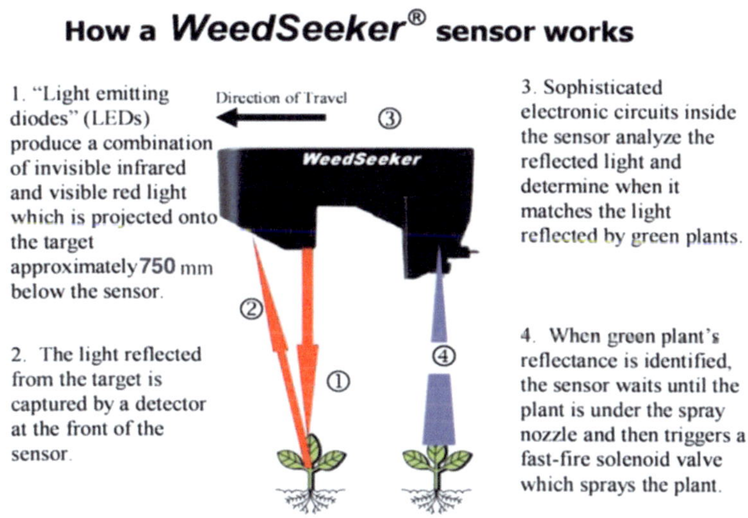

Figure 16.
How a WeedSeeker sensor works. Source: Trimble Agriculture [35].

Figure 17.
WeedSeeker spray nozzles applying herbicide only to weeds. Source: Trimble Agriculture [35].

4.3 Robots for variable rate application (VRA) in postemergence (POST)

Use of autonomous agricultural robots has an interesting potential as a valuable technological tool for precision agriculture, bringing the advantage of being able to make use of the various theories in robotic control, already grounded and consolidated for applications in several other areas [36]. The main characteristic that differentiates an agricultural robot from a simple machine or implement is the freedom degree and autonomy possessed by the robot, including the need for human operation. As agricultural robots must have a high degree of autonomy, tools are necessary so that they can distinguish targets and culture in the field, as well as to orient themselves spatially during movement. The way the distinction is made is through sensors. The main sensors used are GPS real-time kinematic (RTK), cameras, gyroscope, strobe, and proximity [36–38].

The recent trend in the development of mobile robots and autonomous vehicles to perform specific tasks is mainly guided by improving efficiency and leading to operating gains (reduces soil compaction, absence of operator) when compared to the use of large machines [39]. Although much smaller than conventional agricultural machines, they can act cooperatively and perform tasks such as spraying pesticides that pose risks to humans [40]. Sprayers coupled to robots can direct spray nozzles to weeds through a computer vision system. Some models use photovoltaic plates to take advantage of solar energy and reduce or eliminate fossil fuel consumption. With all the advantages related to the autonomy and efficiency of agricultural robots, the farmer can direct his time and efforts toward other agricultural activities such as negotiating sales contracts and making investment decisions.

Robots provide precision spraying, realizing the collection of weed position and incidence information in real time and transmitting them to an atomizer or sprayer that regulates the need for more or less herbicide. Despite having many advantages, the use of robots still has points to be improved, among them are the following:

a. Low autonomy compared to conventional machinery

b. Operational limitations in adverse field conditions

c. State of technological development

The current limitations present in agricultural robots are being resolved with the evolution of the available technology, since the optimization of sensors and algorithms occurs constantly, while in a few years, these limitations can be overcome. Artificial intelligence used in agricultural robots is a way of recognizing patterns so that the computer can identify weeds, pests, disease symptoms, nutritional deficiency, degree of maturation, and cut-off point in the harvest, among others. In a simplified way, artificial intelligence consists of providing the machine with as many examples of situations and decisions as possible, whether historical or simulated based on existing knowledge, so that when faced with similar circumstances, it can make a decision [37, 38]. There are several examples of robots currently used in VRA, two of which are described below.

4.3.1 Robot for Intelligent Perception and Precision Application (RIPPA)

The Robot for Intelligent Perception and Precision Application (RIPPA) is an autonomous system developed by the University of Sydney for detecting weeds and applying herbicides in microdoses (**Figure 18**) [41].

Figure 18.
RIPPA robot model. Source: Sukkarieh [42].

The system has infrared and monochromatic sensors working with neural networks that make it possible to differentiate between crop and weed. In this way, the application and efficiency of the system are much more accurate. Due to its small size and high precision, the system is suitable for smaller areas, such as horticulture. The RIPPA is powered by solar energy through solar panels on the top of the machine. The system also has a sensor for collecting moisture and soil temperature, which makes data collection a little more complete, generating .XLS files so that the producer can create a database with information from his area. **Table 3** contains some additional information from RIPPA [41, 42].

4.3.2 BoniRob

With characteristics similar to RIPPA, BoniRob (**Figure 19**) was developed by the partnership between the companies BOSCH and AMAZONE, in Germany. It is slightly larger than RIPPA, but it is still smaller than a small car and is capable of applying localized pesticides, collecting soil samples, and analyzing to obtain real-time characteristics such as pH and phosphorus levels [43].

Specification description	Value
Track width	1.52 m
Max crop height	0.6 m (adjustable)
IP rating	IP65
Mass (no payload)	Approx. 275 kg
Max payload	100 kg at max operating grade (12°)
Charge-time from empty	> 2 hours (dependent on charger)
Idle discharge time (no solar)	43 hours
Driving discharge time	(0.5 m/s, no solar, no payload) 21.5 hours
Max area traversed per charge (no solar)	8 hectares (~10 hours at 1.6 m/s)

Source: Sukkarieh [42]

Table 3.
Specification description of RIPPA.

Figure 19.
BoniRob model. Source: Sellmann et al. [44].

To ensure its operation, BoniRob has a set of cameras and sensors (**Figure 20**) that work as follows: camera "a" points to the top of the plant with the function of detecting and locating it; camera "b" is positioned to obtain a side view of the plants looking for overlapping plants. In "c," we have a set of light-emitting diodes (LEDs) that are responsible for emitting red and infrared light to assist the cameras when capturing photos. There is also a third camera, which has a high frame rate and resolution (higher than cameras "a" and "b") attached to the sensor responsible for spraying. This sensor, to maintain accuracy in capturing images and also during the application of pesticides, has a strobe that allows, even with variations in the terrain, the camera and the spray tip to remain in the desired position [44].

When it comes to artificial intelligence, based on the culture and species of plants you want to work with and control, machine learning takes place through the developed algorithm and is trained based on obtaining images (millions of them) that allow you to characterize the plants according to their shape, size, and color, among other parameters, allowing them to be recognized and distinguished in the face of a possible action such as spraying it or not [44].

As mentioned earlier, the versatility of agricultural robots is essential, since in the field, the conditions are highly heterogeneous. For this reason, many of these machines allow the installation of modules that perform different functions. In the case of BoniRob, we have a module for phenotypic recognition, a penetrometer, and a localized spraying mode already developed, but there are numerous other possibilities for adaptation and creation based on the particular characteristics to which the use of the machine is intended [44]. Other models of agricultural robots are being developed and gradually made available on the market. A good example is Ecorobotix (**Figure 21**), which applies microdoses of herbicide and

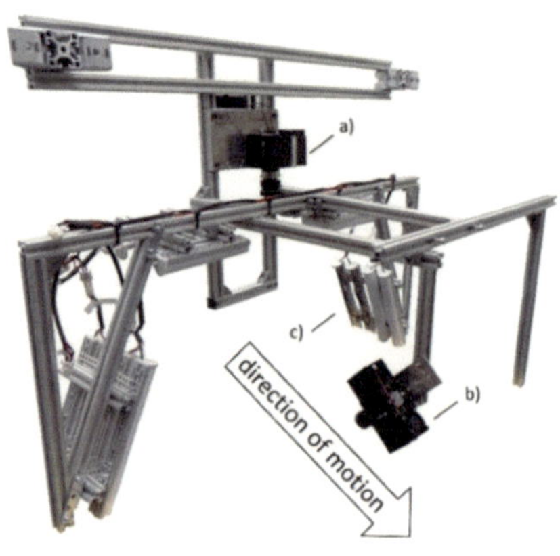

Figure 20.
BoniRob components. Source: Sellmann et al. [44].

Figure 21.
Ecorobotix components. (1) Photovoltaic panels, (2) camera and artificial vision for steering and detection, (3) navigation by GPS and sensors, (4) electrical drive system, (5) rapid robotic arms with sprayers, and (6) tanks for two different products. Source: Ecorobotix [45].

works completely autonomously. Its use is recommended after an initial standard application of herbicide, in order to replace subsequent applications and thus save an important amount of herbicide [45].

The market robots for herbicide application are still at the beginning of its development and consolidation, but it represents a new way of interacting with agriculture, revolutionizing the relationship between man and the field.

5. Variable rate controllers

In order for the VRA to happen efficiently, it is necessary to have a high control in the spraying system responsible for the application of the herbicide. Controllers can act by modifying the pressure at the spray nozzles, or they can change the herbicide concentrations and the water flow in real time. Some of these systems are more complex, while others are simpler. The main controllers will be discussed below.

5.1 Flow-based control systems

In flow-based control system, only the flow and pressure are changed. There is no manipulation of the herbicide concentrations. The system has a flow meter, a speed sensor on the ground, and a servovalve with an electronic controller to apply the desired rate of the tank mixture (**Figure 22**). A microprocessor uses information about the width of the sprayer and the recommendation of the spray volume per hectare to calculate the flow rate appropriate for the current speed of the soil. The servovalve is opened or closed until equal amounts of herbicides are applied regardless of the speed of the machinery. If the controller can be integrated with a recommendation map system, a VRA can be done. These systems have the advantage of being reasonably simple. They are also able to make rate changes across the bar in 3 to 5 seconds [7, 46].

Depending on the speed, problems with drift can occur, as the flow sensor and servovalve control the flow of the tank mixture, allowing variable pressure rates to be delivered to the spray nozzles. Thus, high speeds can represent an increase in the pressure of the nozzles and a consequent decrease in the droplet spectrum.

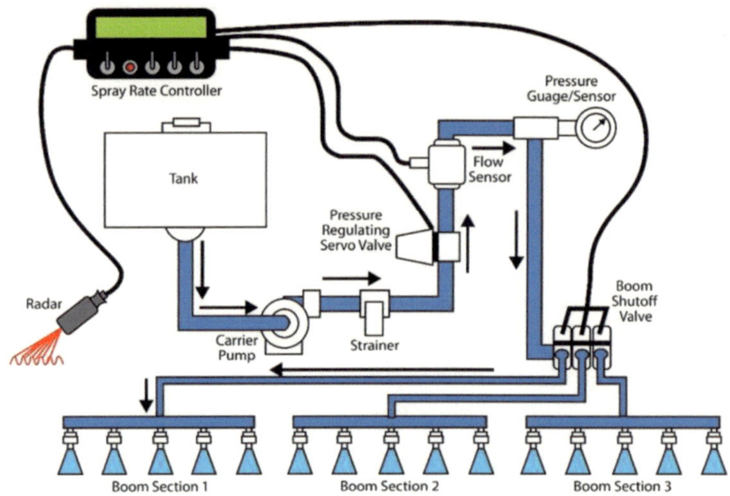

Figure 22.
VRA spraying system that is a flow-based system of application rate. Source: Grisso et al. [7].

5.2 Chemical direct injection systems

In this system, the mixture is prepared with direct injection of the chemical in a flow of water. This system (**Figure 23**) uses a controller and a pump to manage the chemical injection rate instead of the flow rate of a tank mix [46]. The water flow rate is constant and the herbicide injection rate is varied to accommodate changes in soil speed or changes in the prescribed rate.

With the chemical injection, there is no leftover mixture and the direct contact of the operator with toxic products is reduced [10]. The system allows you to control the desired size and spectrum of droplets, since the variation of the application rate does not depend on the flow and pressure on the spray nozzles. Its main disadvantage is the long transport delay between the chemical injection pump and the discharge nozzles at the ends of the boom.

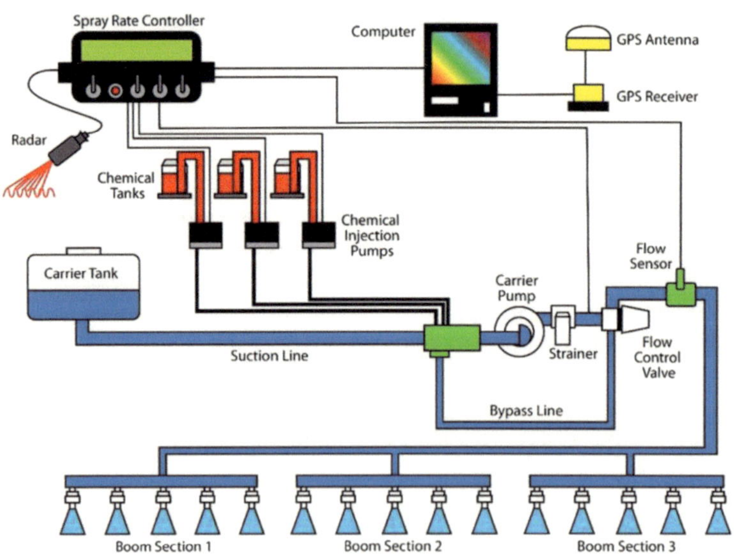

Figure 23.
VRA spraying system that incorporates chemical injection technology. Source: Grisso et al. [7].

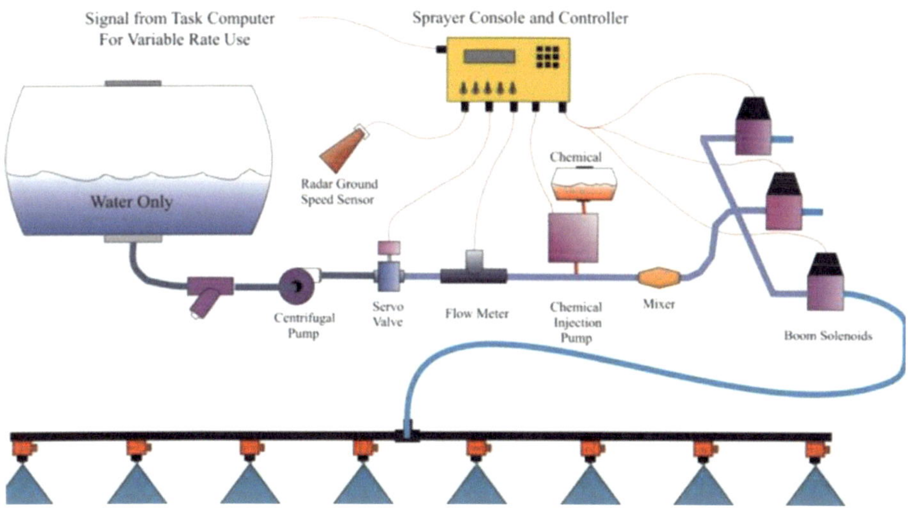

Figure 24.
A direct chemical injection system with carrier control. Source: Rashidi and Mohammadzamani [46].

5.3 Direct chemical injection with carrier control

In this system, there is control of the herbicide injection rate and water flow rate to respond to changes in speed or application rate. A control circuit manages the injection pump, while a second controller operates a servovalve to provide a corresponding water flow (**Figure 24**). Such a system provides a mixture of constant concentration. The system can have many of the advantages of the previous two systems. There is no leftover mixing; the operator is not exposed to chemicals in the tank mixing process; the variation from one rate to another occurs quickly. The disadvantages include related to the complex system, higher initial costs, problem in delivering variable rates of liquid through in the nozzle spray, and modulated spraying nozzle control systems [10, 46].

6. Conclusions

The variable rate application (VRA) of herbicides has great potential for use in agriculture because it allows better control of weeds at lower costs and reduction in the use of inputs and environmental contamination. The main techniques available are based on the generation of application maps and the use of sensors in real time to identify weed infestations, which can be used in the preemergence (PRE) and postemergence (POST) of weeds. Both modalities are equally important in integrated weed management. VRA systems still require relatively high investment, restricting their use. The constant improvement of the VRA should further increase its benefits and reduce the costs of adopting the system, allowing its use by more farmers. The use of precision agriculture in farming systems is a path of no return, in view of the conjuncture of food production needs and scarcity of natural resources. Thus, VRA tends to be used more and more frequently until possible complete replacement of the conventional way of using herbicides in agriculture.

Author details

Alessandro da Costa Lima and Kassio Ferreira Mendes*
Department of Agronomy, Federal University of Viçosa, Viçosa, MG, Brazil

*Address all correspondence to: kfmendes@ufv.br

References

[1] Foley JA, Ramankutty N, Brauman KA, Cassidy ES, Gerber JS, Johnston M, et al. Solutions for a cultivated planet. Nature. 2011;**478**(7369):337-342. DOI: 10.1038/nature10452

[2] Godfray HCJ, Beddington JR, Crute IR, Haddad L, Lawrence D, Muir JF, et al. Food security: The challenge to feeding 9 billion people. Science. 2010;**327**(5957):812-818. DOI: 10.1126/science.1185383

[3] Geberrs R, Adamchuk VI. Precision agriculture and food security. Science. 2010;**327**(5967):828-831. DOI: 10.1126/science.1183899

[4] Molin JP, Amaral LR, Colaço AF. Agricultura de precisão. Oficina de Textos: São Paulo; 2015. p. 223

[5] Gibbons G. Turning a Farm Art into Science - An Overview of Precision Farming [Internet]. 2000. Available from: http://www.precisionfarming.com [Accessed: 12 May 2020]

[6] Soares Filho R, Cunha JPAR. Agricultura de precisão: particularidades de sua adoção no sudeste de Goiás – Brasil. Engenharia Agrícola. 2015;**35**(4): 689-698. DOI: 10.1590/1809-4430-Eng. Agric.v35n4p689-698/2015

[7] Grisso R, Alley M, Thomason W, Holshouser D, Roberson GT. Precision farming tools: Variable-rate application. Virginia Cooperative Extension. 2011;**442**(505):1-16

[8] Sökefeld M. Variable rate technology for herbicide application. In: Oerke EC, Gerhards R, Menz G, Sikora R, editors. Precision Crop Protection - The Challenge and Use of Heterogeneity. Cham: Springer; 2010. pp. 335-347. DOI: 10.1007/978-90-481-9277 9_21

[9] Kempenaar C, Been T. Advances in variable rate technology application in potato in The Netherlands. Potato Research. 2018;**60**(1):295-305. DOI: 10.1007/s11540-018-9357-4

[10] Ahmad L, Mahdi SS. Variable rate technology and variable rate application. In: Ahmad L, Mahdi SS, editors. Satellite Farming. Cham: Springer; 2018. pp. 67-80. DOI: 10.1007/978-3-030-03448-1

[11] Carrara M, Comparetti A, Febo P, Orlando S. Spatially variable rate herbicide application on durum wheat in Sicily. Biosystems Engineering. 2004;**87**(4):387-392. DOI: 10.1016/j.biosystemseng.2004.01.004

[12] Arsego IB. Sorção dos herbicidas diuron e hexazinone em solos de texturas contrastantes. 66 f. Dissertação (thesis). Piracicaba: Escola Superior de Agricultura "Luiz de Queiroz" - Universidade de São Paulo; 2009. DOI: 10.11606/D.11.2009.tde-09092009-110016

[13] Prata F, Lavorenti A. Retenção e mobilidade de defensivos agrícolas no solo. In: Alleoni LRF, Regitano JB, editors. Simpósio Sobre Dinâmica de Defensivos Agrícolas no Solo: Aspectos Práticos e Ambientais. Piracicaba: LSN, ESALQ/USP; 2002. pp. 58-69

[14] Gerstl Z. An update on the $K_{(oc)}$ concept in regard to regional scale management. Crop Protection. 2000;**19**(810):643-648. DOI: 10.1016/s0261-2194(00)00085-5

[15] Mohammadzamani DM, Minaei Z, Alimardani R, Almassi M, Rashid M, Norouzpour H. Variable rate herbicide application using the global positioning system for generating a digital management map. International Journal of Agriculture and Biology. 2009;**11**(2):178-182

[16] Mendes KF, Martins BAB, Reis FC, Dias ACR, Tornisielo VL. Methodologies to study the behavior of herbicides on

plants and the soil using radioisotopes. Planta Daninha. 2017;**35**(1):1-21. DOI: 10.1590/S0100-83582017350100049

[17] Nandula VK, Vencill WK. Herbicide absorption and translocation in plants using radioisotopes. Weed Science. 2015;**63**(1):140-151. DOI: 10.1614/WS-D-13-00107.1

[18] Mendonça CG, Tornisielo VL, Victoria Filho R, Lacerda ALS. Absorption and translocation of 2,4-D in plants of *Memora peregrine*. Journal of Environmental Science and Health - Part B. 2005;**40**(1):137-143. DOI: 10.1081/PFC-200034280

[19] Lund ED, Colin P, Christy D, Drummond PE. Applying soil electrical conductivity technology to precision agriculture. In: Robert P, Rust R, Larson W, editors. Proceedings of the Fourth International Conference on Precision Agriculture. Minnesota: American Society of Agronomy; 1999. pp. 1089-1100. DOI: 10.2134/1999.precisionagproc4.c12b

[20] Johnson CK, Doran JW, Duke HR, Wienhold BJ, Eskridge KM, Shanahan JF. Field-scale electrical conductivity mapping for delineating soil condition. Soil Science Society of America Journal. 2001;**65**(6):1829-1837. DOI: 10.2136/sssaj2001.1829

[21] Machado PLOA, Bernardi ACC, Valencia LIO, Molin JP, Gimenez LM, Silva CA, et al. Mapeamento da condutividade elétrica e relação com a argila de Latossolo sob plantio direto. Pesquisa Agropecuária Brasileira. 2006;**41**(6):1023-1031. DOI: 10.1590/S0100-204X2006000600019

[22] APagri. HTV herbicida em taxa variável [Internet]. 2020. Available from: http://apagri.com.br/htv-herbicida-em-taxa-variavel/ [Accessed: 12 May 2020]

[23] Schuster I, Nordmeyer H, Rath T. Comparison of vision-based and manual weed mapping in sugar beet. Biosystems Engineering. 2007;**98**(1):17-25. DOI: 10.1016/j.biosystemseng.2007.06.009

[24] Sa I, Popović M, Khanna R, Chen Z, Lottes P, Liebisch F, et al. WeedMap: A large-scale semantic weed mapping framework using aerial multispectral imaging and deep neural network for precision farming. Remote Sensing. 2018;**10**(9):e1423. DOI: 10.3390/rs10091423

[25] Tamouridou AA, Alexandridis TK, Pantazi XE, Lagopodi AL, Kashefi J, Kasampalis D, et al. Application of multilayer perceptron with automatic relevance determination on weed mapping using UAV multispectral imagery. Sensors. 2017;**17**(10):e2307. DOI: 10.3390/s17102307

[26] Tsouros DC, Bibi S, Sarigiannidis PG. A review on UAV-based applications for precision agriculture. Information. 2019;**10**(11):349-375. DOI: 10.3390/info10110349

[27] Pajares G. Overview and current status of remote sensing applications based on unmanned aerial vehicles (UAVs). Photogrammetric Engineering & Remote Sensing. 2015;**81**(4):281-329. DOI: 10.14358/PERS.81.4.281

[28] Yang G, Liu J, Zhao C, Li Z, Huang Y, Yu H, et al. Unmanned aerial vehicle remote sensing for field-based crop phenotyping: Current status and perspectives. Fronties in Plant Science. 2017;**8**:1-26. DOI: 10.3389/fpls.2017.01111

[29] Calderón R, Navas-Cortés JA, Lucena C, Zarco-Tejada PJ. High-resolution airborne hyperspectral and thermal imagery for early detection of Verticillium wilt of olive using fluorescence, temperature and narrow-band spectral indices. Remote Sensing of Environment. 2013;**139**:231-245. DOI: 10.1016/j.rse.2013.07.031

[30] Colomina I, Molina P. Unmanned aerial systems for photogrammetry and remote sensing: A review. ISPRS

Journal of Photogrammetry and Remote Sensing. 2014;**92**:79-97. DOI: 10.1016/j.isprsjprs.2014.02.013

[31] Peña JM, Torres-Sánchez J, Castro AI, Kelly M, López-Granados F. Weed mapping in early season maize fields using object-based analysis of unmanned aerial vehicle (UAV) images. PLoS One. 2013;**8**(10):e77151. DOI: 10.1371/journal.pone.0077151

[32] Tian L. Development of a sensor-based precision herbicide application system. Computers and Electronics in Agriculture. 2002;**36**(2-3):133-149. DOI: 10.1016/s0168-1699(02)00097-2

[33] Shiratsuchi LS, Christoffoleti PJ, Fontes JRA. Aplicação localizada de herbicidas. Embrapa Cerrados – Documentos. 2003;**91**:1-18

[34] SmartSensing. WEED-IT Quadro [Internet]. 2020. Available from: http://smartsensingbrasil.com.br/ [Accessed: 13 May 2020]

[35] Trimble Agriculture. WeedSeeker Spot Spray System. 2020. Available from: https://agriculture.trimble.com/product/weedseeker-spot-spray-system/ [Accessed: 13 May 2020]

[36] Kassler M. Agricultural automation in the new millennium. Computers and Electronics in Agriculture. 2001;**20**(1-3):237-240. DOI: 10.1016/S0168-1699(00)00167-8

[37] Weiss U, Biber P. Plant detection and mapping for agricultural robots using a 3D LIDAR sensor. Robotics and Autonomous Systems. 2011;**59**(5):265-273. DOI: 10.1016/j.robot.2011.02.011

[38] Bechar A, Vigneault C. Agricultural robots for field operations: Concepts and components. Biosystems Engineering. 2016;**149**:94-111. DOI: 10.1016/j.biosystemseng.2016.06.014

[39] Pedersen SM, Fountas S, Have H, Blackmore BS. Agricultural robots—System analysis and economic feasibility. Precision Agriculture. 2006;**7**(4):295-308. DOI: 10.1007/s11119-006-9014-9

[40] Grift T. Robotics in crop production. In: Heldman DR, Moraru CI, editors. Encyclopedia of Agricultural, Food, and Biological Engineering. 2nd ed. New York: CRC Press; 2010. pp. 260-262. DOI: 10.1081/E-EAFE-120043046

[41] Hollick V. RIPPA Robot Takes Farms Forward to the Future [Internet]. 2015. Available from: https://sydney.edu.au/news-opinion/news/2015/10/21/rippa-robot-takes-farms-forward-to-the-future-.html [Accessed: 12 April 2020]

[42] Sukkarieh S. An Intelligent Farm Robot for the Vegetable Industry [Internet]. 2016. Available from: https://www.horticulture.com.au/globalassets/laserfiche/assets/project-reports/vg12104/vg12014---final-report-complete.pdf [Accessed: 12 April 2020]

[43] King A. The future of agriculture. Nature. 2017;**540**:21-23. DOI: 10.1038/544S21a

[44] Sellmann F, Bangert W, Grzonka S, Hänsel M. RemoteFarming. 1: Human-machine interaction for a field-robot-based weed control application in organic farming. In: 4th International Conference on Machine Control & Guidance; 19-20 March 2014. Germany: Technische Universität Braunschweig; 2014. pp. 36-42

[45] Ecorobotix. Switch to Smartweeding with Ecorobotix [Internet]. 2020. Available from: http://www.ecorobotix.com/en/ [Accessed: 27 May 2020]

[46] Rashidi M, Mohammadzamani D. Variable rate herbicide application using GPS and generating a digital management map. In: Larramendy ML, Soloneski S, editors. Herbicides, Theory and Applications. London: IntechOpen; 2011. pp. 127-144. DOI: 10.5772/1320

Chapter 11

Effects of Harvest Aids on Sesame (*Sesamum indicum* L.) Drydown and Maturity

William James Grichar, Peter A. Dotray and Derald Ray Langham

Abstract

Harvest aids are traditionally used to desiccate weeds to improve crop quality and harvest efficiency. Field studies were conducted in Texas to determine the effect of harvest aids (glyphosate, diquat-dibromide, glufosinate-ammonium, and carfentrazone-ethyl) on sesame drydown and yield. The objective was to identify one or more harvest aids that could (1) accelerate drydown, (2) burn-down green weeds, (3) even up a field with varying levels of drydown, (4) stop regrowth, (5) stop vivipary, and (6) prepare to plant a new crop. Other than diquat-dibromide, the herbicides were chosen based on the effect on weeds in other crops. The plan was to apply the herbicides 1 week before physiological maturity (PM), at PM, and 1 week after PM. However, sesame maturity is very sensitive to ground moisture, ambient temperature, and relative humidity. The weather was different in all trials and some stages could not be completed. In two cases, the trials had to be abandoned; however, certain patterns emerged. All the herbicides accelerated drydown compared to the untreated check. Diquat-dibromide and glufosinate-ammonium dried sesame faster than glyphosate and carfentrazone-ethyl. The higher rates of the herbicide dried down the sesame faster than the low rate. Although there were some differences in yields across the three application periods, there was no consistent pattern.

Keywords: carfentrazone-ethyl, diquat-dibromide, glyphosate, glufosinate-ammonium, and sesame yield

1. Introduction

After a crop has matured and completed seed fill, there is no weather event that can increase yield, but there are many weather events that can decrease yield and crop quality. Getting the crop out of the field as soon as possible is a critical part of recovering what has taken months to nurture and develop. Harvest aids tradition ally are used to desiccate weeds to improve crop quality and harvest efficiency [1].

In recent years, the use of harvest aids has become important in production of early maturing soybean (*Glycine max* L.) in the states of Kentucky, Mississippi, Missouri, and Tennessee. Soybean leaf retention and presence of green stems and/or green pods in fields where soybean seed are mature (green plant malady) can delay or prevent harvest [2]. Philbrook and Oplinger [3] reported that soybean seed yield loss increased linearly at a rate of 0.2% per day as harvest was delayed out to

42 days. Therefore, harvest aids play an important role in desiccating the crop and accelerating harvest.

Fromme et al. [4] reported using harvest aids to harvest sorghum [*Sorghum bicolor* (L.) Moench] at or after maturity. Harvest aids provide the following advantages: accelerates harvest minimizing weather related damage; prevents or stops seed sprouting; kills the sorghum plant, which is a perennial; provides more efficient and faster threshing; dries out nonproductive suckers and tillers which can delay harvest; reduces differences in harvest maturity across a field; kills late-season weeds; and reduces weed matter in the grain. The application must wait until physiological maturity (PM) to avoid sacrificing yield and reducing test weight. There are three products labelled for use in sorghum: sodium chlorate, glyphosate, and carfentrazone-ethyl. Sodium chlorate provides leaf desiccation but will not kill the plant. Harvest must be timely to avoid regrowth. Desiccation is slowed with low temperatures. Glyphosate is a systemic that will kill the plant and weeds but not accelerate maturity. Carfentrazone-ethyl kills weeds, particularly *Ipomoea* spp. Trostle and McGinty [5] reported good results with the same harvest aids plus diquat-dibromide, which is used only for seed hybrid grain sorghum. Bean [6] reported sodium chlorate is also used for harvesting seed grain sorghum. In North Texas, frost will kill plants without using a harvest aid.

Armstrong [7] reported metsulfuron, 2,4-D, dicamba, glyphosate, and carfentrazone-ethyl and combinations are used as harvest aids for wheat (*Triticum aestivum* L.) to help reduce the amount of green weeds running through the combine. The application must be done after the wheat is mature and different harvest intervals are necessary to allow some of the seed from the weeds to dry so they can be separated by the combine fan. Johnson et al. [8] reported dicamba, glyphosate, and 2,4-D could be used on wheat; however, dicamba and 2,4-D may affect a subsequent double crop. Glyphosate should not be used for seed wheat.

In peanuts (*Arachis hypogaea* L.), Jordan [9] recommended applying carfentrazone-ethyl within 7 days of optimum pod maturity and digging and vine inversion to control *Ipomoea,* which can be a problem at harvest. Chaudhari and Jordan [10] reported carfentrazone-ethyl and pyraflufen-ethyl applied 2 weeks before digging did not reduce the yield or quality of peanuts and controlled *Ipomoea*. Grichar et al. [11] reported 7–52% injury and 4–26% stunting of peanut when carfentrazone-ethyl and pyraflufen-ethyl were applied 35–56 days after planting. They also reported peanut tolerance to carfentrazone-ethyl and pyraflufen-ethyl was cultivar dependent.

Hardke [12] reported sodium chlorate is used as a harvest aid on rice (*Oryza sativa* L.) to accelerate drying in order to accelerate harvest time. Diquat-dibromide, glyphosate, and saflufenacil are used to desiccate canola (*Brassica napus* L.) in Canada [13]. The crop is swathed or harvested direct by using harvest aids to avoid killing frosts that damage the quality of the canola. As opposed to swathing, the advantages of direct harvest are reduced labor and equipment requirements. Diquat-dibromide is a desiccant that will accelerate drydown (4–7 days) but will not control weeds. Glyphosate can dry down canola in warm, sunny days (1–3 weeks), but the primary function is to control weeds. Saflufenacil has some of the functions of diquat-dibromide and glyphosate in terms of contact and systemic activity but none of the harvest aids will hasten maturity and they need to be applied at or after maturity.

Dodds et al. [14] reported harvest aids are used for cotton (*Gossypium hirsutum* L.) defoliation, weed control, and desiccation. All are dependent on the field conditions, weather conditions, and require multiple applications. Defoliation provides the following benefits: removing leaves; eliminating the main source of stain and trash; better lint grades; preventing boll rot; faster and more efficient picker operation; managing maturity, allowing earlier harvest; increased air movement through the crop canopy, which facilitates quicker drying to allow picking to begin earlier in the day; reducing moisture; and improving storage in modules. There are many products used for defoliation in cotton including tribufos, carfentrazone-ethyl, carfentrazone-ethyl + fluthiacet-methyl, pyraflufen-ethyl, saflufenacil, thidiazuron, thidiazuron + diuron, and ethephon + cyclanilide. Glyphosate is used for killing weeds in the crop while paraquat-dichloride and sodium chlorate are used for desiccation.

Zotarelli et al. [15] reported carfentrazone-ethyl, diquat-dibromide, glufosinate-ammonium, pyraflufen-ethyl, and pelargonic acid are used for vine killing in potato (*Solanum tuberosum* L.) harvest. Killing the vine can improve the quality of some potatoes, but used incorrectly, the harvest aids may reduce the quality. Fleury [16] reported diquat-dibromide, glyphosate, carfentrazone-ethyl, and carfentrazone-ethyl + glyphosate are used to desiccate chickpeas (*Cicer arietinum* L.) in Canada. Proper size and color are critical for marketing chickpeas; thus, the crop needs to be mature before using harvest aids. Direct harvest is preferred over swathing because the peas will not cure well in the swath. The cooler nights and shorter days increase the number of days to drydown. Frosts may desiccate the crop without the use of harvest aids. Diquat-dibromide and carfentrazone-ethyl will accelerate drydown while glyphosate will not accelerate drydown. Glyphosate and carfentrazone-ethyl should not be used on chickpea that is to be used for planting seed since they will affect the germination.

In reviewing the information on other crops, there are certain points that pertain to using harvest aids or dessicants that have been studied for sesame (*Sesamum indicum* L.): glyphosate, paraquat-dichloride, diquat-dibromide, glufosinate-ammonium, carfentrazone-ethyl, and pyraflufen-ethyl.

- Harvest aids accelerate the harvest of a crop thereby reducing losses from inclement weather.

- Harvest aids should be applied at or after PM to avoid crop and/or quality loss. Harvest aids will not mature immature seed. The authors have observed that the harvest aids stop all growth on the sesame plant and appear to freeze the seed fill; however, this has not been confirmed in controlled experiments. Although the seed will not increase in weight, it is a living organism and may put on seed color and dry the placenta attachment.

- Harvest aids work better in warm, dry weather than in cold, cloudy, rainy weather. Depending on the temperatures and length of time in subfreezing weather, frosts will kill the sesame and not require harvest aids. Frost generally does not harm seed quality; however, hard freezes may affect the quality. M.L. Kinman (personal communication, 1982) related that in Nebraska a hard freeze killed sesame. When the seed was harvested, it appeared normal, but within a few days the free fatty acids increased quickly rendering the crop unmarketable. The effects of the cold are dependent on the length of time with freezing temperatures, the temperature itself, and the moisture in the seed.

- Harvest aids can help kill weeds facilitating combining and producing a drier crop. In the USA, harvesting sesame during 6% moisture is critical for storage in silos. Sesame has approximately 50% oil, and under high moisture, like other oilseeds, can catch fire in silos. Most of the world can harvest sesame at higher moisture levels because the seed is stored in jute bags that breathe. In harvesting sesame seeds that are 6% moisture, if there are moist weed stems, leaves, and/or seeds in the bin, the moisture will transfer from the weeds to the sesame [17]. Extensive work in Venezuela to dry sesame by passing it through dryers showed that it was too expensive and that if more than 1% of moisture was removed per pass through the machines, the seed quality deteriorated to an unmarketable level (MAVESA representatives, personal communication, 1983).

- Glyphosate, carfentrazone-ethyl, and pyraflufen-ethyl are non-selective herbicides that will kill the sesame and the weeds; however, they are not desiccants. Among these, glyphosate is a better weed killer. Glyphosate should not be used for seed sesame because it can affect the germination.

- Paraquat-dichloride and diquat-dibromide are desiccants and may not kill weeds. They are based on contact with the sesame surface.

- Glufosinate-ammonium is a non-selective herbicide, but also will dry the sesame down at a comparable rate to the desiccants.

- Tolerance to some herbicides is cultivar dependent.

2. Phenology of sesame

Sesame is a survivor crop. For 5500 years it has been planted by subsistence growers in areas that will not support the growth of other crops or under very difficult growing conditions with drought and/or high heat. In some countries, it is grown after the monsoon season on residual moisture with no rains during the production stage while in other areas it is grown during the monsoon season and subject to daily rains during the growing season. In several countries, it is the last crop that can be grown at the edge of deserts where no other crops grow. Very little sesame is grown under high input conditions [18].

There are four phases in the phenology of sesame: vegetative, reproductive, ripening, and drying and there is a tremendous amount of variability in these phases [19]. Sesame is an indeterminate species, and thus, there is an overlap between the reproductive, ripening, and drying phases [18]. Since this chapter deals with sesame desiccation only the ripening and drying phases will be discussed.

Technically, as an indeterminate species, sesame is in the ripening stage from the mid-bloom stage through the full maturity stage. Sesame starts self-defoliation in late bloom stage and leaves have mostly fallen off by the initial drydown stage. As the plant stops flowering and matures, the leaves will drop starting at the bottom of the plant. In some fields, the upper leaves can remain attached providing some photosynthesis for seed fill in the upper capsules. Generally, leaves will turn yellowish green before dropping [18].

The character of leaf drop is very important for mechanical harvest and for using harvest aids. There is world germplasm where the leaves do not drop and can remain green even with a dry, open capsule in the leaf axil. Through plant breeding, the ability to self-defoliate has been incorporated into germplasm in the Americas starting in Venezuela from the mid-1940s [20].

At PM, 75% of the capsules on the main stem have seed with final color (darker than the milky white of the immature seed) as shown in **Figure 1**. Most of the seed grown for edible purposes is a light color, but there are cultivars with brown and black seed. The darker colors at PM are easy to distinguish from the milky white immature seed. Physiological mature seed also will have a brown tip where the

Figure 1.
The capsule on the bottom has seed that is milky white and is not mature. The capsule on the top has seed with final color (Photo: J. Riney).

placenta attachment has dried, and in many varieties, the capsules also may have a dark seed line on one side as shown in **Figure 2**. However, there is world germplasm where the seed line is present but not visible without a magnifying glass.

There are cultivars where the yellowish green color of the capsule will indicate that the seed inside the capsule is at PM. However, there are other cultivars where the capsules are dark green with PM seed inside, and still other cultivars where the capsules are a pale yellow with immature seed inside. Before using capsule color to tell PM, the grower must be familiar with the cultivar.

The concept of PM in sesame was developed in the 1950s (M.L. Kinman, personal communication, 1982) to determine the earliest date that the plants could be cut and still harvest over 95% of the potential yield. When the seed has the final color, the seed can germinate. If the sesame is cut at PM, most of the seed with a greater than 75% darker color will continue maturing sufficiently for germination

Figure 2.
PM seeds have a brown tip (red arrow) where the placenta attachment has dried and on one side of the seed, a seed line is visible (blue arrow) (Photo: D.R. Langham).

but may be lighter in weight. Even in a fully mature plant, the seed weight produced at the top of the plant is lower; however, this loss of seed weight does not seriously affect the potential seed yield of the plant. Physiological maturity is important in the northern US crop where there is a potential for an early frost or freeze. After PM, most of the potential yield can be harvested, even if the plants were terminated by cold temperatures. In south Texas, the rule of thumb is that PM moves up 6–7 node pairs per week below the 75% PM level, and 4–5 node pairs per week above the 75% level. At higher latitudes where night temperatures are cooler, the progress of PM is 3–4 node pairs per week below the 75% level and 1–2 node pairs above it. The threshold temperature for growing degree days of sesame is 16°C. When night temperatures go below the threshold, it takes longer for the crop to mature. Physiological maturity also delineates the earliest time that harvest aids should be applied; applying them earlier will reduce potential yield.

The drying phase is divided into three stages: full maturity, initial drydown, and late drydown. Full maturity occurs from PM until 90% of the plants have all seeds mature. This stage usually occurs from 107 to 112 days after planting and lasts for approximately 1 week. With direct harvest, without the use of harvest aids, this stage is not important. With harvest aids, the plants will be killed, and the seeds will no longer fill. At the end of this stage, the plants will have the highest potential yield and can be terminated to accelerate drydown. However, since the capsules in the top 2–3 node pairs contribute little seed, the practical time to apply harvest aids may be at some point between PM and when all seeds are mature.

The initial drydown stage occurs from the time all seeds are mature until the sesame plants have one dry capsule. This stage typically occurs 113–126 days after planting and can last for up to 2 weeks. This is a unique stage for most sesame that is grown in the Americas. In the world germplasm, a few cultivars have dry capsules with a green leaf in the same leaf axil, and many cultivars have a dry capsule when the top of the plant is still flowering. As a result, growers cut the plants to prevent seed loss and create a situation where some of the seed is mature and other seed is immature. In the Americas, the capsules do not drydown and open until flowering is complete and the majority of seed fill is complete. This character is described as "delayed shattering" (W. Wongyai, personal communication, 1998). Since sesame is basically indeterminate, the ability to stop flowering is difficult. There is germplasm that will stop flowering based on daylength; others that will stop based on cold nights; but most stop when they run out of moisture and/or fertility. The only problem with the latter germplasm is potential regrowth, which will be discussed later. A potential problem with delayed shattering is vivipary, which also will be discussed later.

The main stem will generally have dry capsules before the branches; however, the branches will generally drydown before the main stem. The lower capsules dry first with the top capsules drying last. There are some cultivars where the bottom capsules at the 2–3 node pairs drydown late even though the seed is at PM. Parts of the stem will dry before all the capsules are dry.

The late dry-down stage occurs from the time of the first capsule drydown until enough dry-down has occurred to produce 6% or less moisture seed. The first capsule drydown usually occurs 127 to 146 days after planting and the late dry-down stage can last up to 3 wks. If the reproductive phase is shorter because of a lack of fertility, the first capsule drydown will occur a shorter time from planting but will not necessarily change the length of time of the late-drydown stage. However, if the reproductive phase is shorter because of a lack of moisture, the first capsule drydown will have a shorter time from planting and a shorter length of time of the late dry-down stage.

Phase	Days from planting		Length of phase, days	
	Range	Mean	Range	Mean
Vegetative	29–59	42	29–59	42
Reproductive	56–116	89	16–70	47
Ripening	77–140	108	(14)[a]–54	11
Drying	102–181	150	11–57	38

[a]*In some lines, there are dry capsules above green leaves while the upper part of the plant is still flowering creating a negative range.*

Table 1.
Range of days in phases for world germplasm.

The above paragraphs show nominal days for one cultivar in the USA. There is a range in the world germplasm planted in the USA (**Table 1**).

Generally, if a cultivar starts flowering early, it will also stop flowering, mature, and dry early. The point is that if a crop is left in the field to drydown without harvest aids, it takes about 5–6 weeks to be dry enough for a combine to harvest. If it is in the rainy season, it could take longer.

3. Shattering nature of sesame

Due to the shattering nature of the capsules, in most of the world, the sesame needs to be cut when green and shocked (stacking in bundles) so less seed will fall out as the plants dry (**Figures 3** and **4**).

Growers will cut the plants before the capsules start opening. Over 99% of the sesame harvest in the world involves some or total manual labor to cut the sesame, and no plants are left standing to dry in the field. Therefore, most of the sesame is harvested during or at the end of the ripening phase. Once shocked, the capsules will start drying and opening. Since the sesame is stacked and in a shock, which does not bend in the wind, most of the seed will stay in the capsules with some falling out of the top. In a commercial field in Venezuela, the author examined shocks that

Figure 3.
The plants are cut before the capsules start drying even though they have leaves and are not completely mature to the top (Photo: D.R. Langham).

Figure 4.
The plants are placed in shocks to dry. The leaves shrivel quickly, but it will take about 2–3 weeks to dry, particularly in center of shock (Photo: D.R. Langham).

were about 75% dry and found that about 10–15% of the seed had fallen out. That evening, it rained about 10 mm, and two days later the outer bundles in the shock had lost 20–40% of the seed, while the inside bundles which had not been wet were still at about 10–15% loss. **Figure 5** shows a similar Mexican commercial variety after a rain. Shattering is necessary for a manual harvest (**Figure 6**) where minimal force is used to get the seed out of the capsules.

Since the 1940s, the goal in Venezuela and the USA has been to completely mechanize the sesame harvest. Leaving a shattering cultivar in the field to harvest

Figure 5.
More seed shattered out than usual with some capsules still green because of rain (Photo: D.R. Langham).

Figure 6.
Shattering is essential to release seed as in this photo. Many growers hold the sesame plant upside down and hit with an implement. Some seed is lost while drying in the shock; other seed is lost in handling the sesame from the shock to the threshing area but upon threshing 95–100% of the remaining seed is collected (Photo: N. Smith).

direct may lead to as much as 90% loss of seed as discovered in 1978–1980 with attempts to harvest shattering varieties left to dry in the field.

In the 1940s, D.G. Langham [20] harvested sesame by using a binder to cut and bind the sesame at maturity. The sesame was manually shocked and when dry, the bundles were thrown into the combine header. Langham [21–23] discussed the history of improving shatter resistance in the USA. Basically by 1982, there was enough shatter resistance to be able to swath sesame into a windrow and leave it on

Figure 7.
Improved non-dehiscent capsule. These capsules hold their seed better and will still release it in the combine. This photo was taken on Dec 17 when the crop could have been combined the first week of Oct. The seed is still in the capsules 76 days after the plants were dry enough for combining (Photo: D.R. Langham).

the ground instead of shocking it manually. The windrows were harvested with a combine equipped with a pickup attachment. By 1988, the shatter resistance had improved to the point where the sesame could be left in the field to dry and then cut. The goal had been to have the seeds stay in the capsules until the combines arrived and then release the seed in the combine with a minimum of force. Improvements continued with the development of non-dehiscent sesame [24, 25] and later improved non-dehiscent sesame [25]. The seed would stay in the capsules through wind and rain even after the plants were dry enough to combine (**Figure 7**).

4. History of sesame harvest aids

Gollifer and Radley [26] studied the use of diquat-dibromide (a desiccant) at 3.5 L ha^{-1} and endothal (a defoliant) at 8.2 L ha^{-1} using the Venezuela 51 sesame variety in Trinidad at 78 and 88 days after planting. These applications reduced yield, moisture, and oil content. For the first application there were reductions in sesame yield (35%), seed weight (12%) and oil content (14%). Also, endothal reduced germination 16%. There were no appreciable differences at the 88 day after planting application between the two herbicides. The diquat-dibromide was a good desiccant while the endothal was not an effective defoliant.

Perez and Gonzalez [27] evaluated 1% paraquat-dichloride as a harvest aid compared to the untreated check using one variety (Morada indehiscente) at three timings in Venezuela. They concluded that there was no difference in yield, but there was in oil content.

Urdaneta and Mazzani [28] studied the effect of diquat-dibromide at 1.5 and 2 L ha^{-1} with different surfactants (JF-4825, JF-4826, and Agral 90) applied 96 and 103 days after emergence (DAE), using one variety (Aceitera R) in Venezuela. They concluded that diquat-dibromide at 1.5 L ha^{-1} resulted in a higher yield than at 2.0 L ha^{-1}; yield from the 103 DAE application was greater than the 96 DAE application; there were differences in the surfactant used; harvesting plots without a harvest aid at the same time as the application of harvest aids yielded less than the application of the harvest aids; and there were no differences in germination of the seed harvested from treated and untreated plants.

Lee [29] sprayed two concentrations (0.3 and 0.5%, v/v) of diquat-dibromide on one variety in South Korea. Both concentrations reduced sesame moisture content and increased capsule dehiscence. He concluded diquat-dibromide could be used as a harvest aid to accelerate desiccation up to 2 weeks from normal field conditions.

In South Korea, Han et al. [30] applied diquat-dibromide and paraquat-dichloride at 250, 500, and 1000 ppm on sesame 3 days or 3 h before cutting the plants. The moisture content decreased rapidly while yield and germination were not different from untreated check. There were no detectable diquat-dibromide or paraquat-dichloride residues when sprayed at 250 ppm 3 days before cutting but there were higher residues for all other concentrations and for the 3 h before cutting application. There was 25% less labor required in threshing the diquat-dibromide and paraquat-dichloride treated plots as opposed to the untreated and 100% of the seed was recovered in 9 days as opposed to 12 days for the untreated.

In Australia, Bennett [31] studied the effects of diquat-dibromide applied 81, 86, and 89 days after planting (DAP) and rates (1.2, 2.2, and 4.5 L ha^{-1}) in 1992 and 1993 using the sesame variety, Yori 77. He concluded that diquat-dibromide dried the sesame quicker than natural drydown and stem moisture was higher than the seed moisture. The greener stem could be a problem in combining in that the sap from the stems could coat the seeds and give them a bad flavor. There was less sap when the stems reached 10% moisture; however, it could take an additional 7 days for the stem to reach this moisture level leading to more seed loss in the field. Also, desiccation did not have an effect on defoliation, and the ability to harvest earlier through desiccation increased seed yield. In a grower book, Bennett et al. [32] recommended a diquat-dibromide application of 3 L ha^{-1}. This application rate reduced shattering but seed loss was still evident even while waiting for sufficient drydown to spray (**Figure 8**). Also, there was still more loss when the cutter bar struck the sesame and the sesame was being moved by the auger into the feeder housing of the combine.

Bennett and Routley [33] studied diquat-dibromide at 0, 1.4, 2.6, and 5.3 L ha^{-1} and time of application (97 DAP with 40% green capsules, 103 DAP with 18% green capsules, 109 DAP with 0% green capsules, and untreated) in the 1993 and 1994 growing season using the sesame variety, Y1:44. Results indicated that application rates of 2.0 L ha^{-1} were cost effective and that time of application should be between 20 and 40% green capsules. They also concluded that desiccation was a risk management tool. Sesame naturally drying down can be exposed for 30–40 days of various weather conditions, including rains and high winds, while a desiccated crop has a drydown of 7–10 days and, therefore, less of a chance of being exposed to inclement weather conditions. Also, desiccation allows the grower to plan for contract harvesting and trucks.

Mazzani [34] summarizing sesame research in Venezuela stated after almost 30 years of research and grower experience, the recommendations were to use diquat-dibromide at 1 kg ha^{-1} at normal cutting time, which resulted in only a 15% lower yield over the untreated. Applying diquat-dibromide at 2 kg ha^{-1} a week earlier than normal cutting time resulted in 97% seed loss due to immature seeds. The desiccation system is 4–8% of the cost of traditional system of cutting, binding, and shocking. Mazzani (personal communication, 1999) said that the timing of the herbicide application required grower experience of the sesame maturity stage based on capsule color and variety instead of number of days from emergence. Also, combining after using diquat-dibromide could be a bit earlier than the traditional method. Using the traditional shocking method, the inner bundles in the shocks take longer to drydown resulting in high risk, since the start of the monsoon season can occur while the sesame is still in shocks. One of the major advantages of using desiccants is that there is less manual labor available at harvest for cutting and shocking. On the negative side, the use of desiccants may produce less yield than the traditional method, particularly if there is a wind during drydown, which may cause

Figure 8.
Sesame on the ground even before combining from using dehiscent varieties (Photo: M. Bennett).

Figure 9.
(a) and (b) Spraying of mature sesame with diquat-dibromide to drydown the sesame for direct harvest (Photos: S. Araujo).

Figure 10.
Combining sesame in various countries. US (Photo: J. Riney), China (Photo: H.Y. Zhang), Brazil (Photo: V. Queroga). (a) USA; (b) China; (c) Brazil.

some seed to shatter or in the case of severe winds most of the seed to shatter. Some growers (L. Jimenez, personal communication, 1998) who have large plantings use both methods since all the fields may be ready for harvest at the same time and there is not enough labor to cut all the sesame at the same time. However, in years when there is asynchronous maturity and labor is available, they prefer cutting instead of applying a harvest aid.

Currently in Venezuela, Araujo [35] reported that two applications of diquat-dibromide at 2 kg ha^{-1} are applied a week apart (**Figure 9a** and **b**).

In the USA, glyphosate has been approved for use as a harvest aid by the Environmental Protection Agency (EPA) and it is used extensively for harvest in the southern USA (**Figure 10a**). In the northern latitudes, glyphosate is not as effective because of possible low temperatures at harvest. In addition, in the northern areas, there are normally frosts or freezes that accelerate drydown.

P. Bazyar (personal communication, 2018) reported diquat-dibromide is currently used as a harvest aid in Iran. Diquat-dibromide is sprayed when 65–75% of the capsules have changed color from dark green to bright green and the crop is harvested with a wheat combine. H.M. Miao (personal communication, 2019) reported either paraquat-dichloride or ethylene (1.5 kg ha^{-1}) are used as a harvest aid in direct harvest sesame in Xinjiang Province in China (**Figure 10b**). V. Queiroga and N. Arriel (personal communication, 2019) reported paraquat-dichloride or diquat-dibromide are used as a harvest aids in direct harvest sesame in Mato Grosso, Brazil (**Figure 10c**).

5. Rationale for harvest aids in the USA

In the USA, all the sesame is mechanically harvested with a combine and the use of a harvest aid can help facilitate harvest in most cases. In northern latitudes, with cool evenings, harvest aids are slower acting and a frost or freeze will dry down the sesame just as effectively as a harvest aid. There are six reasons to consider a harvest aid for use in sesame:

1. **Accelerate drydown.** This is basically using the idea that no weather event can increase yield after the crop is mature, but there are many weather events that can reduce yield. Natural drydown averages 5–6 weeks and inclement weather can prevent a combine from entering a field for as many as 8–9 weeks. With harvest aids, the drydown is 1–2 weeks depending on heat and humidity. In the USA, sesame is grown from the southern-most tip of Texas to the Kansas border. Planting starts in late March in the south and ends in early July in the north while harvest starts in late August in the south and ends in December in the north. Into the fall and early winter, the daylength shortens and temperatures cool with a pattern in most areas of increased rainfall. In south Texas, harvest aids will accelerate the drydown and avoid poor weather (particularly the threat of hurricanes along the Texas coast). In the northern part of Texas, Oklahoma, or Kansas, frost will generally kill sesame without having to use harvest aids. There is a gray zone in between where in some years, harvest aids are used and in other years the growers will wait for the frost, particularly on late planted sesame.

2. **Burn-down green weeds.** Green weeds can add moisture to the sesame seed in the combine bin. As stated above, moisture needs to be around 6% at harvest. Sesame can be dry and yet in the combine bin and the truck, moisture from the weeds can be absorbed by the sesame seed. The two major weed problems are *Amaranthus* spp. and *Echinochloa crus-galli*. *Amaranthus* spp. has a lot of moisture in the stems while green *E. crus-galli* seeds will not blow out the back of the combine and will go into the bin with a remarkable amount of moisture. Certain weeds can be a serious problem in processing sesame. A large percentage of any weed seed can be a problem when cleaning sesame even though it may be as small as *Amaranthus* spp. or as large as the seed of *Ipomoea* spp. Certain weed seeds such *Sorghum halepense*, *Solanum rostratum*, *Coreopsis* spp., *Salsola tragus, Salvia reflexa, Chenopodium album*, and *Kochia scoparia* are more of a problem, even in small quantities. No one wants any seed, other than sesame, on top of their hamburger bun. Although much of *Sorghum halepense* seed is a different size compared to sesame, the elongated seeds of *Sorghum halepense* can go through the round holes of the cleaner head first. Once through the cleaner, they have a specific gravity close to sesame seed and cannot be easily separated on the gravity table. Finally, a portion will be close enough in color to sesame that it will not be separated by the color sorter. *Solanum rostratum* is the same size and specific gravity as sesame and the seed may not be cleaned out of the sesame. *Salvia reflexa* is a special problem in that when it contacts water, the seed coat will swell into a sticky gelatinous substance that will clump sesame seeds together. As mentioned by Bennett [31], sap from some weed stems can coat the seeds and give them a bad flavor.

3. **Even up a field with varying levels of drydown.** Many fields have low spots that can still be green while high spots are drying. The difference between the stage of the driest and the stage of the greenest and the proportion between the dry and the green must be considered. In most cases, the "green portions" are 1–2 weeks behind the 'dry portions". In this case, the dry portions should go past PM until the capsules start drying down to allow the green portions to mature further increasing the yield potential. However, if the differences are

more than 2 weeks, the field should be harvested at different times. On the other hand, if the proportion of green is very small, it is not practical to harvest at different times, and the yield loss in the green portion should be acceptable.

There is no set formula to determine the solution. Like most farming decisions, it is up to the judgement of the grower. In some cases, some seed maybe planted in dry soil and only germinate after a rain resulting in two growth stages. The decision to harvest is therefore not clear. A rain 1–2 weeks after planting is different from a rain 4–6 weeks after planting. As stated above, the proportion is also important. If most of the field is of late germination, the early plants should be ignored. In a higher proportion of early germination, the older plants will have a more aggressive root system, shading the younger plants, and therefore the younger plants will act like weeds. The younger plants will not produce a substantial amount of seed and should be ignored in the decision as to when to use harvest aids. Using glyphosate on uneven fields will kill all the sesame and the stage of seed fill will be frozen. Killing sesame while leaves are still attached does cause some problems with dry leaves (particularly the petioles) being broken up and entering the combine bin as foreign matter.

4. **Stop regrowth.** Certain sesame varieties have a propensity for restarting growth and flowering after the main stem has stopped flowering. Regrowth usually occurs in areas where conditions are such that the plants have run out of moisture and/or fertility. If there is rain, some varieties will form branches at the bottom of the plant and these will flower and set capsules while the main stem and the older branches will not start flowering again (**Figures 11** and **12**). There are three types of regrowth: top (restarts at the tops of the main stems), middle (branches emerge from the middle of the main stem), and bottom (branches start in the axils of other branches or below the branches). There are varieties that show spontaneous branching whereby branches start in the middle of the capsule zone under capsules. However, this is not considered regrowth because spontaneous branches begin during flowering and stop flowering before the top of the main stem stops flowering. In regrowth,

Figure 11.
Bottom regrowth (red arrow) below the lowest capsule (yellow arrow).

Figure 12.
Middle regrowth (red arrow) in the middle of the capsule zone. Note: the sesame without regrowth is dry. Both of these are minor regrowth which has already stopped flowering. In some cases, there may be over 20 capsules on each branch.

capsules on the original plant will dry down while it may take another 60 days for the regrowth capsules to mature and dry out. Since most of the yield is in the old growth, it can shatter while waiting for the regrowth to dry to be combined. Combining when the original plant is dry and the regrowth is green is similar to the weed situation in that it will introduce moisture into the combine bin. Using glyphosate as a harvest aid will kill the plant and stop the regrowth.

Figure 13.
In many areas of the world, seed will germinate inside the green capsule when the temperatures are high. Even if only the top seeds germinate, the roots from the seedlings will trap the seeds below, preventing them from being harvested (Photo: D.R. Langham).

5. **Stopping vivipary.** Under some conditions, there is vivipary in sesame—the seeds will germinate in the capsules (**Figure 13**). Not only are the germinated seeds lost, but the roots of the seedlings bind the rest of the seed and keep it in the capsule when combined.

   Many growers have felt that the opening of the capsules allows water to enter and germinate the seed. The opposite occurs. Seeds in open capsules do not germinate because the moisture will evaporate out of the capsule before the seed can germinate. Vivipary occurs in closed capsules. It is believed that this is a dispersal mechanism to open the capsule and allow the seed to fall out. The exception in dry capsule is if the tip of the capsule has a minimum opening. Not only can there be vivipary, but there is also the danger of mold forming inside the dry capsules. When there is vivipary, the seedlings persist as the capsules continue to dry. Vivipary is controlled genetically through seed dormancy with some varieties having a greater propensity than others. Vivipary is rare in the USA because normally at harvest the night temperatures are below 15°C, the minimum germination temperature. Glyphosate will kill the sesame plants and stop vivipary.

6. **Prepare to plant a new crop**. In many areas of the USA, wheat and sesame are double cropped. The earlier the sesame is harvested, the earlier the wheat can be planted.

6. Harvest aid research in the USA

The first testing in the USA was done by D.M. Yermanos (personal communication, 1982) in California in the mid-1970s. He used paraquat-dichloride; however, the varieties at the time were shattering and leaving the plants standing even for 2 weeks longer resulted in lower yields than cutting and drying in shocks. In addition, a paraquat-dichloride application at harvest would not be acceptable in the US food market since there would be the potential to get the paraquat-dichloride on the seed in an open capsule. Paraquat-dichloride, a restricted use pesticide, is considered to have oral, dermal, and inhalation toxicity to humans (skull and cross bone symbol) [36]. In 1975, St. Andre et al. [37] reported that diquat-dibromide, either alone or in combination with products called 'Bolls Eye' or 'Dinitro", had potential to serve as dessicants in sesame.

There was no additional harvest aid research until 2003 after the 2002 harvest season showed that weather had significantly affected sesame yield in some areas. Getting the crop out of the field earlier became a priority. Many felt that sesame should follow the cotton harvest pattern of first defoliating the plants and then drying the sesame down. In 2003, test strips in a nursery were sprayed with paraquat-dichloride, diquat-dibromide, glyphosate, carfentrazone-ethyl, and pyraflufen-ethyl when the plants had stopped flowering but still had their leaves. Paraquat-dichloride and diquat-dibromide caused spotting on the leaves where the drops contacted the plants; however, this did not kill the sesame. The amount of spotting was greater at the top of the plant than at the bottom. Glyphosate killed the sesame but retained its leaves and did not mature the seed in the upper capsules. Carfentrazone-ethyl and pyraflufen-ethyl produced no visible effect.

At this point in time, it was thought that the sesame plant had to have the leaves to allow the herbicide (particularly glyphosate) to be absorbed and translocated to the roots. However, sesame is different from cotton in that there is sesame germplasm that will allow plants to self-defoliate as the capsules mature. Parallel to this, a grower sprayed glyphosate on a sesame crop with no leaves to eliminate volunteer sorghum and weeds within the crop and the glyphosate also killed the sesame without leaves. It was postulated that the capsules and stems absorbed the glyphosate and it translocated to the roots. In addition, the lack of leaves allowed the glyphosate to reach the lower stem and capsules. Subsequent testing was done at PM when the leaves had dropped naturally.

With the above field observations, research was conducted to answer some questions. Previous experience, when sesame was swathed at maturity, had shown that if the sesame was cut earlier than PM, yield and quality of the seed would be reduced. Other experience had shown that even though sesame had been bred with improved non-dehiscence, weather could reduce the yield. Therefore, the objective of this research was to identify herbicides that may help promote desiccation of sesame and the effect of these herbicides on sesame development and also determine the optimum application timing to determine the amount of loss if the harvest aid was applied too early or too late.

7. Field studies

The weather was variable and it was very difficult to predict when PM would occur to schedule the sprayings. As mentioned previously, PM proceeds between 1 and 7 node pairs per week depending on the weather. The variability can be exacerbated when planting sesame under pivots. In several years of successive droughts, the moisture below 30 cm was depleted and the soil was very dry; therefore, no roots could penetrate the soil. Water in Texas can be scarce and when irrigating with an overhead irrigation system pivot, the minimum amount necessary is used. The results are roots that are often in the top 20–30 cm of soil. Irrigation is terminated when the crop stops flowering. If there is hot weather with low humidity, the sesame can go to drydown in less than 2 weeks instead of the normal 5–6 weeks. One harvest aid experiment in Uvalde was cancelled because the plants were in the late drydown stage at the predicted 1 week before PM. In another study, near Lorenzo, Texas, the crop was sprayed at least 2 weeks before PM because a cold front had delayed PM and the plants still had their leaves.

7.1 Materials and methods

7.1.1 Research sites

Field studies were conducted from 2006 through 2008 near Uvalde (29.468° N, 99.7061° W) in the south Texas sesame growing region and near Lorenzo (33.6684° N, 101.5354° W) in the Texas High Plains sesame growing region to evaluate sesame response to harvest aids. Soil type at Uvalde was a Winterhaven silty clay loam (fine-silty, carbonatic, hyperthermic Fluventic Ustochrepts) with less than 1.0% organic matter and pH 7.8. Soil type at Lorenzo was an Amarillo sandy clay loam (fine loamy, mixed, thermic Aridic Paleustalf) with 0.8% organic matter and pH 7.8.

7.1.2 Herbicides and application

A randomized complete-block experimental design was used, and treatments were replicated three times. Treatments consisted of a factorial arrangement of four herbicide treatments (carfentrazone-ethyl, diquat-dibromide, glufosinate ammonium, and glyphosate) at two rates. A non-treated control was included for comparison. Herbicides included glyphosate (Durango® Herbicide,

Dow AgroSciences, 9330 Zionsville Rd., Indianapolis, IN 46268 in south Texas and Roundup Weathermax, Bayer CropScience, 800 N. Lindberg Blvd., St. Louis, MO 63167 in the Texas High Plains) at 0.63 and 0.84 kg ae ha^{-1}; diquat-dibromide (Reglone® Herbicide, Syngenta Crop Protection, Inc., P.O. Box 18,300, Greensboro, NC 27409) at 0.28 and 0.56 kg ai ha^{-1}; glufosinate-ammonium (Bayer CropScience) at 0.47 and 0.58 kg ai ha^{-1}; and carfentrazone-ethyl (FMC, 1735 Market St., Philadelphia, PA 19103) at 0.018 and 0.035 kg ai ha^{-1}. Carfentrazone-ethyl was only evaluated in 2006 and 2008 at Lorenzo.

These herbicides were applied prior to physiological maturity (PRE PM), at physiological maturity (PM), or after physiological maturity (POST PM). The PRE PM treatment was not applied in 2006 at Lorenzo or 2007 at Uvalde due to sesame development that was farther alone than anticipated. At Lorenzo, the PRE PM treatments were applied 99 and 103 DAP in 2007 and 2008, respectively; PM treatments applied 97, 105, and 112 DAP in 2006, 2007, and 2008, respectively; and POST PM treatments applied 113, 112, and 121 DAP in 2006, 2007, and 2008, respectively. At Uvalde, the 2006 treatments were cancelled because drydown developed much faster than anticipated. The PRE PM treatments in 2008 were applied 104 DAP; PM treatments were applied 102 and 110 DAP in 2007 and 2008, respectively; and POST PM treatments were applied 114 and 119 DAP in 2007 and 2008, respectively.

In the original concept in designing the experiment, herbicide sprays would be done within 2 weeks: the PRE PM followed by PM a week later followed by POST PM a week later. However, scheduling and weather problems prevented the exact 1 week intervals. Sprayings were postponed when the sesame was wet or rains were imminent.

Plot size was five rows (76 cm apart) by 9.1 m in south Texas (Uvalde) and four rows (101 cm apart) by 7.3 m in the Texas High Plains (Lorenzo). Only the two middle rows were sprayed and the other rows were untreated and served as buffers.

Herbicides were applied in water using a CO_2-pressurized backpack sprayer with either Teejet 11002 DG, TeeJet® 11002DG, or Turbotee 110015 flat fan spray tips (Teejet Spraying Systems Co., P.O. Box 7900, Wheaton, IL 60188) nozzles calibrated to deliver 190 L/ha at 180 kPa at Uvalde and Turbotee 110015 calibrated to deliver 140 L/ha at 207 kPa at Lorenzo.

7.1.3 Sesame plantings, observations, and harvest

The sesame variety 'S29' was planted at Lorenzo in 2006; however, in 2007 and 2008 at both locations 'S32' was grown since it was the main variety used in all sesame growing areas of the US at that time [17]. Planting dates at the Uvalde location were June 2 in 2007 and May 30 in 2008 while at the Lubbock location, sesame was planted May 26 in 2006, June 18 in 2007 and June 26 in 2008. Each sesame cultivar was seeded approximately 2.0 cm deep at 9 kg/ha at both locations.

Sesame drydown was evaluated 7–14 days after herbicide application based on a scale of 0 (no drydown, plants still green) to 100 (complete drydown, no green tissue). Due to the extremely dry conditions at Uvalde in 2008, an accurate assessment of drydown could not be obtained. Plants in each plot were hand-cut, bagged, and threshed with a harvester to obtain sesame yield. Sesame yields were not obtained from the PRE PM and PM diquat-dibromide and glufosinate-ammonium treatments at Uvalde in 2008 because they had been dry long before the other treatments and comparisons were invalid. One of the problems with the methodology was that all plots were harvested at the same time regardless of treatment drydown (other than 2008 as stated above).

7.1.4 Data analysis

Data for percentage of sesame drydown were transformed to the arcsine square root prior to analysis; however, non-transformed means are presented because arcsine transformation did not affect interpretation of the data. Data were subjected to analysis of variance (ANOVA) and analyzed using SAS PROC MIXED with locations and years designated as random effects in the model. A mixed model was chosen because the actual environments experienced at both locations in Texas where the experiments were conducted are unlikely to occur again in the future. Allowing the four environments to be random allows estimates of treatment responses to be made over a range of environments. Treatment means were separated using Fisher's Protected LSD at $P \leq 0.05$. The untreated check was used for yield comparison and a visual comparison for sesame drydown and was included in the drydown and yield data analysis.

8. Results

Since weather conditions and sesame development varied from year-to-year and across locations no attempt was made to combine data across years or locations. Also, in some years (2006 at Lorenzo, 2007 at Uvalde), the PRE PM stage occurred sooner than anticipated and a timely herbicide application was not made. Since drydown in 2008 at Uvalde occurred so quickly due to the extreme drought conditions, an accurate assessment of sesame drydown could not be obtained. An herbicide treatment by application timing interaction was noted in each year for drydown and sesame yield; therefore, data are presented separately for herbicide treatment and application timing.

8.1 Sesame drydown

8.1.1 Lorenzo

In 2006, when the PM applications were compared, diquat-dibromide and glufosinate-ammonium accelerated drydown more than glyphosate or carfentrazone-ethyl (**Table 2**). With the POST PM timing, a rate effect was noted with carfentrazone-ethyl, diquat-dibromide, and glyphosate but not glufosinate-ammonium. The high rates of glyphosate and diquat-dibromide improved drydown 44–54% while the high rate of carfentrazone-ethyl improved drydown 29% when compared with the lower rates while virtually no difference was noted between glufosinate-ammonium rates. The high rate of glyphosate resulted in the greater drydown than other herbicides. All treatments improved drydown over the untreated check. In 2007 at the PRE PM and PM stage, diquat-dibromide at 0.28 and 0.56 kg ai ha^{-1} accelerated sesame drydown greater then glufosinate-ammonium at 0.47 and 0.58 kg ai ha^{-1}, which was greater than glyphosate at 0.63 and 0.84 kg ae ha^{-1}; however, the high rate of glyphosate and both rates of glufosinate ammonium performed similarly at the POST PM stage (**Table 2**). Diquat-dibromide at either rate applied POST PM provided ≥87% sesame drydown. Sesame drydown with the untreated check was ≤39% at all application stages.

In 2008 at the PRE PM stage, diquat-dibromide at 0.28 and 0.56 kg ai ha^{-1} and glufosinate-ammonium at 0.47 kg ha^{-1} produced 63–80% drydown while both rates of glyphosate and the untreated check resulted in <10% drydown (**Table 2**). Carfentrazone-ethyl at 0.018 and 0.035 kg ai ha^{-1} was intermediate in sesame

	Carfentrazone-ethyl (kg ai ha^{-1})		Diquat-dibromide (kg ai ha^{-1})		Glufosinate-ammonium (kg ai ha^{-1})		Glyphosate (kg ae ha^{-1})		
Stage[a]	0.018	0.035	0.28	0.56	0.47	0.58	0.63	0.84	Check
					%				
Lorenzo									
2006									
PRE PM	—	—	—	—	—	—	—	—	—
PM	12	12	24	28	26	26	15	15	0
POST PM	28	36	26	40	44	46	36	52	16
LSD (0.05)					6				
2007									
PRE PM	—	—	78	84	65	73	12	7	4
PM	—	—	67	74	62	66	13	13	11
POST PM	—	—	87	93	60	59	38	56	39
LSD (0.05)					8				
2008									
PRE PM	14	18	67	80	63	57	4	7	1
PM	18	16	63	70	45	48	14	16	10
POST PM	88	94	97	98	96	98	86	94	28
LSD (0.05)					6				
Uvalde									
2007									
PRE PM	—	—	—	—	—	—	—	—	—
PM	—	—	67	67	70	73	53	57	50
POST PM	—	—	80	80	80	77	67	70	73
LSD (0.05)					7				

[a]*PRE PM, prior to physiological maturity; PM, physiological maturity; POST PM, after physiological maturity.*

Table 2.
Sesame drydown (7–14 d after treatment) at Lorenzo and Uvalde as influenced by herbicides.

drydown. At the PM stage, diquat-dibromide at either rate produced the greatest drydown (63–70%), glufosinate-ammonium was intermediate (45–48%) and carfentrazone-ethyl and glyphosate resulted in ≤18% drydown. At the POST PM stage, the high rates of carfentrazone-ethyl and glyphosate and both rates of diquat-dibromide and glufosinate-ammonium resulted in ≥94% drydown while the untreated check showed only 28% drydown.

8.1.2 Uvalde

At the PM stage, diquat-dibromide and glufosinate-ammonium resulted in 67–73% sesame drydown while glyphosate drydown was no more than 57% and no better than the untreated check (**Table 2**). At POST PM no herbicide treatment was better than the untreated check.

8.2 Sesame yield

8.2.1 Lorenzo

In 2006 at the PM stage, diquat-dibromide at 0.28 kg ai ha^{-1} and glufosinate-ammonium at 0.58 kg ai ha^{-1} produced yields that were lower than the untreated check (**Table 3**). None of the other herbicide treatments produced yields that were different from the untreated check. At POST PM, all sesame yields were lower than the untreated check and no differences were noted between herbicide treatments.

	Carfentrazone-ethyl (kg ai ha^{-1})		Diquat-dibromide (kg ai ha^{-1})		Glufosinate-ammonium (kg ai ha^{-1})		Glyphosate (kg ae ha^{-1})		
Stage[a]	0.018	0.035	0.28	0.56	0.47	0.58	0.63	0.84	Check
Lorenzo									
(kg ha^{-1})									
2006									
PRE PM	—	—	—	—	—	—	—	—	
PM	650	583	466	525	554	408	670	637	
POST PM	451	412	247	448	448	417	433	327	717
LSD (0.05)					250				
2007									
PRE PM	—	—	1282	1345	1364	1298	1058	1132	
PM	—	—	1357	1383	1409	1361	1245	1139	
POST PM	—	—	1195	1325	1151	1158	1280	1049	1091
LSD (0.05)					191				
2008									
PRE PM	785	890	818	890	946	773	670	818	
PM	753	795	998	964	773	868	762	841	
POST PM	863	829	751	1002	860	836	874	778	851
LSD (0.05)					197				
Uvalde									
2007									
PRE PM	—	—	—	—	—	—	—	—	
PM	—	—	1535	1657	1562	1602	1645	1444	
POST PM	—	—	1473	1506	1585	1475	1547	1506	1645
LSD (0.05)					NS				
2008									
PRE PM	—	—	—	—	—	—	1344	1330	
PM	—	—	—	—	—	—	1406	1352	
POST PM	—	—	1310	1304	1350	1390	1432	1562	1430
LSD (0.05)					217				

[a]*PRE PM, prior to physiological maturity; PM, physiological maturity; POST PM, after physiological maturity.*

Table 3.
Sesame yield at Lorenzo and Uvalde as influenced by drydown herbicides.

In 2007 at the PRE PM stage, diquat-dibromide at 0.56 kg ha^{-1} and both rates of glufosinate-ammonium produced yields that were greater than the untreated check (**Table 3**). At the PM stage, both diquat-dibromide and glufosinate-ammonium increased yield over the untreated check while glyphosate was not different than the untreated check. At the POST PM stage, only diquat-dibromide at 0.56 kg ha^{-1} increased yield over the untreated check.

In 2008 at any sesame development stage, none of the herbicides improved yields over the untreated check. However, at the POST PM stage, diquat-dibromide at 0.56 kg ha^{-1} improved sesame yield 29–34% over diquat-dibromide at 0.28 kg ha^{-1} and glyphosate at 0.84 kg ha^{-1} (**Table 3**).

8.2.2 Uvalde

In 2007 or 2008 no differences in sesame yield was noted between the untreated check and any herbicide treatment (**Table 3**). In 2008, glyphosate at 0.84 kg ae ha^{-1} applied POST PM did result in greater yield than glyphosate at either rate applied PRE PM or diquat-dibromide at both rates applied POST PM.

9. Discussion

Since the 1950s study by M. Kinman (personal communication, 1982), there have been no studies to try to define PM. Those studies were based on cultivars that had dry, open capsules at the bottom of the plant while the top was not mature. His criterion was to find the point of maximum yield knowing there is an offset of the weight gained by more seed fill versus the weight loss by shattering. Current cultivars do not have dry, open capsules until after the tops of the plants have completed seed fill. Testing should show the amount of yield lost by spraying 1, 2, 3, and 4 weeks ahead of PM. There is no credible way to show the loss by spraying 1, 2, 3, and 4 weeks after PM because the weather cannot be controlled. Rain, fog, dews, temperature, relative humidity, and wind affect the amount of shattering and lodging. In doing this type of study it is important to test all the phenotypes. Sesame phenotypes can be classified by the amount of branching (no branching, few branches, or many branches) and the number of capsules per leaf axil (single or triple capsule). Over 30 lines were segmented into lower main stem, middle main stem, upper main stem, lower branches, and upper branches with the percentage of seed weight in each of the segments depending on phenotype (**Table 4**). The branches complete flowering and seed fill before the main stem [18]. If one phenotype is chosen for the study and the phenotype does not have branches, there will

Branch	Caps	LMS[1]	MMS	UMS	MS	LBR	UBR	BR
				%				
None	1	31.2	37.1	31.7	100.0			
Few	1	27.1	28.8	23.7	79.6	9.9	10.5	20.4
Many	1	22.4	23.5	17.7	63.6	15.9	20.5	36.4
None	3	25.5	41.8	32.7	100.0			
Few	3	19.1	32.5	26.9	78.5	10.6	10.9	21.5

[1]*Abbreviations: BR, branches; LMS, lower main stem; MMS, middle main stem, UMS, upper middle stem; MS, main stem; LBR, lower branches; UBR, upper branches.*

Table 4.
Percentage of seed weight in segments of the plants.

Reasons for harvest aid	Glyphosate (GLY)	Glufosinate-ammonium (GLU)	Carfentrazone-ethyl (CAR)	Diquat-dibromide (DIQ)	Untreated (UNT)
Accelerate drydown	DIQ > GLU > GLY > CAR				No
Burn-down green weeds[a]	GLY > GLU > CAR			No	No
Evening up a field with varying levels of drydown	Yes	Not tested, probably yes	Needs to be tested	No	No
Stopping regrowth	Yes	Not tested			No
Stopping vivipary	Yes	Not tested			No
Preparing to plant a new crop	Yes	Yes	Not sure	Yes	No

[a]*There is an exception with herbicide tolerant weeds. There are also some weeds that are killed more effectively by one herbicide, e.g.* Ipomoea *spp. with carfentrazone-ethyl.*

Table 5.
Performance of herbicides on sesame harvest aids requirements.

probably be a shorter cutting window before there is substantial yield loss by cutting early as opposed to a phenotype with many branches.

These studies led to the US EPA approval of glyphosate as a harvest aid. As a result, glyphosate has been used as a harvest aid on hundreds of thousands of hectares in the last 10 years. Past experience plus these experiments provide an idea of how these herbicides satisfy the requirements for harvest aids for sesame (**Table 5**).

10. Conclusions

The use of herbicides in sesame did accelerate drydown although this did not always result in an increase in yield over the untreated check. However, there was no weather event such as rain, fog, dews, or strong winds during the experiment which would have an impact on the results. Diquat-dibromide and glufosinate-ammonium dried the sesame faster than glyphosate and carfentrazone-ethyl.

With all herbicides, the higher rate dried the sesame down faster.

Efforts should be continued to try to persuade manufacturers to support for registration of diquat-dibromide and glufosinate-ammonium. Even though carfentrazone-ethyl is not as effective as glyphosate, with the present legal battles with glyphosate as a carcinogen, efforts should be made to register carfentrazone-ethyl. Stopping regrowth and vivipary is not an experiment that can be set up because they do not occur in most of the fields or nurseries. However, when a field is found with regrowth and/or vivipary, glufosinate-ammonium, carfentrazone-ethyl, and diquat-dibromide should be tested and glyphosate should be added for comparison purposes.

Acknowledgements

Sesaco Corporation provided funding for this research.

Conflict of interest

There is no conflict of interest with any of the authors.

Author details

William James Grichar[1]*, Peter A. Dotray[2] and Derald Ray Langham[3]

1 Texas A&M AgriLife Research, Corpus Christi, TX, USA

2 Texas A&M AgriLife Research, Lubbock, TX, USA

3 Sesame Research LLC, San Antonio, TX, USA

*Address all correspondence to: w-grichar@tamu.edu

References

[1] Griffin JL, Boudreaux JM, Miller DK. Herbicides as harvest aids. Weed Technology. 2010;**58**:355-358

[2] Boudreaux JM, Griffin JL, Etheredge LM Jr. Utility of harvest aids in indeterminate and determinate soybeans. Proceedings of the Southern Weed Science Society. 2007;**60**:91

[3] Philbrook BD, Oplinger ES. Soybean field losses as influenced by harvest delays. Agronomy Journal. 1989;**81**: 251-258

[4] Fromme D, Price OPIII, Stephenson DO. Utilizing Harvest Aids in Grain Sorghum. Louisiana State University Ag Center; 2017. Available from: http://www.lsuagcenter.com/profiles/aiverson/articles/page1503673878799 [Accessed: 22 September 2018]

[5] Trostle C, McGinty J. Texas Grain Sorghum Weed Control & Harvest Desiccation Guide. Texas AgriLife Extension; 2016. Available from: texassorghum.org/2016-texas-grain-sorghum-weed-control-harvest-desiccation-guide.html [Accessed: 08 October 2019]

[6] Bean B. Sorghum and the use of a harvest-aid product. In: United Sorghum Checkoff Program. 2017. Available from: http://www.sorghumcheckoff.com/news-and-media/newsroom/2017/09/15/sorghum-and-the-use-of-a-harvest-aid-product/ [Accessed: 22 September 2018]

[7] Armstrong J. Harvest aid weed control in wheat. In: Production Technology Report. PT 2009-2 21(2). Oklahoma State University Extension. 2009. Available from: http://weedscience.okstate.edu/4-h/small-grains/PT2009-2-%20Harvest%20Aid%20Weed%20Control%20in%20Wheat.pdf/ [Accessed: 23 September 2018]

[8] Johnson B, Nice G, Bauman T. Harvest aid herbicides for winter wheat. Purdue Weed Science. 2004. Available from: ag.purdue.edu/btny/weedscience/documents/harwheat03.pdf [Accessed: 07 October 2019]

[9] Jordan D. Peanut weed management. In: Peanut Information Handbook. University of North Carolina Extension; 2018. Available from: content.ces.ncsu.edu/peanut-information/peanut-weed-management [Accessed: 07 October 2019]

[10] Chaudhari S, Jordan D, Jennings K. Peanut (*Arachis hypogaea* L.) response to carfentrazone-ethyl and pyraflufen-ethyl applied close to harvest. Peanut Science. 2017;**44**:47-52

[11] Grichar WJ, Dotray PA, Baughman TA. Peanut variety response to postemergence applications of carfentrazone-ethyl and pyraflufen-ethyl. Crop Protection. 2010;**29**: 1034-1038

[12] Hardke J. Harvest Aids on Hybrid Rice. University of Arkansas Research and/Extension; 2018. Available from: http://www.arkansas-crops.com/2018/09/06/harvest-aids-hybrid/ [Accessed: 25 September 2018]

[13] Anonymous. Harvest management. In: Canola Encyclopedia. 2018. Available from: https://www.canolacouncil.org/canola-encyclopedia/managing-harvest/harvest-management/ [Accessed: 21 September 2018]

[14] Dodds DM, Reynolds DB, Barber LT, Raper TB. Mid-South Cotton Defoliation Guide. Mississippi State University Extension. University of Arkansas Extension and University of Tennessee Extension; 2016. Available from: https://extension.tennessee.edu/publications/Documents/W376.pdf [Accessed: 23 September 2018]

[15] Zotarelli L, Sargent S, Dittmar P, Makani M. Potato Vine Killing or Desiccation. University of Florida Extension HS925; 2016. Available from: https://edis.ifas.ufl.edu/pdffiles/HS/HS18100.pdf [Accessed: 23 September 2018]

[16] Fleury D. Chickpea harvest management. In: Saskatchewan Pulse Growers. Undated. Available from: https://saskpulse.com/files/general/150729_CHICKPEA_harvest_management.pdf [Accessed: 21 September 2018]

[17] Langham DR, Riney J, Smith G, Wiemers T. Sesame Growers Guide. 2008. Available from: baylor.agrilife.org/files/2011/05sesamegrowers guide2008.pdf [Accessed: 08 October 2019]

[18] Langham DR. Growth and Development of Sesame. American Sesame Grower Association; 2008. Available from: researchgate.net/publication/265308920_GROWTH_AND_DEVELOPMENT_OF-SESAME [Accessed: 07 October 2019]

[19] Langham DR. Phenology of sesame. In: Janick J, Whipkey A, editors. Issues in New Crops and New Uses. Alexandria, VA: ASHS Press; 2007. pp. 144-182

[20] Langham DG, Rodriguez M. El ajonjoli (*Sesamum indicum* L.): su cultivo, explotacion, y mejoramiento. (Spanish) [Sesame: Its cultivation, exploitation, and improvement]. Publicacion del Ministerio de Agricultura y Cria. 1945;**2**:132

[21] Langham DR. Shatter resistance in sesame. In: Zanten LV, editor. Sesame Improvements by Induced Mutations, Proceedings Final FAO/IAEA Co-ord. Research Meeting. IAEA. TECDOC-1195. 2001. pp. 51-61

[22] Langham DR, Wiemers T. Progress in mechanizing sesame in the US through breeding. In: Janick J, editor. Trends in New Crops and New Uses. Alexandria, VA: ASHS Press; 2002. pp. 157-173

[23] Langham DR. Mechanization of sesame (*Sesamum indicum* L.)—A family's life journey. ResearchGate.net. 2019:122

[24] Langham DR. Method for making non-dehiscent sesame. In: United States Patent 6,100,452. 2000. p. 22

[25] Langham DR. Non-dehiscent sesame. In: United States Patent 8,080,707. 2011. p. 21

[26] Gollifer DE, Radley RW. Experiments to determine the effect of defoliation and desiccation on pod maturation, seed yield, and oil content of sesame. In: Annual Report. Department of Crop Production, University of the West Indies; 1967, 1967

[27] Perez PN, Gonzalez G. Uso de desecantes en ajonjolí indehiscente (*Sesamum indicum*). (Spanish). Use of desiccants on indehiscent sesame. Jornadas Agronómicas. 1970;**1**(7)

[28] Urdaneta R, Mazzani B. Efecto de la aplicacion de Reglone sobre el secado y rendimiento del ajonjoli. (Spanish). Effects of Reglone application on the drydown and yield of sesame. Agronomia Tropical Maracay. 1978; **28**(5):463-471

[29] Lee HJ. Sesame seed maturity, determination of harvesting time, and effect of chemical desiccation. In: Agronomy Abstracts. 72nd annual meeting, American Society of Agronomy. 1980. p. 101

[30] Han SS, Park KH, Yoo CH, Lee YK. Investigation on the use of paraquat and diquat as a desiccant for sesame harvest-aid. Korean Journal of Weed Science. 1993;**313**(1):7-13

[31] Bennett MR. Effect of time and rate of Reglone application on desiccation of sesame. In: Sesame Research Report, 1992–1993 Wet Season. Technical Report No. 218. Katherine: Department of Primary industry and Fisheries; 1994, 1994. pp. 38-45

[32] Bennett MR, Imrie BC, Raymond L, Wood IM. Sesame Growers Guide. 1st Australian Workshop. Northern Territory Department of Primary Industry and Fisheries; 1995. p. 24

[33] Bennett MR, Routley G. Effect of time and rate of Reglone application on desiccation of sesame. In: Sesame Research Report, 1993–1994 wet season. Technical Report No. 232. Katherine: Department of Primary Industry and Fisheries; 1995. pp. 18-37

[34] Mazzani B. Investigacion y tecnologia del cultivo del ajonjoli en Venezuela. Caracas, Venezuela: Ediciones del Consejo Nacional de Investigaciones Cientificos y Tecnologicas (CONICIT); 1999. p. 76 (Spanish)

[35] Araujo S. El ajonjolí, la "semilla milagrosa" en Venezuela. In: Periodismo Agropecuario (Spanish). 2017. 6 pp

[36] Anonymous. Gramoxone Safety Data Sheet. Greensboro, NC: Syngenta Crop Protection, LLC; 2015. 8 pp

[37] St. Andre J, Yermanos DM, Taylor C. Sesame harvesting trial, 1975. In: Progress Report-1975, Fresno County. 1975. 11 pp

Section 3

Animal Pests and Diseases

Chapter 12

Pathological Changes Associated with Natural Outbreak of Swine Pasteurellosis

Mamta Choudhary, Binod Kumar Choudhary and Ratan Chandra Ghosh

Abstract

Swine pasteurellosis is usually observed in descript as well as nondescript pigs imparting in huge economic losses to the pig producers. The disease is characterized by pyrexia, dullness, staggering gait, anorexia, serous nasal discharge and dyspnoea. Case fatality rate may as high as 95% in adult animals and 100% in piglets. Typical lesions of oedematous swellings may remarkably visible in the pharyngeal region, these swellings spread to the ventral cervical region and brisket of pigs. Gross lesions include severe pneumonia and haemorrhages in lungs, petechial haemorrhages on serous membranes and other visceral organs. Lymph nodes usually get enlarged, oedematous and haemorrhagic. The blood smears from heart blood and tissue impression smears reveal teaming numbers of bipolar organisms indicating the presence of *Pasteurella* spp., the etiological organism. The bacteriological isolation and characterization of causative agent should be ruled out to identify by Gram' staining for purity and bipolar morphology and biochemical characterization of the organisms. Molecular characterization necessitates to confirm *Pasteurella multocida* along with capsular types of the organism. Histopathological examination of lungs usually reveals typical fibrinous bronchopneumonia, multifocal suppuration and pleural thickening. Heart of some pigs may show presence of thrombi, haemorrhages and necrosed myocardium.

Keywords: Swine Pasteurellosis, *Pasteurella multocida*, Haemorrhagic septicaemia, capsular types, fibrinous bronchopneumonia

1. Introduction

Pasteurella multocida is of substantial economic significance in the livestock industry [1]. Infections by *Pasteurella multocida* have been reported in all the animals and fowls [2]. It is an important principal animal pathogen for over a century and is becoming crucial as human pathogen [3] leading to a disease process termed Pasteurellosis. *Pasteurella multocida* B:2, which causes haemorrhagic septicaemia (HS) of ruminants, is believed to enter the host via respiratory and oral routes. While the role of respiratory route of infection has been established, *Pasteurella multocida* is one of the most fascinating Gram-negative bacteria and is a commensal of the upper respiratory tract of many animal species as the organism is also a primary or secondary pathogen and responsible for a wide range of economically important diseases in domesticated animals throughout the world. Pasteurellosis is an infection of cattle, buffalo, swine and other species of animals caused by Gram-negative coccobacillary bipolar organism, *Pasteurella multocida*. It is OIE list B disease of ruminants in the tropical countries. *Pasteurella multocida* strains express a polysaccharide capsule on their cell surfaces and the antigenic specificity of the capsule determines the serogroups: A, B, D, E or F [4]. It has long been recognized that there is relationship exist between the capsular type and disease predilection [5], which suggests that the capsular polysaccharide type plays a role in host and disease specificity. For example, the majority of cases of fowl cholera are caused by capsular type A strains. Progressive atrophic rhinitis (PAR) of pigs is associated predominantly with capsular type D isolates, bovine and porcine pneumonia are associated mainly with capsular type A strains and haemorrhagic septicaemia of cattle and water buffaloes is caused exclusively by capsular type B and E isolates [6].

Capsular types A and D cause economic losses in swine because of their association with progressive atrophic rhinitis and enzootic pneumonia [7]. Its association with acute septicemic pasteurellosis in pigs has been recognized. *Pasteurella multocida,* a part of the commensal flora in the upper respiratory tract of pigs is shown to appear intermittently in the nasopharynx and subsequently shed in nasal secretions [8]. During this period, the carrier animal act to become a source of infection for in-contact susceptible animals. The role of pig as a reservoir of *Pasteurella multocida* for the transmission of the disease between pigs and cattle has been suggested. *Pasteurella multocida* is an important pathogen of pigs. It causes pneumonic pasteurellosis and is characterized by pneumonia, purulent bronchopneumonia and pleurisy. Affected pigs may have fever of up to 106^{0}F, are anorectic and disinclined to move. They show significant respiratory distress, often breathing through the mouth. Death is common after a clinical course of 4–7 days. There is a marked tendency of the disease to become chronic, resulting in reduced weight gains and frequent relapses. On post mortem examination there is a chronic bronchopneumonia with abscessation. Pleuritis is common and there may also be pericarditis. Peracute cases show an acute necrotizing fibrinous bronchopneumonia. Septicaemic disease with death asymptomatic acute deaths may occur within 12 hours in piglets. In India, it is associated with infection by capsular serotype B. The disease occurs in all ages of pigs including adults and is manifested by fever, dyspnoea and congestion on serosal surfaces.

Yet, Pasteurellae have been shown to be a common microflora of the upper respiratory tract in normal animals [9]. The organisms more often than not act as secondary invaders in animals with concurrent diseases or suffering from debilitating stressful conditions. HS is a peracute disease and is considered to be one of the most economically important diseases in Asia particularly in South and South East Asia leading to huge economic loss in livestock industry. *Pasteurella multocida* type B:2 assumed to be transmitted between the animals by aerosol infection and ingestion of contaminated river water or material with *P. multocida* especially during the HS outbreak. The clinical indication of this disease is often characterized by rapid course of high fever, respiratory distress, dullness, depression followed by death [10]. Pathogenesis of *P. multocida* is a complex interaction between host specific factors and specific bacterial virulence factors; therefore, understanding the disease pathogenesis is complex and depends on the bacterial strain, the animal model and their interactions. The key virulence factors identified in *Pasteurella multocida* include capsule, lipopolysaccharides, surface adhesions, iron regulated and iron acquisition proteins [11].

2. Epidemiology

The scope of epidemiology in modern animal husbandry practice is continuously widening. Epidemiological data provide information on various diseases which are pre-requisite for planning, execution and monitoring of disease control programmes. It is an important requirement for assessing economic impact of a disease and also for developing disease forecasting system. The disease is usually associated with wet, humid weather and increased incidence is recorded during wet, humid weather and during wet seasons. In countries where systemic epidemiological studies have been carried out, it has become evident that outbreaks do occur throughout the year but those occurring during wet seasons tend to spread presumably due to the longer survival of the organism under moist conditions [8]. Zhao *et al.* [12] examined one hundred and sixty-four clinical isolates of *Pasteurella multocida* recovered from two swine herds in Minnesota. The isolates were characterized by restriction endonuclease analysis (REA) and rRNA gene restriction fragment length patterns. They concluded that these genomic fingerprinting techniques were highly discriminatory and that capsular serotyping in combination with REA or ribotyping was an appropriate technique for epidemiological studies of *Pasteurella multocida* of swine origin.

3. Cultural, Phenotypic and Biochemical Characterization

Pasteurella multocida is one of the most fascinating bacterial pathogens. It is a small, Gram-negative rod or coccobacillary, non-motile, non-spore forming, facultative anaerobe belonging to the family *Pasteurellaceae.* The organism can be identified on the bases of cultural, morphological and biochemical characteristics. The organism is a Gram-negative rod with bipolar staining characteristics, which is non-haemolytic on sheep blood agar, aerobic to facultatively anaerobic and produces indole, oxidase, catalase and produce indole and ferment carbohydrates with slight gas production [13–15].

4. Genotypic Chracterization of *Pasteurella multocida*

Since the initial development of the PCR in 1985, the basic principle of *in vitro* nucleic acid amplification through repetitive cycling has had extensive application in all aspects of fundamental and applied clinical sciences [16]. The application of PCR technology for *Pasteurella multocida* identification was first reported in 1994 when primers constructed from the sequence of the toxA gene (encoding the dermonecrotic toxin implicated in progressive atrophic rhinitis) were used to detect toxigenic *Pasteurella multocida* strains. PCR techniques play a critical role in the clinical laboratory diagnosis as rapid and specific detection of microorganism. It has provided remarkable advances in the diagnosis of infectious agents, particularly in cases where the presence of organism is having significance. Lichtensteiger *et al.* [17] investigated the feasibility of PCR for accurate, rapid detection of toxigenic *Pasteurella multocida* from swabs. They developed a PCR protocol which resulted into amplification of an 846-nucleotide segment of the toxA gene. They developed a concordance of PCR results with (i) detection of toxA gene with colony blot hybridization, (ii) detection of toxA protein with colony immunoblot analysis, and (iii) lethal toxicity of sonicate in mice in a test set of 40 swine diagnostic isolates. Results of an enzyme-linked immunosorbent assay for toxA agreed with the other assays except for a negative reaction in one of the 19 isolates that the other assays identified as toxigenic. They suggested that PCR detection of toxigenic *Pasteurella multocida* directly from clinical swab specimens should be feasible.

5. Gross pathological lesions

On post-mortem examination of dead pigs from natural outbreaks, the gross pathological lesions may be marked by congestion and petechial haemorrhages on all over the serous membranes. Widespread petechial haemorrhages in the wall of thoracic cavity is the hallmark of the disease. Hydrothorax with presence of straw-coloured fluid in thoracic cavity can be seen (**Figure 1**). All the visceral organs may exhibit petechial to ecchymotic haemorrhages on the serosal surfaces. In some animals, hydrothorax, pleurisy and hydropericardium can also be prominent [18, 19].

The lungs usually show congestions with varying degrees of consolidation and with a marked thickening of the interlobular septa, pleura and rubbery consistency of lungs. There may be petechiae over the lungs (**Figure 2**). In acute cases the lungs may be severely consolidated with liver-like consistency. Whereas, subacute to chronic infection manifest grossly by marbled appearance of lungs (**Figure 3**), rubbery consistency and thickening of pleura (**Figure 4**), and emphysematous changes in lungs [20–22]. Heart may be severely congested and there may be presence of petechial as well as haemorrhagic streaks and necrotic foci which can be visible upon removal of pericardium (**Figure 5**). Rounding of heart and haemorrhages were also observed by Kapoor *et al.* [23]. The liver is one of the severely affected organs in this disease. The lesions may be characterized by congestion, petechiae and multiple necrotic foci on the surface of liver [21, 22]. Splenomegaly is a constant lesion seen in all the cases. There may be haemorrhagic enteritis in pigs died of swine pasteurellosis.

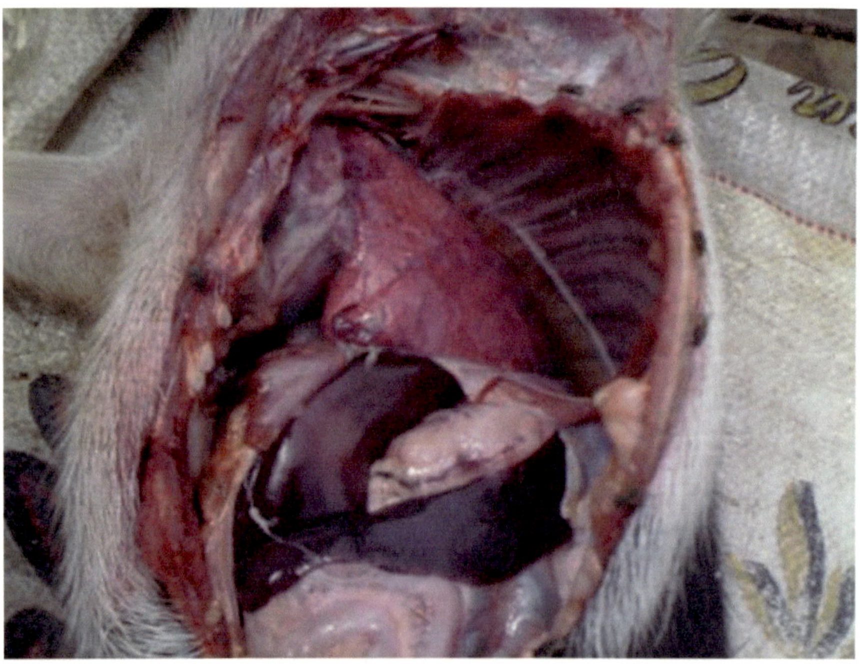

Figure 1.
Consolidation of lungs, hydrothorax, congested liver and splenomegaly in a pig died of swine pasteurellosis.

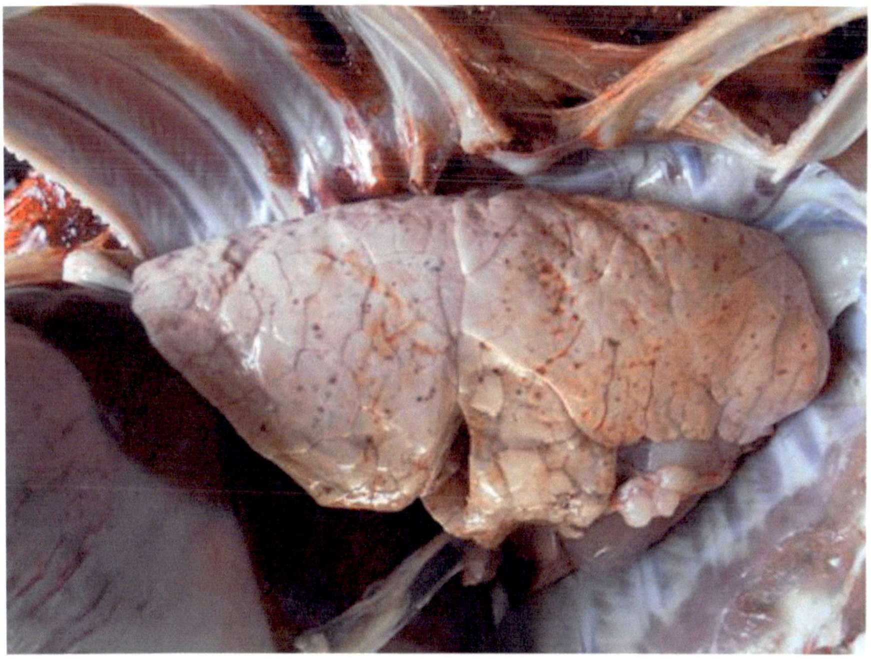

Figure 2.
Petechiae over lungs of a pig died of swine pasteurellosis.

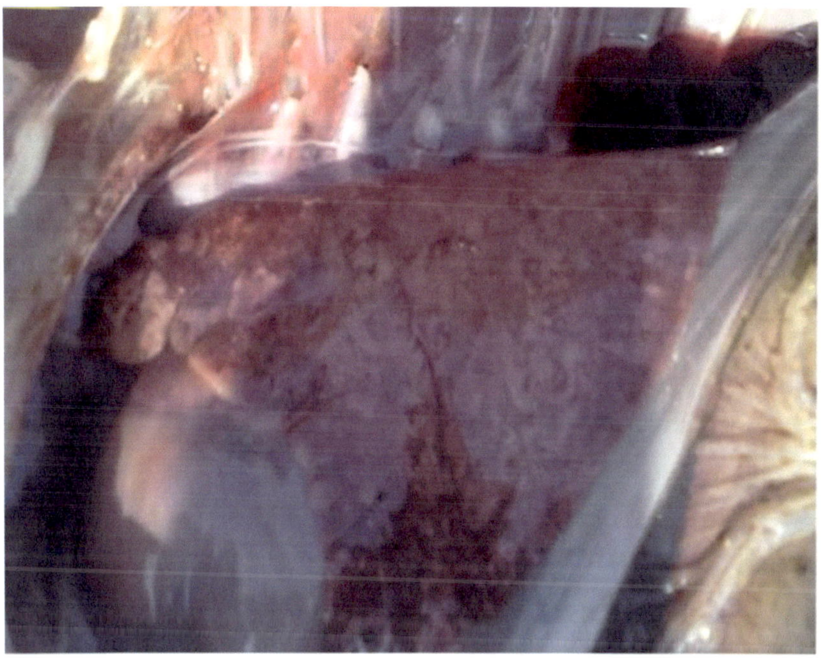

Figure 3.
Marbled appearance of lungs of a pig died of swine pasteurellosis.

6. Histopathological lesions

Lungs, the primarily affected organ, microscopically shows a variety of lesions from congestion of capillaries with thickened interlobular septa and atelectasis to severe lesions of perivascular and bronchial infiltration of inflammatory cells. There may be

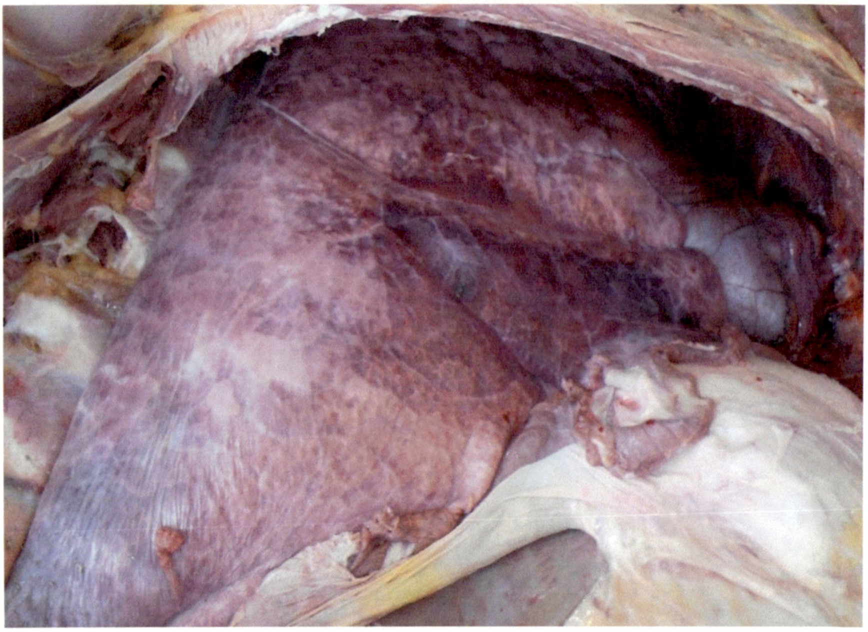

Figure 4.
Rubbery consistency of lungs and pleural thickening in a pig died of swine pasteurellosis.

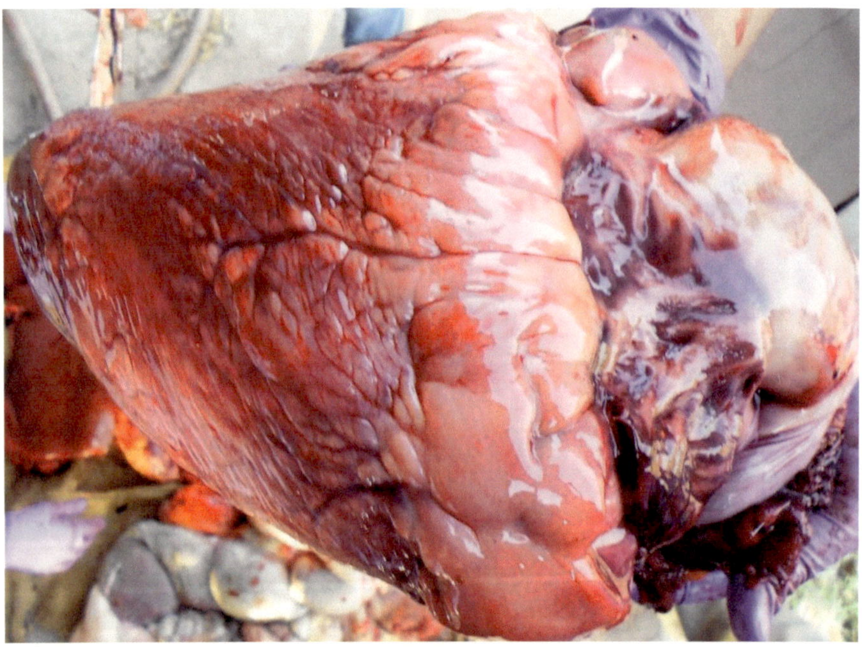

Figure 5.
Severe congestion and presence of haemorrhagic streaks in heart of a pig.

presence of oedema in lungs. The pneumonic lesions microscopically characterized by fibrinous pneumonia (**Figure 6**), necrotizing fibrinohaemorrhagic pneumonia, (**Figure 7**), interstitial pneumonia (**Figure 8**) and purulent bronchopneumonia (**Figure 9**). The acute fibrinous pneumonia characterized by serofibrinous exudation and infiltration with polymorphonuclear cells, macrophages and erythrocytes may

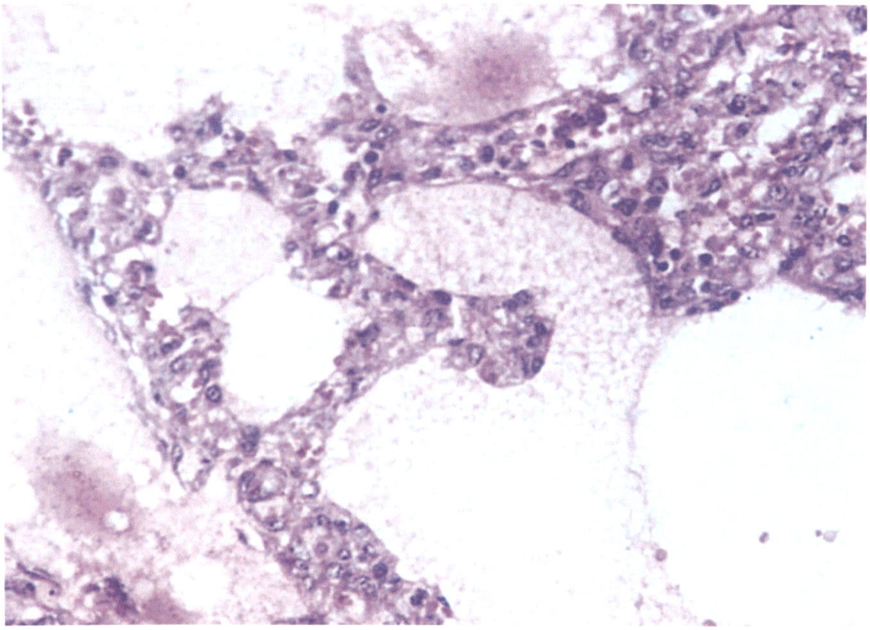

Figure 6.
Photomicrograph showing Fibrinous exudate in alveoli (H&EX400).

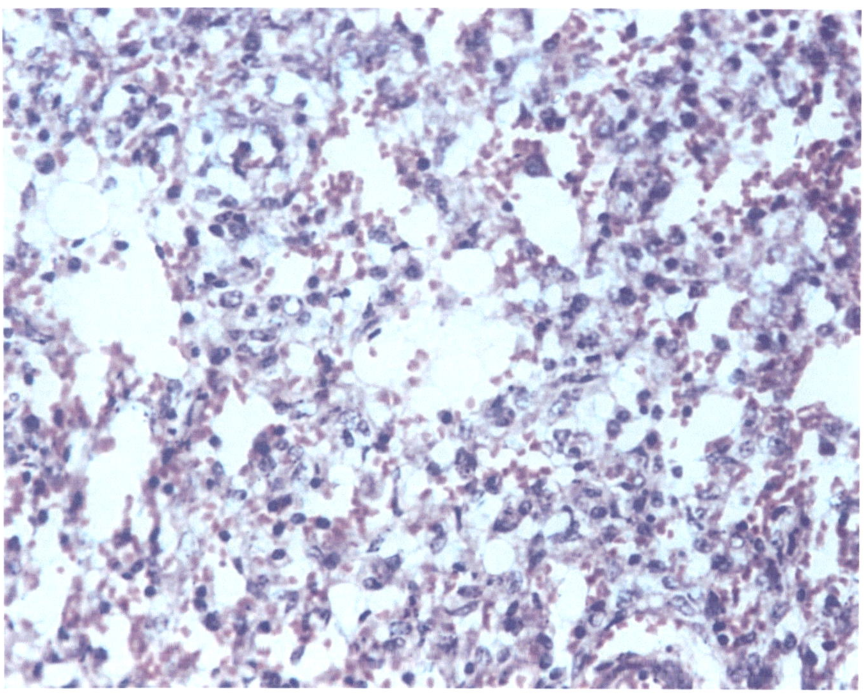

Figure 7.
Photomicrograph showing necrotizing fibrinohaemorrhagic pneumonia (H&E X400).

be observed. The bronchial and alveolar lumen usually filled with infiltrated erythrocytes, polymorphonuclear cells and macrophages (**Figure 10**). The alveoli showed variable changes from congestion to severe haemorrhages. Pleura and alveolar septa get thickened with fibrin, oedema and infiltration of polymorphonuclear cells. [23–26].

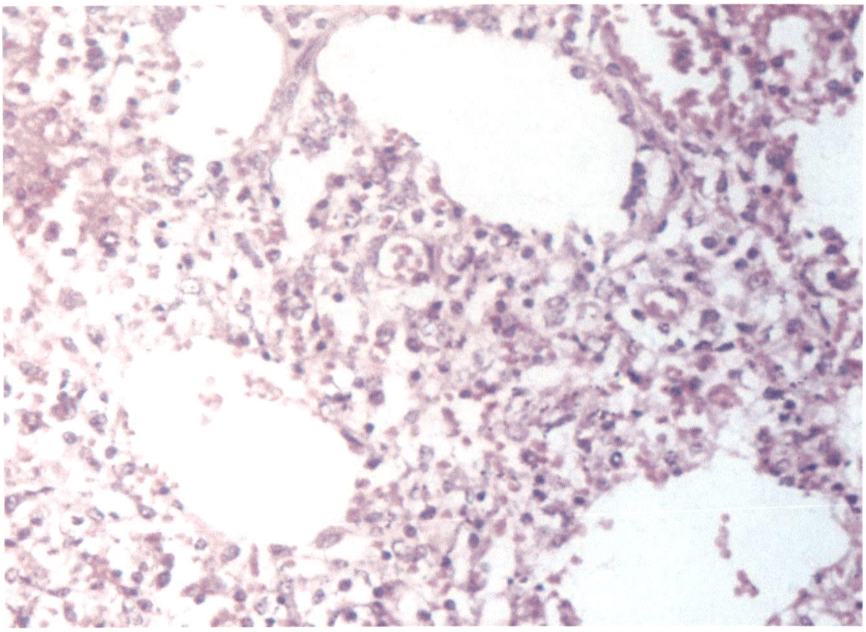

Figure 8.
Photomicrograph showing interstitial pneumonia (H&E X400).

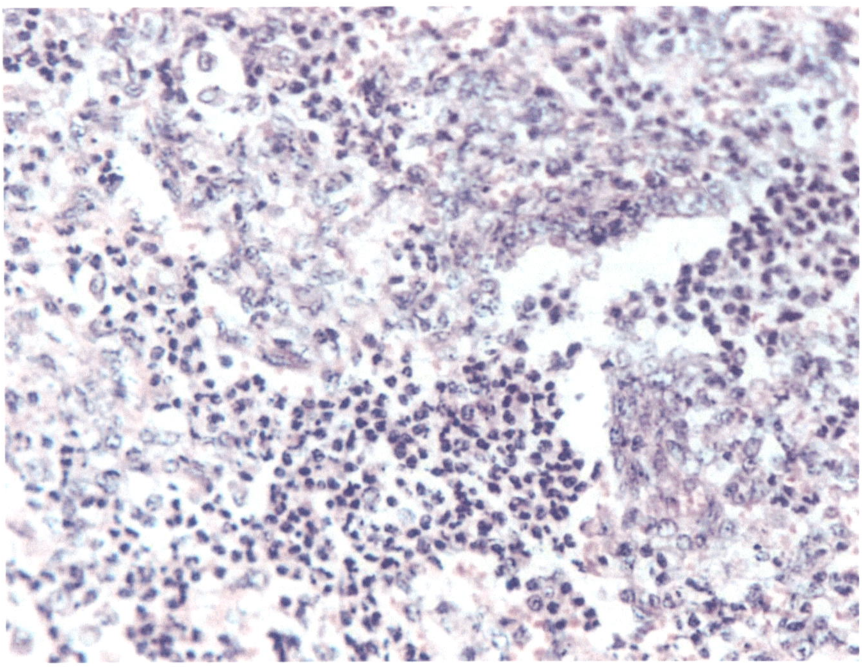

Figure 9.
Photomicrograph showing flooding of polymorphonuclear cells in alveoli (H&E X400).

Haemorrhages and necrosis may be evident in heart. Sub-pericardial haemorrhage and presence of erythrocytes in between the myocardial fibers may be remarkably noted. Myocardial necrosis can be marked as loss of striations of muscle fibers (**Figure 11**). There may be presence of thrombi in the blood vessels and fibrinous pericarditis in heart as a common finding. Liver is the consistently

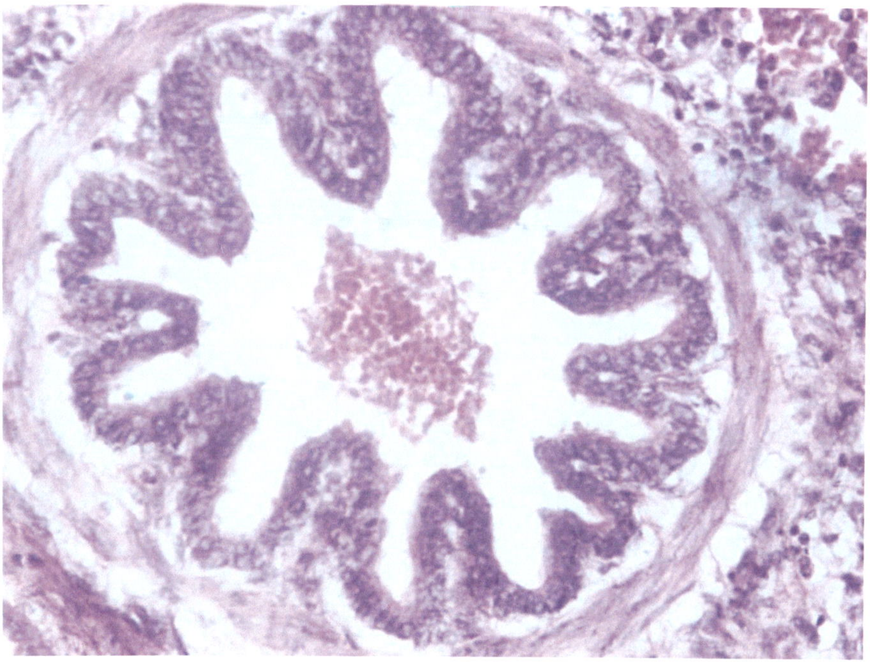

Figure 10.
Photomicrograph showing bronchiolar lumen containing exudate composed of erythrocytes (H&E X400).

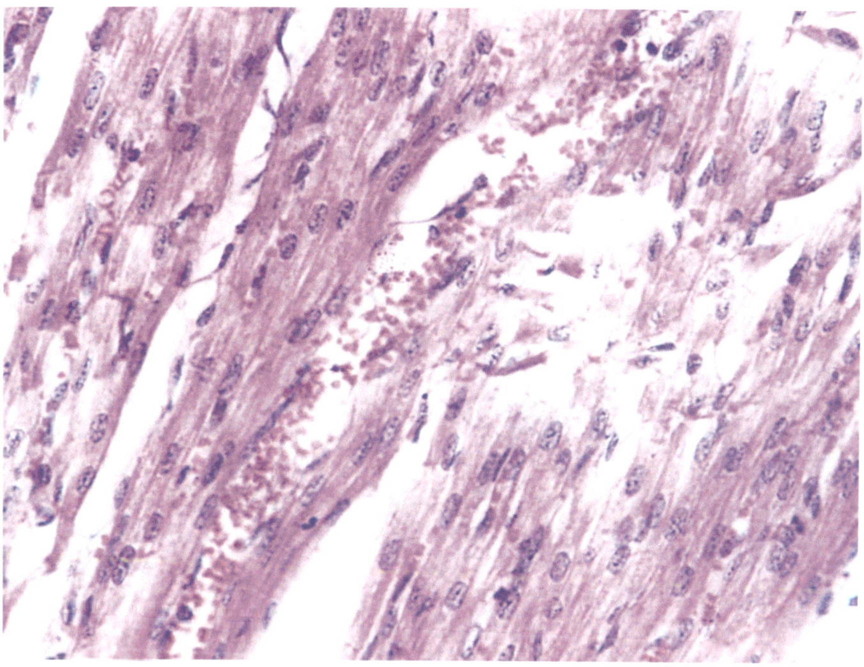

Figure 11.
Photomicrograph showing necrosis of myocardium and haemorrhages (H&E X400).

affected organ in this disease and the lesions reveal as invariably dilated and engorged blood vasculature and sinusoids. Focal areas of haemorrhages are usually seen. There will be hepatocytic swelling and increased activity of Kupffer cells in the parenchyma and focal areas of degenerative changes and hepatocytic necrosis.

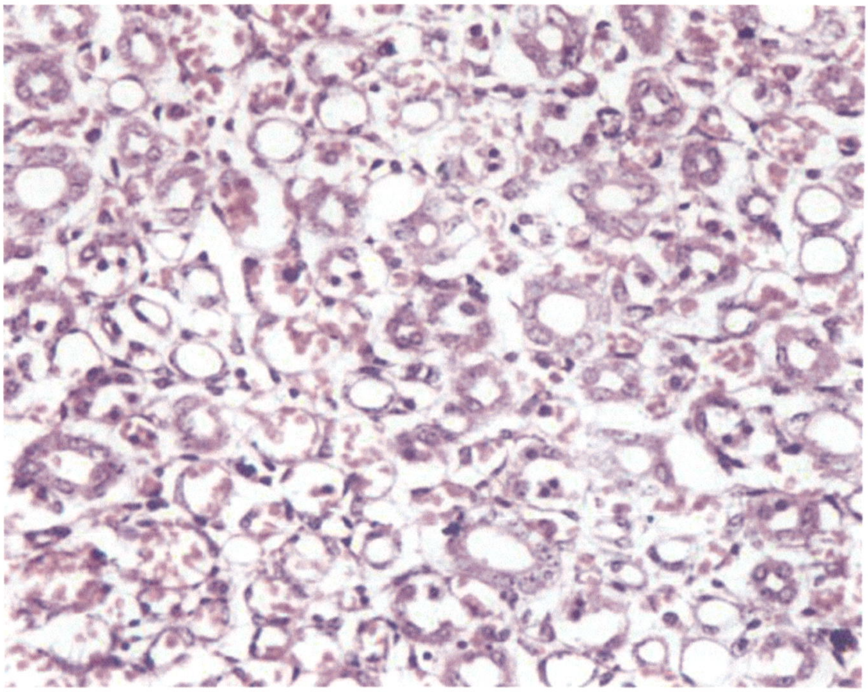

Figure 12.
Photomicrograph of kidney showing haemorrhages and sloughing of lining epithelium of renal tubules (H&E X400).

The hydropic degeneration of hepatocytes are characterized by cytoplasmic vacuolations and areas of hepatocytic necrosis with cellular infiltration. There may be dilatation of sinuses and disruption of hepatic cords seen in affected animals [20, 22, 27]. Lesions in Kidneys reveal as vascular congestion and focal areas of haemorrhages. Haemorrhages used to be intertubular (**Figure 12**). Cortical tubular epithelium may invariably swollen or degenerated with increased cytoplasmic granularity. The degenerative and necrotic changes of tubular epithelium will be diffuse in nature. Generalized degenerative and necrotic changes in the tubular epithelial cells may also be seen with variable severity from mild to high [20, 28].

There will be depletion of lymphocytes from germinal centre of the spleen and widespread necrosis can also be seen Spleen reveals as variably dilated and engorged vasculature, haemosiderosis, necrosis of lymphoid elements and infiltration of inflammatory cells mainly neutrophils [29, 30]. The microscopic lesions in the intestine are characterized by haemorrhages, necrosis of villi epithelium and desquamation of lining epithelium with focal infiltration of mononuclear cells in lamina propria and increase number of goblet cells [25].

7. Conclusions

Swine Pasteurellosis is an acute infection in swine caused by members of the *Pasteurella multocida*. Swine throughout the world are affected by pasteurellosis. *Pasteurella multocida* of swine can be isolated from natural field cases. Field isolate of *Pasteurella multocida* on sheep blood agar yield non-hemolytic, round, grayish, smooth or mucoid colonies. The isolates are Gram negative, cocco-bacilli in morphology and non-motile facultative anaerobe. Biochemically the isolates are positive for oxidase, catalase, indole production, reduction of nitrate, glucose and sorbitol fermentation, but fail to ferment lactose, arabinose and adonitol. The isolate also found negative for citrate, lysine decarboxylase, urease, phenylalaline deamination and H_2S production. The genomic DNA of test isolates and organisms upon PCR using the primer pair KMT1SP6 and KMT1T7 produce an amplified product of approximately 460 bp size. On post-mortem examination of dead pigs, the gross pathological lesions observed in different organs marked by congestion, petechial to ecchymotic haemorrhages on serosal surfaces of all the visceral organs. The histo-pathological examination reveals as acute fibrinous pneumonia with variable degree of haemorrheges; mild to severe congestion and focal haemorrhages in heart, with loss of striation of heart muscle, thrombus formation in cardiac blood vessels, hydropic degeneration in liver, haemorrhages, hepatocytic necrosis and increased activity of Kupffer cells in the hepatic parenchyma and mild to severe necrosis in renal tubules with presence of focal haemorrhages.

Acknowledgements

Authors are thankful to the Dean, Veterinary College, Anjora, Durg, Chhattisgarh and Indian council of Agricultural Research for providing the necessary facilities to carry out the research.

Conflict of interest

The authors declare no conflict of interest.

Author details

Mamta Choudhary[1]*, Binod Kumar Choudhary[1] and Ratan Chandra Ghosh[2]

1 ICAR-National Institute of Biotic Stress Management, Raipur, Chhattisgarh, India

2 Department of Veterinary Pathology, Chhattisgarh Kamdhenu University, Durg, India

*Address all correspondence to: chiyamum@gmail.com

References

[1] Collins, F M. Mechanisms of acquired resistance to Pasteurella multocida infection: a review. Cornell Vet. J.1977: 67: 103-138

[2] Soltys, M A. Pasteurella, Francisella and Actinobacillus. In: Introduction to Veterinary Microbiology. 1979.National Government Publication, Universiti Pertanian Malaysia, Ipoh, Malaysia, 158- 167

[3] Biberstein, EL. Handbook Series in Zonooses. (Stoenner, H., Torten, M. and Kaplan, W. eds.), CRC Press Inc., Boca Raton, Florida, pp 495-514 (1079)

[4] Rhoades, K R and Rimler, R B. Capsular groups of *Pasteurella multocida* isolated from avian hosts. Avian Dis.1987: 31: 895-898

[5] Boyce JD, Adler B. A capsular *Pasteurella multocida* B: 2 can stimulate protective immunity against pasteurellosis. Infect. Immun. 2001;**69**:1943-1946

[6] Carter, GR and De Alwis, MCL. Haemorrhagic septicemia, 1989. pp.131-160.In (Eds) C. Adlam and J.M. Rutier Pasteurella and Pasteurellosis. Academic Press Ltd. London, England

[7] Djordjenic SP, Eamens GJ, Ha HW, MJ and Chin, JC. Demonstration of Australian *Pasteurella multocida* isolates from sporadic outbreaks of porcine pneumoniaare non toxigenic (toxA) and display heterogeneous DNA restricrion enzyme profiles compared with toxigenic isolates from herds with progressive atrophic rhinitis. J. Med. Microbiol. 1998;**47**(8):679-688

[8] De Alwis, MCL. Haemorrhagic septicaemia. ACIAR Monograph 1990. No.57,p.36

[9] Campbell, RSF. Pasteurellosis. In: Veterinary Epidemiology. Melbourne, Australian Universities' International Development Program, 1983. 113-115

[10] Boyce, JD, M Harper, IW Wilkie and Adler, B. Pasteurella. In: Pathogenesis of Bacterial Infections in Animals (Gyles, C.L., Prescott, J.F., Songer, G. and Thoen, C.O. eds.)., Wiley-Blackwell, Ames, 2010. p 664-664.

[11] Hawari, AD, Obeidat, M, Awaisheh, SS, AlDaghistani HI and Al-Abbadi, AA. Am. J. Anim. Vet. Sci., 2014:9: 116-121

[12] Zhao, G, Pijoan, C, Murtaugh, M P Molitor, T W Use of restriction endonuclease analysis and ribotyping to study epidemiology of *Pasteurella multocida* in closed swine herds. Infect Immun. 1992: 60:1401-1405.

[13] Bergey, DH, Harrison, FC, Breed, RS, Hammer, BW and Huntoon, FM. Bergey's Manual of determinative bacteriology 9th edn., 1994. Williams and Wilkins, Baltimore

[14] Chawak, MM, Verma, KC, Kataria, JM and Kamar, AA. Characterization of indigenous isolates of avian *P. multocida.* Indian Comp. Microbiol. Immunol. Infec. Dis. 2000: 21:111-114

[15] Anupama, M, Venkatesha, MD, Yasmeen, N and Gowda, RNS. Evaluation of polymerase chain reaction (PCR) for identification of virulent *P. multocida* and its comparison with animal inoculation test. Indian J. Anim. Sci. 2003: 73:166-167

[16] Rapley R, Theophilus BOM, Bevan IS, Walkar MR. Fundamentals of polymerase chain reaction: a future in clinical diagnostics. Med. Lab. Sci. 1992;**49**:119-128

[17] Lichtensteiger, CA, Steenbergen, SM, Lee, R.M, Polson, DD and Vimr, ER. Direct PCR analysis for toxigenic

Pasteurella multocida. J. Clin. Microbiol. 1996: 34: 3035-3039

[18] Kumar, H, Sharma, S, Mahajan, V, Verma, S, Arora, AK, Kaur, P and Sandhu, KS. Pathology and PCR based confirmation of haemorrhagic septicemia outbreaks in bovines. Indian J. Vet. Pathol. 2006:30(1):5-8

[19] Abdullah, FFJ, Khaleel, MM, Adamu, L, Osman, AY and Haron, AW Polymerase chain reaction detection of *Pasteurella multocida* type B: 2 in mice infected with contaminated river water. Am. J. Anim. Vet. Sci. 2013: 8: 146-151

[20] Kapoor, V, Katoch, RC, Sharma, M and Asrani, RK. Pathogenicity test of *P. multocida* in mice A:1. Indian J. Anim. Sci. 2004: 74:495-496

[21] Srinivas VMV, Parthiban S. Mukhopadhyay, HK and Antony, PX *Pasteurella multocida* infection in Emu (*Dromaius novaehollandiae*). Indian Vet. J. 2014;**91**(6):09-12

[22] Bhat MA, Darzi NM, Wari SA, Willayat M. Pathology of spontaneously occurring *P. multocida* infections in pigeons. Indian J. Anim. Sci. 2002;**72**:384-385

[23] Choudhary M, Choudhary BK, Ghosh RC, Bhoyar S, Chaudhari S, Barbuddhe SB. Cultivable microbiota and pulmonary lesions in polymicrobial bovine pneumonia. Microbial Pathogenesis. 2019;**134**:103577

[24] Praveena, P E, Periasamy, S, Kumar, A A and Singh, N. Pathology of Experimental Infection by *Pasteurella Multocida* Serotype A1 in Buffalo Calves. Vet. Pathol. 2014: doi:10.1177/0300985813516647

[25] Sujatha K, Srilatha CH, Ahmed NN. An outbreak of pasteurellosis in pigs. Indian Vet. J. 2003;**80**:341-343

[26] Kumar H, Mahajan V, Sharma S. Concurrent pasteurellosis and classical swine fever in Indian pigs. J. Swine Health Prod. 2007;**15**(5):279-283

[27] Das, SS and Bhagwan PSK. Isolation and characterization of *P. multocida* from ovine pneumonia. Indian J. Anim. Sci. 1997:67:29-30

[28] Ali S, Mahmoud MZ. Screening picture on pasteurellosis in a flock of sheep at Assiut Governorate. Assiut Vet. Med. J. 1993;**29**(57):164-176

[29] Pilla AG, Katiyar AK, Awadhiya RP, Vegad JL. An outbreak of pasteurellosis in swine. Indian Vet. J. 1986;**63**:527-529

[30] Katoch S, Sharma M, Patil, RD, Kumar, S and Verma, S. *In vitro* and *in vivo* pathogenicity studies of *Pasteurella multocida* strains harbouring different ompA. Vet. Res. Commun. 2014: DOI 10.1007/s11259-014-9601-6

Chapter 13

Identification of Ticks in Dogs with Ehrlichiosis

Koperumselvan Karthika

Abstract

Examination of ticks collected from ehrlichiosis positive dogs revealed the occurrence of *Rhipicephalus sanguineus*. The distribution of ehrlichiosis in dogs is related to the spreading of vectors. *Ehrlichia canis* is the etiologic agent of canine monocytic ehrlichiosis (CME) and recognized as the most prevalent tick-borne disease affecting dogs and is transmitted by the brown dog tick *Rhipicephalus sanguineus* with an expanding global distribution. Infection of the vertebrate host occurred when an infected tick ingested a blood meal which in turn contaminated the feeding site with its salivary secretion. Blood transfusions from infected donors can also transmit the organisms. Hence, identification of ticks is necessary to detect the disease affecting dogs.

Keywords: ticks, ehrlichiosis, Monocytosis, canine, *Rhipicephalus sanguineus*

1. Introduction

Ehrlichial diseases have emerged as significant problems for human and animals over the past two decades [1–3]. In 1935, *Ehrlichia canis* was first discovered in dog in Algeria [4]. Before the outbreak in military working dogs in Southeast Asia in 1967, canine ehrlichiosis was considered to be a mild disease characterized by fever, vomiting and naso-ocular discharge [5]. Since then, the disease in dogs has spread worldwide [6] and caused serious effects.

Canine ehrlichiosis or tropical pancytopenia is an acute, subacute or chronic tick borne disease caused by *E. canis* which is a Gram-negative intracellular bacterium [7] which occurs particularly in tropical and subtropical regions due to its geographical distribution of its vector tick *Rhipicephalus sanguineus* [8]. This disease is characterized clinically by anorexia, fever, vomiting, loss of weight, enlargement of the liver, spleen and lymph nodes, epistaxis, superficial bleeding and thrombocytopenia [9]. Dogs with canine monocytic ehrlichiosis may die due to hemorrhage and/or secondary infection [10]. Hematological changes in dogs affected with *E. canis* of all stages of infection include a reduction in hematocrit, hemoglobin concentration, blood cell count, thrombocytopenia and leucopenia. However, poor reticulocyte response (non-regenerative anemia) is associated with chronic ehrlichiosis [11]. Diagnosis of ehrlichiosis can be made based on clinical signs, demonstration of morulae in the monocytes, serological testing with the detection of antibodies against *E. canis* and polymerase chain reaction (PCR). Demonstration of morulae inclusions in blood smears of dogs in the subclinical and chronic stages of the disease was often difficult or impossible and has a low sensitivity rate as this organism is usually present in very low concentrations and hence cannot be used for diagnosis of the disease [10]. The evolutionary thesis suggests that both ixodid and argasid ticks have been in existence since the late Paleozoic to early Mesozoic eras [12]. Antecedent forms evolved as obligate ectoparasites of smooth-skinned reptiles during the late Paleozoic era [5].

Ticks are the most important ectoparasites in tropical and sub-tropical areas. They are also responsible for severe economic losses either through direct effects of blood sucking or indirectly as vectors of pathogens and toxins. Ticks (Acari: Ixodida) are blood feeding ectoparasites acts as vectors of human diseases next to mosquitoes, but comparatively more important as vectors of animal diseases [13–15]. Ticks belong to

Phylum: Arthropoda
Class: Arachnida
Subclass: Acari
Order: Parasitiformes
Suborder: Ixodida

Ixodida contains three families: Argasidae (soft ticks having dorsum without chitin), Ixodidae (hard ticks having dorsum totally or partially covered with chitin) and Nuttalliellidae (an ill known monotypic family represented by Nuttalliella namaqua), among which Argasidae and Ixodidae are more important. In turn, according to morphological characters, the family Ixodidae is subdivided into the Prostriata group (genus Ixodes) and Metastriata group (all other genera in Ixodidae).

Traditionally, classifications and phylogenetics inferences for Ixodida were based on morphological, biological and ecological characteristics [16–21]. Tick classification largely based on morphological characteristics, and the value given to differences and similarities among groups of ticks, resulting in non-homogeneous tick arrangements. The molecular taxonomy associated with conventional morphological cataloging will be useful to obtain a more homogeneous and independent criterion for classification, although in the short term this may not be obvious.

1.1 Importance of tick identification in dogs

Many ticks are responsible for causing various diseases. Among which the tick *Rhipicephalus sanguineus* (brown dog tick) plays vital role because it causes ehrlichiosis in dogs which is life threatening disease in dogs which causes symptoms similar to that of dengue in human beings. Reduction in platelet count and multi-organ failure are the major detrimental things in case of dogs in this specific ehrlichiosis disease. That is why identification of ticks is very important to rule out the disease and early identification will help in saving the life of the animal by giving appropriate treatment. Presence of ticks itself will help in identifying subacute cases so that life loss can be avoided. Due to its veterinary and public health relevance, *Rhipicephalus sanguineus* is one of the most studied ticks.

Ticks able to survive in adverse conditions too as they have heavy protective, chitinous covering and can withstand long periods of starvation and also have wide host range. They can deposit large number of eggs at a time and are relatively free from natural enemies and are tenacious blood suckers.

Medical and veterinary importance of ticks based on their capability of disease transmission. The important diseases transmitted by ticks are Lyme borreliosis (*Borrelia burgdorferi*), *Canine babesiosis* (*babesia* sp.), Ehrlichiosis (*Ehrlichia* sp.), Anaplasmosis (*Anaplasma* sp.), Hepatozoonosis (*Hepatozoon* sp.).

2. Materials and methods

The dogs presented to Teaching Veterinary Clinical Campus that were diagnosed for ehrlichiosis by nPCR were utilized for the study. Around 3 or 4 ticks collected from different sites of the affected dogs were fixed in a 70% ethanol solution. It was further processed and was identified as per the morphology described by [14]

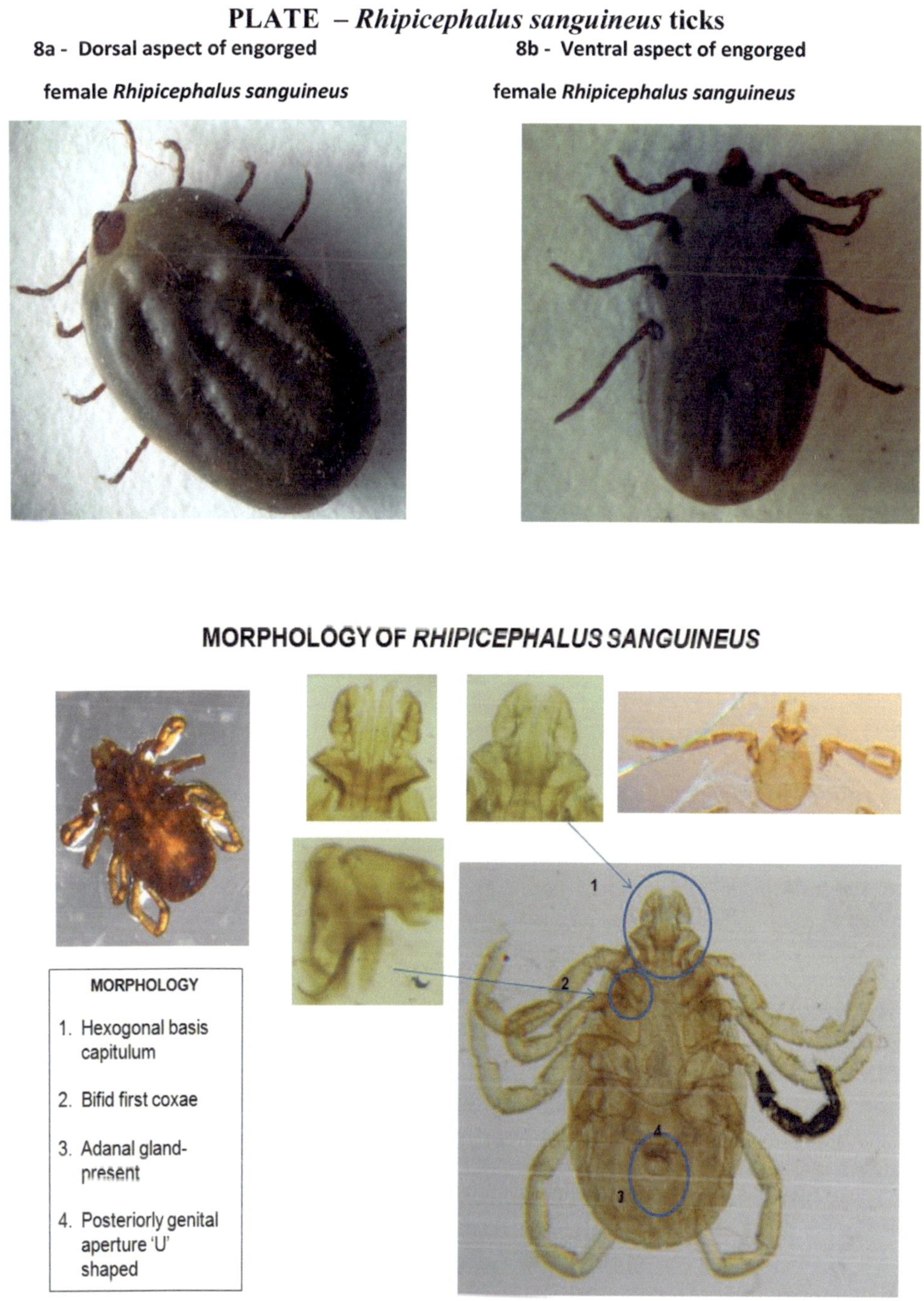

Figure 1.
Rhipicephalus sanguineus *ticks.*

using stereomicroscope and magnifying lens. The stereomicroscope was used at a low magnification of 10× and magnifying lens was used at 40× magnification for identification of specific morphological features.

2.1 Collection of ticks

Unengorged/engorged male and female ticks were collected from dogs either by gently plucking from the body of the dog by hand manipulation or with the help of blunt pointed forceps without damaging their mouth parts. The specimen collected in a plastic container with ventilated cap was labeled appropriately as per host and sites of attachment. Label must contain information about date and place of collection, host, age and site of collection. These samples were transported to the laboratory for further studies.

2.2 Tick identification

These ticks were identified using standard keys [22, 23].

The ticks in the present study were identified as *R. sanguineus* (**Figure 1**). Sen and Fletcher [24] reported that *R. sanguineus* was the only tick that infested dogs in India. Bashir et al. [25] from Pakistan reported 96.8% of the ticks were identified as *R. sanguineus* and the remaining identified as *Dermacentor* and *Haemaphysalis* species. In the present study, all the ticks were identified as *R. sanguineus* and concurred with the findings of [24]. Krogt [26] demonstrated that *R. sanguineus* ticks were able to transmit *E. canis* from a naturally infected dog to an uninfected dog via the bite of the infected tick. Filippova [27] from Japan reported that *E. canis* developed in the salivary glands of *R. sanguineus.* Though, *R. sanguineus* seems to be the vector for *E. canis* in Puducherry, definite studies regarding tick transmission of ehrlichiosis caused by *E. canis* in India is lacking. Hence, transmission studies needs to be undertaken to determine its vector potentiality.

2.3 Tick control measures

Economic losses can be reduced by adopting tick control measures like chemical acaricides [7]. The major reason to control ticks includes disease transmission, tick paralysis or toxicosis by *Rhipicephalus* sp. [7] and physical damage caused by ticks. Keeping animals away from tick-prone areas is the most effective way to control exposure.

3. Results and discussion

Out of 46 dogs found positive for ehrlichiosis, 35 dogs (76.10%) were infested with ticks (**Figure 2**). The ticks collected from different sites of the dogs suffering from ehrlichiosis were identified as *R. sanguineus* based on specific morphological features viz. the reddish brown scutum and conscutum, slightly convex shaped eyes, hexagonal basis capitulum, bifid first coxae, posterior "U" shaped genital aperture and the presence of adanal glands [9]. Bashir et al. [25] from Pakistan reported 96.8% of the ticks were identified as *R. sanguineus* and the remaining identified as *Dermacentor* and *Haemaphysalis* species. *R. sanguineus* was the most commonly encountered tick in India as reported by [28]. In the present study, all the ticks were identified as *R. sanguineus* which concurred with the findings of [24] who reported that *R. sanguineus* was the only tick that infested dogs in India [29]. Filippova [27] reported that *E. canis* developed in the salivary glands of *R. sanguineus* and were able

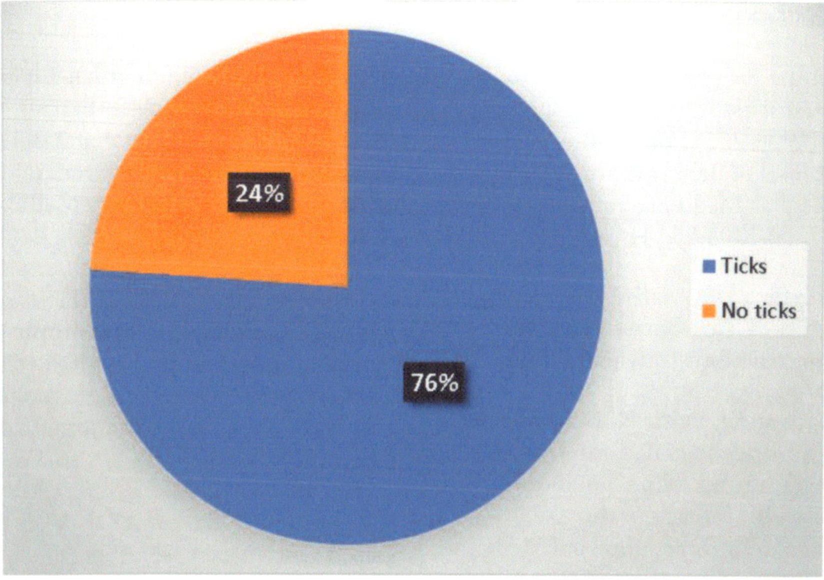

Figure 2.
Representation of ticks in dogs with ehrlichiosis.

to transmit *E. canis* from a naturally infected dog to an uninfected dog via the bite of an infected tick. Although, *R. sanguineus* seems to be the vector for spread of *E. canis*, definite studies regarding tick transmission of the disease is lacking in India. Hence, transmission studies needs to be undertaken to determine its vector potentiality.

4. Summary

The present study on ticks collected from 35 dogs affected with *Ehrlichia canis* were identified as *R. sanguineus* based on the typical morphological features which included hexagonal basis capitulum, bifid first coxae, presence of adanal shields, posterior "U" shaped genital aperture and the presence of adanal glands. Hence, it is concluded that *R. sanguineus* ticks were responsible for transmitting *E. canis* infection in dogs of Puducherry.

Author details

Koperumselvan Karthika
Indian Veterinary Research Institute, Izatnagar, India

*Address all correspondence to: karthika1988.6@gmail.com

References

[1] Moreira SM, Bastos CV, Araujo RB, Santos M, Passos LMF. Retrospective study (1998-2001) on canine ehrlichiosis in Belo Horizonte. Arquivo Brasileiro de Medicina Veterinária e Zootecnia. 2003;**55**:141-147

[2] Soneshine DE, Mather TN. Ecological Dynamics of Tick-borne Zoonoses. New York: Oxford University Press; 1994

[3] Wardrop KJ, Reine N, Birkenheuer A, Hale A, Hohenhaus A, Crawford C, et al. Canine and feline blood donor screening for infectious disease. Journal of Veterinary Internal Medicine. 2005;**19**:135-142

[4] Donatien A, Lestoquard F. Existence en Algérie d'une Rickettsia du chien. Le Bulletin de la Société de Pathologie Exotique. 1935;**28**:418-419

[5] Hoogstraal H, Aeschlimann A. Tick-host specificity. Bulletin de la Société Entomologique Suisse. 1982;**55**:5-32

[6] Regendanz P, Muniz J. O *Rhipicephalus sanguineus* como transmissor da piroplasmose canina no Brasil. Memórias do Instituto Oswaldo Cruz. 1936;**31**:81-84

[7] Drummond RO. Tick-borne livestock diseases and their vectors. Chemical control of ticks. World Animal Review (FAO). 1983;**36**:28-33

[8] Andereg PI, Passos LMF. *E. canis*—Revisao. La Clinica Veterinaria. 1999;**19**:31-39

[9] Karthika K, Vijayalakshmi P, Sreekrishnan R, Das SS, Antony PX. Identification of ticks recovered from dogs affected with ehrlichiosis in Puducherry. The Indian Veterinary Journal. 2014;**91**(11):95-96

[10] Jongejan F, Uilenberg G. The global importance of ticks. Parasitology. 2004;**129**:S1-S12

[11] Walker JS, Rundquist JD, Taylor R, Wilson BL, Andrews MR, Barck J, et al. Clinical and clinicopathologic findings in tropical canine pancytopenia. Journal of the American Veterinary Medical Association. 1970;**157**:43-55

[12] Hoogstraal H, Kim KC. Tick and mammal coevolution, with emphasis on Haemaphysalis. In: Kim KC, editor. Coevolution of Parasitic Arthropods and Mammals. New York: John Wiley & Sons; 1985. pp. 505-568

[13] Dumler JS, Barbet AF, Bekker CP, Dasch GA, Palmer GH, Ray SC, et al. Reorganization of genera in the families Rickettsiaceae and Anaplasmataceae in the order Rickettsiales: unification of some species of *Ehrlichia* with *Anaplasma*, *Cowdria* with *Ehrlichia* and *Ehrlichia* with *Neorickettsia*, descriptions of six new species combinations and designation of *Ehrlichia equi* and 'HGE agent' as subjective synonyms of *Ehrlichia phagocytophila*. International Journal of Systematic and Evolutionary Microbiology. 2001;**51**:2145-2165

[14] Huxsoll DL. The historical background and global importance of ehrlichiosis. In: Williams JC, Kakoma I, editors. Ehrlichiosis: A Vector-Borne Disease of Animals and Humans. Dordrecht: Kluwer Academic; 1990. p. 164

[15] Saito TB, Cunha-Filho NA, Pacheco RC, Ferreira F, Pappen FG, Farias NAR, et al. Canine infection by Rickettsiae and Ehrlichiae in Southern Brazil. American Journal of Tropical Medicine and Hygiene. 2008;**79**:102-108

[16] Camicas JL, Morel PC. Position systématique et classification des tiques (Acarida: Ixodida). Acarologia. 1977;**18**:410-420

[17] Clifford CM, Kohls GM, Sonenshine DE. The systematics of the

subfamily Ornithodorinae (Acarina: Argasidae) I. The genera and subgenera. Annals of the Entomological Society of America. 1964;**57**:429-437

[18] Fivaz B, Petney T, Horak I. Tick Vector Biology: Medical & Veterinary Aspects. Brimingham, England: Wrox Press, Computer Book Publisher, Springer Verlag; 1992

[19] Higuchi S, Fujimori M, Hoshi F, Kawamura S, Yasuda Y. Development of *Babesia gibsoni* in the salivary glands of the larval tick *Rhipicephalus sanguineus*. The Journal of Veterinary Medical Science. 1995;**57**:117-119

[20] Jeyathilakan N, Kowsigan A, Kiruthika K, Abdul Basith S. Occurrence of tick infestation in a wild bison (*Bos gaurus indicus*) in Tamilnadu. The Indian Veterinary Journal. 2012;**89**:116

[21] Klompen JSH, Oliver JH. Systematic relationships in the soft ticks (Acari: Ixodida: Argasidae). Systematic Entomology. 1993;**18**:313-331

[22] Pospelova-Shtrom MV. On the system of classification of ticks of the family Argasidae Can., 1890. Acarologia. 1969;**11**:1-22

[23] Shaw SE, Day MJ, Birtles RJ, Breitschwerdt EB. Tick-borne infectious diseases of dogs. Trends in Parasitology. 2001;**17**:74-80

[24] Sen SK, Fletcher TB. Veterinary Entomology and Acarology for India. New Delhi: Indian Council of Agricultural Research; 1962. p. 457

[25] Bashir IN, Chaudhry ZI, Ahmed S, Saeed MA. Epidemiological and vector identification studies on canine babesiosis. Pakistan Veterinary Journal. 2009;**29**:51-54

[26] van der Krogt JS. An overview of the clinical picture and current diagnostics and therapies. In: Jongejan F, editor. *Ehrlichia canis* infection on the Island of Curacao. A Research Report. February 2010

[27] Filippova NA. Classification of the subfamily Amblyomminae (Ixodidae) in connection with a reinvestigation of the chaetotaxy of the anal valves. Parazitologiya. 1994;**28**:3-12

[28] Abd Rani PAM, Irwin PJ, Coleman GT, Gatne M, Traub RJ. A survey of canine tick-borne diseases in India. Parasites and Vectors. 2011;**4**:141

[29] Tilak R, Gupta KKD, Verma AK. Vector data bank in the Indian Armed Forces. Medical Journal Armed Forces India. 2008;**64**:36-39